Lecture Notes in Economics and Mathematical Systems 419

Toru Maruyama   Wataru Takahashi (Eds.)

Nonlinear and Convex Analysis in Economic Theory

Springer

Editors

Toru Maruyama
Director of the Conference
Department of Economics
Keio University
2-15-45, Mita, Minato-ku
Tokyo 108, Japan

Wataru Takahashi
Chief director of the Conference
Department of Information Science
Tokyo Institute of Technology
2-12-1, Ohokayama, Meguro-ku
Tokyo 152, Japan

Library of Congress Cataloging-in-Publication Data.

Nonlinear and convex analysis in economic theory / Toru Maruyama, Wataru Takahashi (eds.). p. cm. – (Lecture notes in economics and mathematical systems; 419) "Contributions to T. I. Tech/K.E.S. Conference on Nonlinear and convex Analysis in Economic Theory ... Keio University, July 2-4, 1993"–CIP pref. Includes bibliographical references.

ISBN 978-3-540-58767-5 ISBN 978-3-642-48719-4 (eBook)
DOI 10.1007/978-3-642-48719-4

1. Economics, Mathematical-Congressses. 2. Spatial analysis (Statistics)–Congresses. 3. Business cycles–Congresses. I. Maruyama, Tōru. II. Takahashi, Wataru, 1944- . III. Keiō Gijuku Keizai Gakkai. IV. Tōkyō Kōgyō Daigaku. V. T. I. Tech/K.E.S. Conference on Nonlinear and Convex Analysis in Economic theory (1993): Keio University) VI. Series. HB 135.N65 1995 330' .01'–dc20 94-39352 CIP

Typesetting: Camera ready by author
SPIN: 10486680 42/3140-543210 - Printed on acid-free paper

Preface

The papers collected in this volume are contributions to T.I.Tech./K.E.S. Conference on Nonlinear and Convex Analysis in Economic Theory, which was held at Keio University, July 2-4, 1993. The conference was organized by Tokyo Institute of Technology (T. I. Tech.) and the Keio Economic Society (K. E. S.) , and supported by Nihon Keizai Shimbun Inc..

A lot of economic problems can be formulated as constrained optimizations and equilibrations of their solutions. Nonlinear-convex analysis has been supplying economists with indispensable mathematical machineries for these problems arising in economic theory.

Conversely, mathematicians working in this discipline of analysis have been stimulated by various mathematical difficulties raised by economic theories.

Although our special emphasis was laid upon "nonlinearity" and "convexity" in relation with economic theories, we also incorporated stochastic aspects of financial economics in our project taking account of the remarkable rapid growth of this discipline during the last decade.

The conference was designed to bring together those mathematicians who were seriously interested in getting new challenging stimuli from economic theories with those economists who were seeking for effective mathematical weapons for their researches.

Thirty invited talks (six of them were plenary talks) given at the conference were roughly classified under the following six headings :

1) Nonlinear Dynamical Systems and Business Fluctuations,
2) Fixed Point Theory,
3) Convex Analysis and Optimization,
4) Eigenvalue of Positive Operators,
5) Stochastic Analysis and Financial Market,
6) General Equilibrium Analysis.

As we all knew, the Japanese economy was in the midst of a serious recession in the summer of 1993. Nevertheless, presidents of many companies and organizations had kindly offered us generous financial supports. On behalf of the programme committee, we would like to extend our cordial gratitude to them all for their warm cooperations.

Two hundred mathematicians and economists from ten countries attended the conference. We are indebted to all of the participants for the successful completion of the entire programme in a quite friendly atmosphere. Finally, we appreciate the very skillful assistance of E. Shimada, secretary of the Department of Information Science at Tokyo Institute of Technology, and S.

Atsushiba, Y. Kimura and M. Shigenari, students of the same Department, in the course of preparation of this volume.

August 28, 1994

Toru Maruyama
Wataru Takahashi

Table of Contents

The Bargaining Set in Large Finite NTU Exchange Economies

Robert M. Anderson*

Department of Economics and Department of Mathematics
University of California at Berkeley, Berkeley, CA 94720, U.S.A.

The bargaining set was originally defined by Aumann and Maschler [6]. Several different definitions have been subsequently proposed; the most frequently used definition was proposed by Davis and Maschler [8]. In the exchange economy context we consider here, the core consists of all allocations such that no coalition can propose an alternative set of trades which is feasible for the coalition on its own and which makes all of its members better off. All definitions of the bargaining set restrict the ability of coalitions to block ("object to") an allocation, by taking into account the possibility that a second coalition might propose yet another set of trades ("counterobject") and thereby cause some members to defect from the first coalition. In the Aumann-Maschler and Davis-Maschler definitions, the original objection is proposed by a single individual known as the leader of the objection; any counterobjecting coalition must exclude this leader. An objection is said to be justified if there is no counterobjection; the bargaining set consists of all allocations with no justified objection.

Geanakoplos [15] considered sequences of transferable utility (TU) exchange economies with smooth preferences. He modified the Aumann-Davis-Maschler definition to allow a group of leaders, each of whom had to be excluded from any counterobjection. The group of leaders contains a fixed (but small) fraction of the number of agents in the economy; thus, as the number of agents grows along the sequence of economies, the number of leaders grows proportionately. He showed that this Geanakoplos bargaining set becomes asymptotically competitive as the number of agents grows; the proof, which uses Nonstandard Analysis, is quite lengthy.

*The author is grateful particularly to Walter Trockel, and in addition to Bob Aumann, Larry Ausubel, Truman Bewley, Don Brown, Gérard Debreu, Eddie Dekel-Tabak, Glenn Ellison, Drew Fudenberg, John Geanakoplos, Ted Groves, Bruce Hamilton, Walt Heller, Ehud Kalai, Edi Karni, Ali Khan, Michael Maschler, Eric Maskin, Andreu Mas-Colell, Roger Myerson, Wilhelm Neuefeind, David Pearce, Matthew Rabin, Debraj Ray, Arijit Sen, Chris Shannon, Max Stinchcombe, Rajiv Vohra, Akira Yamazaki, Bill Zame and Lin Zhou for helpful discussions.

Shapley and Shubik [21] showed that the Aumann-Davis-Maschler bargaining set converges in replica sequences of TU exchange economies with smooth preferences.

Mas-Colell [20] considered nontransferable utility (NTU) exchange economies with a continuum of agents but without smooth preferences. His definition differs from the Aumann-Davis-Maschler and Geanakoplos definitions because it does not allow an individual or group to be designated the leader of an objection. Because this increases the number of potential counterobjections, it makes it harder to form a justified objection, and hence the Mas-Colell bargaining set is larger than the Aumann-Davis-Maschler and Geanakoplos bargaining sets. Under hypotheses similar to those of Aumann's core equivalence theorem, he showed that the Mas-Colell bargaining set coincides with the set of Walrasian allocations. Since models with a continuum of agents are thought of as idealizations of large economies, it seemed reasonable to expect that Mas-Colell's bargaining set would become approximately competitive in sequences of finite exchange economies as the number of agents increased. Zhou [23] has proposed additional restrictions on counterobjections; these restrictions are satisfied in the Aumann-Davis-Maschler definition. Since Zhou's additional restrictions make it easier to form a justified objection, the Zhou bargaining set is smaller than the Mas-Colell bargaining set.

We report here on two recent papers: Anderson, Trockel and Zhou [5], which provides a TU (and thus, *a fortiori*, an NTU) nonconvergence example for the Mas-Colell and Zhou bargaining sets, and Anderson [4], which provides NTU convergence theorems for the Aumann-Davis-Maschler and Geanakoplos bargaining sets.

Anderson, Trockel and Zhou [5] show that Mas-Colell's and Zhou's bargaining sets need not converge in replica sequences of economies. In the example, the measure of the set of individually rational Pareto optimal equal-treatment (IRPOET) allocations which are *not* in the Mas-Colell and Zhou bargaining sets tends to zero as the economy is replicated; in particular, the set of IRPOET bargaining set allocations converges in the Hausdorff distance to the set of *all* IRPOET allocations. The replica sequence in the example satisfies the hypotheses of the Debreu-Scarf theorem [11] and of Debreu's rate of convergence theorem for the core [9]. The cooperative game generated by the example has transferable utility and satisfies the assumptions of Shapley and Shubik [21]; thus, the Aumann-Davis-Maschler bargaining set *does* converge.

The essence of the nonconvergence example is easy to describe. It is a replica sequence with two goods and two types of agents. Let f be an IRPOET allocation and let f_n denote the n-fold replica of f. Let the agents of type 2 be relatively favored by f, and the agents of type 1 disfavored. One first shows that if a coalition S can make a justified Mas-Colell or Zhou objection to f, then S must contain all n agents of type 1; if not, one could form a counterobjection by switching one of the type 1 agents omitted from S for one of the type 1 agents included in S. There is an ideal ratio t of the number of type 2 agents to type 1 agents that maximizes the utility that can be provided to type 1 agents. For most n, it will be the case that there exist $n' < n$ and m' such that

$$\left|\frac{m'}{n'} - t\right| < \min_{m \in \mathbf{N}} \left|\frac{m}{n} - t\right|.$$

But then a coalition consisting of n' agents of type 1 and m' agents of type 2 can counterobject. It is only in the rare case that the best approximation to t of the form $\frac{m'}{n'}$ with $n' \leq n$ has $n' = n$ that f_n is *not* in the Mas-Colell and Zhou bargaining sets. Thus, the nonconvergence example is driven entirely by an integer problem.

Anderson [4] presents three theorems showing that the Aumann-Davis-Maschler and Geanakoplos bargaining sets are approximately competitive in large finite *NTU* exchange economies. The first two theorems deal with the Geanakoplos bargaining set. In addition to dropping the assumption of transferable utility, certain other assumptions (notably smoothness of preferences) required in Geanakoplos [15] are weakened. The first theorem requires assumptions similar to those needed for certain core convergence results (see Anderson [3]). Preferences are assumed to be tight, i.e. for every $\sigma > 0$, there is a compact set K of preferences such that the proportion of agents with preferences in K is at least $1 - \sigma$. Note that a compact set of preferences is equimonotone (see Anderson [3]). This first theorem also requires uniformly integrable endowments, though it does not require that the social endowment of each good be bounded away from 0. It establishes no rate of convergence.

The hypotheses of the second theorem are incomparable to those of the first. It is assumed that there is a set P of preferences exhibiting bounded marginal rates of substitution such that the proportion of agents with preferences in P is bounded away from 0. Instead of uniform integrability of endowments, it is only required that the largest individual endowment be small compared to the number of agents; however, the *per capita* social endowment is required to be bounded away from 0 in each commodity. In the event that endowments are uniformly bounded, the rate of convergence is arbitrarily close to the inverse of the number of agents.

The third theorem in Anderson [4] concerns the Aumann-Davis-Maschler bargaining set in sequences of NTU exchange economies with smooth preferences. Anderson [2], Kim [18] and Geller [16] have previously shown that economies with smooth preferences exhibit faster core convergence rates than non-smooth economies. In the bargaining set context, the rate of convergence is expressed in terms of the size of the competitive gap and the size of the leader set. The faster convergence afforded by smoothness allows one to restrict the leader set to be a single individual, and thereby prove a convergence theorem for the Aumann-Davis-Maschler bargaining set. The third theorem generalizes the Shapley-Shubik result to nonreplica sequences of NTU exchange economies with smooth preferences and uniformly bounded endowments.

The Aumann-Davis-Maschler bargaining set differs from the Mas-Colell bargaining set only by the designation of a single individual as the leader of an objection. The nonconvergence example for the Mas-Colell bargaining set in Anderson, Trockel and Zhou [5] occurs in a replica sequence of TU economies with smooth preferences which satisfies all the assumptions of the convergence theorem for the Aumann-Davis-Maschler bargaining set. It is remarkable that the designation of a single leader should make such a profound difference in the resulting bargaining set.

Smoothness of preferences is a strong assumption because it requires that indifference surfaces not cut the boundary of the consumption set; in particular, it implies that at every Walrasian equilibrium, all agents with positive income consume positive amounts of all goods. It may be possible to weaken smoothness to a linkedness condition, as Cheng [7], Propositions 7.4.12 and 7.4.16 of Mas-Colell [19], and Kim [18] have done for the core.

The three main theorems of Anderson [4] are all derived from a main proposition, which adapts Mas-Colell's equivalence proof to the case of a fixed finite economy. The proof of this proposition is conceptually the same as Mas-Colell's proof in the continuum case, with the Shapley-Folkman Theorem substituting for Richter's Theorem, but the derivation of the needed estimates is quite complex. The error terms that arise from this substitution can be absorbed in the Geanakoplos leaders' consumption (and in the Aumann-Davis-Maschler leader's consumption if preferences are smooth) but they are sufficient to destroy the Mas-Colell and Zhou objections. The derivation of the second theorem on the Geanakoplos bargaining set from the main proposition is quite straightforward, while the derivation of the first theorem is rather indirect. The derivation of the convergence theorem for the Aumann-Davis-Maschler bargaining set is similarly indirect; in addition, it incorporates the central idea in the fast core convergence theorem of Anderson [2].

The discrepancy between the behavior of Mas-Colell's bargaining set in the continuum and its behavior in sequences of large finite economies gives reason to be cautious in accepting the continuum as the proper idealization of a "large" economy. In a game with an atomless measure space of agents, an individual player makes no difference to the game. Hence, if one interprets the leader as a single individual, there is no difference between the Mas-Colell and Aumann-Davis-Maschler bargaining sets in atomless games. This motivated Geanakoplos to consider leader sets of small positive measure as the appropriate extension of the Aumann-Davis-Maschler bargaining set to continuum games. However, in the nonstandard hyperfinite games considered by Geanakoplos (as well, of course, in large finite games), an individual leader makes *sense.* What is surprising is that, given smooth preferences, an individual leader makes a *difference.* In the light of the nonconvergence example for the Mas-Colell bargaining set in Anderson, Trockel and Zhou [5], which satisfies all the assumptions of the convergence theorem for the Aumann-Davis-Maschler bargaining set, we see that allowing a single leader can shrink the bargaining set from a set containing essentially every individually IRPOET allocation to a set of allocations which are approximately competitive. There appears to be no way to capture the importance of a single leader in a continuum model.

References

[1] Anderson, Robert M., *An Elementary Core Equivalence Theorem,* Econometrica **46** (1978), 1483-1487.

[2] Anderson, Robert M., *Gap-Minimizing Prices and Quadratic Core Convergence,* Journal of Mathematical Economics **16** (1987), 1-15; correction, Journal of Mathematical Economics **20** (1991), 599-601.

[3] Anderson, Robert M., *The Core in Perfectly Competitive Economies,* in Robert J. Aumann and Sergiu Hart (eds.), Handbook of Game Theory, Amsterdam: Elsevier Science Publishers B.V., 1992, 413-457.

[4] Anderson, Robert M., *Convergence of the Aumann-Davis-Maschler and Geanakoplos Bargaining Sets,* Working Paper, Department of Economics, University of California at Berkeley, January 1994.

[5] Anderson, Robert M., Walter Trockel and Lin Zhou, *Nonconvergence of the Mas-Colell and Zhou Bargaining Sets,* Working Paper, Department of Economics, University of California at Berkeley, January 1994.

[6] Aumann, Robert J. and Michael Maschler, *The Bargaining Set for Cooperative Games,* in M. Dresher, Lloyd S. Shapley and Albert W. Tucker (eds.), Advances in Game Theory, Princeton: Princeton University Press, 1964, 443-476.

[7] Cheng, Hsueh-Cheng, *What is the Normal Rate of Convergence of the Core (Part I),* Econometrica **49** (1981), 73-83.

[8] Davis, Morton and Michael Maschler, *Existence of Stable Payoff Configurations for Cooperative Games,* Bulletin of the American Mathematical Society **69** (1963), 106-108.

[9] Debreu, Gérard, *The Rate of Convergence of the Core of an Economy,* Journal of Mathematical Economics, **2** (1975), 1-7.

[10] Debreu, Gérard, *Existence of Competitive Equilibrium,* Chapter 15 in Kenneth J. Arrow and Michael D. Intrilligator (eds.), Handbook of Mathematical Economics, Volume II. Amsterdam: North-Holland Publishing Company, 1982, 697-743.

[11] Debreu, Gérard and Herbert Scarf, *A Limit Theorem on the Core of an Economy,* International Economic Review **4** (1963), 236-246.

[12] Dierker, Egbert *Gains and Losses at Core Allocations,* Journal of Mathematical Economics **2** (1975), 119-128.

[13] Dutta, Bhaskar, Debraj Ray, Kunal Sengupta, and Rajiv Vohra, *A Consistent Bargaining Set,* Journal of Economic Theory **49** (1989), 93-112.

[14] Edgeworth, Francis Y., Mathematical Psychics, 1881. London: Kegan Paul.

[15] Geanakoplos, John, *The Bargaining Set and Nonstandard Analysis,* Chapter 3 of Ph.D. Dissertation, Department of Economics, Harvard University, Cambridge, MA, 1978.

[16] Geller, William, *Almost Quartic Core Convergence,* Presentation to Econometric Society North American Summer Meeting, Berkeley, June 1987.

[17] Hildenbrand, Werner, Core and Equilibria of a Large Economy, 1974. Princeton: Princeton University Press.

[18] Kim, Wan-Jin, *On the Rate of Convergence of the Core,* Working Paper #8617, Department of Economics, University of California at Berkeley, 1986.

[19] Mas-Colell, Andreu (1985), The Theory of General Economic Equilibrium: A Differentiable Approach. Cambridge: Cambridge University Press.

[20] Mas-Colell, Andreu, *An Equivalence Theorem for a Bargaining Set,* Journal of Mathematical Economics **18** (1989), 129-139.

[21] Shapley, Lloyd S. and Martin Shubik, *Convergence of the Bargaining Set for Differentiable Market Games,* Appendix B in Martin Shubik, A Game Theoretic Approach to Political Economy, Cambridge, MA: M.I.T. Press, 1984.

[22] Vohra, Rajiv, *An Existence Theorem for a Bargaining Set,* Journal of Mathematical Economics **20** (1991), 19-34.

[23] Zhou, Lin, *A New Bargaining Set of an N-Person Game and Endogenous Coalition Formation,* preprint, Cowles Foundation, Yale University, April 1992.

Epi-convergence of Integral Functionals Defined on the Space of Measures

Charles Castaing Vincent Jalby
Département de Mathématiques, Université Montpellier II
34095 Montpellier Cedex 05, FRANCE

1. Introduction

This paper is concerned with the epi-convergence of integral functionals defined on the space of vector measures with bounded variation with values in a reflexive separable Banach space.

Section 2 recalls several facts about vector measures with bounded variation. In Section 3, we give some new results for the epi-convergence of integral functionals defined on the space of vector measures with bounded variation, which are refinements of classical results obtained by C. Castaing ([9]) and Y. Reshetnyak ([20]). In Section 4, some useful results about integral functionals defined on the space of measures with bounded variation are summarized.

2. Notations. Definitions

Let T be a Polish space (i.e. T is a separable topological space and there is a complete metric defining the topology of T) and $\mathcal{B}(T)$ be its Borel σ-algebra. Let E be a separable reflexive Banach space with norm $\|\cdot\|$, E' be its strong dual space and $\langle\cdot,\cdot\rangle$ be the duality bilinear form between E and E'. The closed unit ball B' of E' is equipped with the weak* topology $\sigma(E',E)$; it is a metrizable compact space. We denote by $B(x,r)$ the open ball of E with center x and radius r.

We denote by $\mathcal{C}^b(T,E)$ the Banach space of all bounded continuous E-valued functions on T equipped with the uniform norm.

An E'-valued measure on T is a σ-additive set-function m from $\mathcal{B}(T)$ into E'. The variation of the measure m is the non-negative real-valued measure $|m|$ defined on T by

$$|m|(A) = \sup\left\{\sum_{i\in I} \|m(A_i)\| : (A_i)_{i\in I} \text{ finite } \mathcal{B}(T)\text{-partition of } A\right\}, \forall A \in \mathcal{B}(T)\,.$$

We denote by $\mathcal{M}^b(T,E')$ the space of all E'-valued measures m on T with bounded variation (i.e. $|m|$ is a bounded Radon measure on T). We set $\|m\| = |m|(T)$.

For every $m \in \mathcal{M}^b(T,E')$, there is a $|m|$-measurable function $\frac{dm}{d|m|}: T \longrightarrow B'$ such that $m = \frac{dm}{d|m|}|m|$, that is

$$m(A) = \int_A \frac{dm}{d|m|}(t)\, d|m|(t)\,, \qquad \forall A \in \mathcal{B}(T)\,.$$

For every $f \in \mathcal{C}^b(T,E)$ and every $m \in \mathcal{M}^b(T,E')$, we define the integral of f with respect to m by

$$\int f\, dm = \int \langle f(t), \frac{dm}{d|m|}(t)\rangle\, d|m|(t)\,.$$

Thus, the space $\mathcal{M}^b(T,E')$ is identified with a subspace of the topological dual space of $\mathcal{C}^b(T,E)$. It is equipped with the weak* topology $\sigma(\mathcal{M}^b(T,E'), \mathcal{C}^b(T,E))$ usually called *weak* (or *narrow*) topology.

For further details on vector measures, we refer to [19], [13], [16], [12].

A subset $\mathcal{H}$ of $\mathcal{M}^b(T,E')$ is *bounded* if

$$\sup_{m\in\mathcal{H}} \|m\| < +\infty\,.$$

It is *tight* (or it satisfies *Prokhorov's condition*) if for every $\varepsilon > 0$, there is a compact subset K_ε of T such that

$$|m|(T \setminus K_\varepsilon) \le \varepsilon\,, \qquad \forall m \in \mathcal{H}\,.$$

We recall Prokhorov's theorem ([5, §5, Théorème 1], [11, Théorème III.59]):

Theorem 2.1 *Let $\mathcal{H}$ be a bounded tight subset of $\mathcal{M}^b(T,\mathbb{R})$. Then $\mathcal{H}$ is relatively weakly compact in $\mathcal{M}^b(T,\mathbb{R})$.*

All the forthcoming results still hold without tightness assumption, if T is a Polish locally compact space and if $\mathcal{M}^b(T,E')$ is equipped with the *vague* topology (i.e. the topology of the pointwise convergence on the space $\mathcal{C}_c(T,E)$ of all continuous functions from T to E with compact support).

3. Epi-convergence of integral functionals

We first recall several classical notions of epigraphical analysis (see [1]):

Let X be a Hausdorff first countable topological space. Let $(F_n)_n$ be a sequence of functions from X into $[-\infty,+\infty]$. The *epigraphical inferior* (resp. *superior*) *limit* $\mathrm{li_e}\, F_n$ (resp. $\mathrm{ls_e}\, F_n$) of the sequence $(F_n)_n$ is defined by

$$\begin{aligned}\mathrm{li_e}\, F_n(x) &= \min_{\{x_n\to x\}} \liminf_n F_n(x_n)\,, \qquad \forall x\in X\,,\\ \mathrm{ls_e}\, F_n(x) &= \min_{\{x_n\to x\}} \limsup_n F_n(x_n)\,, \qquad \forall x\in X\,.\end{aligned}$$

The sequence $(F_n)_n$ *epi-converges* to F at x if $F(x)=\mathrm{li_e}\, F_n(x)=\mathrm{ls_e}\, F_n(x)$.

The first part of our main result is the following theorem:

Theorem 3.1 *Let Y be a Polish space and $\widehat{\mathbb{N}}=\mathbb{N}\cup\{\infty\}$. Let $\{\phi_k : k\in\widehat{\mathbb{N}}\}$ be lower semi-continuous functions on $T\times B'\times Y$ with values in $[0,+\infty]$ such that for all $(t,y,k)\in T\times Y\times\widehat{\mathbb{N}}$, $\phi_k(t,\cdot,y)$ is convex and positively 1-homogeneous on B', i.e., for all $x\in B'$ and $\lambda\in[0,1]$, $\phi_k(t,\lambda x,y)=\lambda\phi_k(t,x,y)$. We assume that $\mathrm{li_e}\,\phi_k\geq\phi_\infty$ on $T\times B'\times Y$, that is, for every sequence $((t_k,x_k,y_k))_k$ in $T\times B'\times Y$ which converges to (t,x,y), we have*

$$\liminf_k \phi_k(t_k,x_k,y_k)\geq\phi_\infty(t,x,y)\,. \tag{3.1.1}$$

Let $(m_k)_k$ be a bounded tight sequence in $\mathcal{M}^b(T,E')$ which converges weakly to $m\in\mathcal{M}^b(T,E')$. Let $(u_k)_k$ be a sequence of continuous functions on T with values in Y which converges uniformly on every compact subset of T to a continuous function u. Then we have

$$\liminf_k \int_T \phi_k(t,\frac{dm_k}{d|m_k|}(t),u_k(t))\,d|m_k|(t)\geq\int_T\phi_\infty(t,\frac{dm}{d|m|}(t),u(t))\,d|m|(t)\,.$$

To prove this theorem, we will use an extension to Banach valued measures defined on a Polish space of a result due to Y. Reshetnyak (Theorem 4.1) and the two following lemmas concerning vector measures. These results are classical for real-valued measures and can be found in [4, Chap. III, §4]. The proofs in the vector case are slightly the same.

Lemma 3.2 *Let S be a Polish space. Let $m\in\mathcal{M}^b(T,E')$ and $\lambda\in\mathcal{M}^b(S,\mathbb{R})$. Then there exists a unique measure $m\otimes\lambda$ in $\mathcal{M}^b(T\times S,E')$ such that*

$$m\otimes\lambda(A\times B)=m(A)\,\lambda(B)\,,\qquad\forall(A,B)\in\mathcal{B}(T)\times\mathcal{B}(S)\,.$$

Moreover, we have

$$|m\otimes\lambda|=|m|\otimes|\lambda|$$

and

$$\frac{d(m\otimes\lambda)}{d|m\otimes\lambda|}=\frac{dm}{d|m|}\otimes\frac{d\lambda}{d|\lambda|}\,,\qquad |m\otimes\lambda|\text{-}a.e.$$

Lemma 3.3 *Let S be a Polish space. Then the map Π: $\mathcal{M}^b(T,E')\times\mathcal{M}^b(S,\mathbb{R}) \longrightarrow \mathcal{M}^b(T\times S,E'), (m,\lambda)\longmapsto m\otimes\lambda$ is continuous on each subset $\mathcal{H}\times\mathcal{K}$ of $\mathcal{M}^b(T,E')\times\mathcal{M}^b(S,\mathbb{R})$ where $\mathcal{H}$ is a bounded tight subset of $\mathcal{M}^b(T,E')$ and $\mathcal{K}$ is a bounded tight subset of $\mathcal{M}^b(S,\mathbb{R})$. Moreover $\Pi(\mathcal{H}\times\mathcal{K})$ is bounded and tight in $\mathcal{M}^b(T\times S,E')$.*

Proof of Theorem 3.1. For convenience sake, we set $m_\infty = m$ and $u_\infty = u$. Remark that the (Alexandroff) one point compactification $\widehat{\mathbb{N}}$ of $\mathbb{N}$ is a compact metric space. We define a function $\psi: T\times\widehat{\mathbb{N}}\times B' \longrightarrow [0,+\infty]$, by $\psi(t,k,x) = \phi_k(t,x,u_k(t))$. Let us show that ψ is lower semi-continuous on $T\times\widehat{\mathbb{N}}\times B'$. Let $(t_k,p_k,x_k)_k$ be a sequence in $T\times\widehat{\mathbb{N}}\times B'$ converging to (t,p,x). We set $a = \liminf_k \psi(t_k,p_k,x_k)$. If $p\in\mathbb{N}$, for k large enough, $p_k = p$. Since u_p is continuous, the sequence $(u_p(t_k))_k$ converges to $u_p(t)$ in Y. It follows from the lower semi-continuity of ϕ_p that

$$\begin{aligned}\liminf_k \phi(t_k,p_k,x_k) &= \liminf_k \phi_{p_k}(t_k,x_k,u_{p_k}(t_k)) = \liminf_k \phi_p(t_k,x_k,u_p(t_k))\\ &\geq \phi_p(t,x,u_p(t)) = \psi(t,p,x)\,.\end{aligned}$$

Consider now the case $p=\infty$. Extracting a subsequence, we may assume that $a = \lim_k \psi(t_k,p_k,x_k)$ and that the sequence $(p_k)_k$ is increasing. For each $n\in\mathbb{N}$, we define

$$s_n = \begin{cases} t_1, & \text{if } n < p_1,\\ t_k, & \text{if } p_k \leq n < p_{k+1},\end{cases} \quad\text{and}\quad y_n = \begin{cases} x_1, & \text{if } n < p_1,\\ x_k, & \text{if } p_k \leq n < p_{k+1}.\end{cases}$$

Then, we have $t = \lim_n s_n$, $(s_{p_k})_k = (t_k)_k$ and $x = \lim_n y_n$, $(y_{p_k})_k = (x_k)_k$. Moreover, since u is continuous and $(u_n)_n$ converges uniformly to u on the compact set $\{t, s_n : n\in\mathbb{N}\}$, the sequence $(u_n(s_n))_n$ converges to $u(t)$ in Y. Hence, by (3.1.1), we get

$$\begin{aligned}\liminf_k \psi(t_k,p_k,x_k) &\geq \liminf_n \psi(s_n,n,y_n) = \liminf_n \phi_n(s_n,y_n,u_n(s_n))\\ &\geq \phi_\infty(t,x,u(t)) = \psi(t,\infty,x)\,.\end{aligned}$$

That proves the lower semi-continuity of ψ. Now, for each $k\in\widehat{\mathbb{N}}$, consider ν_k the measure in $\mathcal{M}^b(T\times\widehat{\mathbb{N}},E')$ defined by $\nu_k = m_k\otimes\delta_k$ (where δ_k is the Dirac measure at k in $\mathcal{M}^b(\widehat{\mathbb{N}},\mathbb{R})$). Since the sequence $(\delta_k)_k$ is bounded, obviously tight and converges weakly to δ_∞ in $\mathcal{M}^b(\widehat{\mathbb{N}},\mathbb{R})$, by Lemmas 3.2 and 3.3, we have $|\nu_k| = |m_k|\otimes\delta_k$ and the sequence $(\nu_k)_k$ converges weakly to ν_∞ in $\mathcal{M}^b(T\times\widehat{\mathbb{N}},E')$. Now we apply Reshetnyak's theorem (Theorem 4.1) to the function ψ and the sequence $(\nu_k)_k$. We obtain

$$\liminf_k \int_{T\times\widehat{\mathbb{N}}} \psi(t,n,\frac{d\nu_k}{d|\nu_k|}(t,n))\,d|\nu_k|(t,n) \geq \int_{T\times\widehat{\mathbb{N}}} \psi(t,n,\frac{d\nu_\infty}{d|\nu_\infty|}(t,n))\,d|\nu_\infty|(t,n)\,. \tag{3.3.1}$$

Since for each $k \in \widehat{\mathbb{N}}$, $\nu_k = m_k \otimes \delta_k$, by Lemma 3.2, we have

$$\frac{d\nu_k}{d|\nu_k|} = \frac{d(m_k \otimes \delta_k)}{d(|m_k| \otimes \delta_k)} = \frac{dm_k}{d|m_k|} \otimes 1 .$$

We deduce that for each $k \in \widehat{\mathbb{N}}$,

$$\begin{aligned}
&\int_{T\times\widehat{\mathbb{N}}} \psi(t,n,\frac{d\nu_k}{d|\nu_k|}(t,n))\, d|\nu_k|(t,n) \\
&\quad = \int_T \int_{\widehat{\mathbb{N}}} \psi(t,n,\frac{dm_k}{d|m_k|}(t).1(n))\, d|\delta_k|(n)\, d|m_k|(t) \\
&\quad = \int_T \psi(t,k,\frac{dm_k}{d|m_k|}(t))\, d|m_k|(t) \\
&\quad = \int_T \phi_k(t,\frac{dm_k}{d|m_k|}(t),u_k(t))\, d|m_k|(t) .
\end{aligned}$$

Finally, by (3.3.1), we find

$$\liminf_k \int_T \phi_k(t,\frac{dm_k}{d|m_k|}(t),u_k(t))\, d|m_k|(t) \geq \int_T \phi_\infty(t,\frac{dm}{d|m|}(t),u(t))\, d|m|(t) .$$

And the proof is complete. □

Remarks 3.4 The set $\widehat{\mathbb{N}}$ has also been used by E. Balder ([2]), M. Valadier ([27]).

We can now state the complete result on epi-convergence of integral functionals defined on the space of vector measures with bounded variation.

Theorem 3.5 *Let $\{\phi_k : k \in \widehat{\mathbb{N}}\}$ be lower semi-continuous functions on $T\times B'$ with values in $[0,+\infty]$ such that for all $k \in \widehat{\mathbb{N}}$ and all $t \in T$, $\phi_k(t,\cdot)$ is convex and positively 1-homogeneous on B'. We assume that $\mathrm{li_e}\,\phi_k \geq \phi_\infty$ on $T \times B'$, and for all $t \in T$, $\phi_\infty(t,\cdot) \geq \mathrm{ls_e}\,\phi_k(t,\cdot)$ on B'. That is, for all $(t,x) \in T \times B'$, for every sequence $(t_k,x_k)_k$ in $T \times B'$ which converges to (t,x), we have*

$$\liminf_k \phi_k(t_k,x_k) \geq \phi_\infty(t,x) , \tag{3.5.1}$$

and there exists a sequence $(x_k)_k$ in B' (weakly) converging to x such that*

$$\phi_\infty(t,x) \geq \limsup_k \phi_k(t,x_k) . \tag{3.5.2}$$

Consider the functionals $\{I_{\phi_k} : k \in \widehat{\mathbb{N}}\}$ defined on $\mathcal{M}^b(T,E')$ by

$$I_{\phi_k}(m) = \int_T \phi_k(t,\frac{dm}{d|m|}(t))\, d|m|(t) .$$

Then for every bounded tight sequence $(m_k)_k$ in $\mathcal{M}^b(T, E')$ which converges weakly to $m \in \mathcal{M}^b(T, E')$, we have

$$\liminf_k I_{\phi_k}(m_k) \geq I_{\phi_\infty}(m) ,$$

and for each $m \in \mathcal{M}^b(T, E')$, there exists a bounded tight sequence $(m_k)_k$ in $\mathcal{M}^b(T, E')$ weakly converging to m such that

$$\limsup_k I_{\phi_k}(m_k) \leq I_{\phi_\infty}(m) .$$

That is, $(I_{\phi_k})_k$ epi-converges sequentially on the bounded tight subset of $\mathcal{M}^b(T, E')$.

Proof. By Theorem 3.1, we already know that for every bounded tight sequence $(m_k)_k$ in $\mathcal{H}$ which converges weakly to $m \in \mathcal{M}^b(T, E')$, we have

$$\liminf_k I_{\phi_k}(m_k) \geq I_{\phi_\infty}(m) .$$

Let us show the second inequality. We adapt a proof of A. Salvadori ([24, Proof of Theorem 3.1]): fix $m \in \mathcal{M}^b(T, E')$. We set $v = \frac{dm}{d|m|} \in L^1_{E'}(T, \mathcal{B}(T), d|m|)$. If $I_{\phi_\infty}(m) = +\infty$, there is nothing to prove. So we can assume that $\phi_\infty(\cdot, v(\cdot)) \in L^1_{\mathbb{R}}(T, \mathcal{B}(T), d|m|)$. Let $\psi_k: T \times B' \longrightarrow [0, +\infty]$ be the function defined by

$$\psi_k(t, x) = [\phi_k(t, x) - \phi_\infty(t, v(t))]^+ .$$

It is clear that ψ_k is measurable on $T \times B'$ and that $\psi_k(t, \cdot)$ is lower semi-continuous on the compact set B'. Consider now the multifunction $\Gamma_k: T \rightrightarrows B'$ defined by

$$\Gamma_k(t) = \{y \in B' : d(v(t), y) + \psi_k(t, y) = \min_{x \in B'}\{d(v(t), x) + \psi_k(t, x)\}\} ,$$

where d denotes a metric on B' (equipped with the weak* topology). Since B' is a compact set, it is clear that for all $t \in T$, $\Gamma_k(t)$ is non empty. Then, by [10, III.39], there exists a selection $v_k \in L^1_{E'}(T, \mathcal{B}(T), d|m|)$ of Γ_k. For every $t \in T$, by (3.5.2), there exists a sequence $(x_k)_k$ in B' (weakly*) converging to $v(t)$ such that $\lim_k \psi_k(t, x_k) = 0$. Since,

$$d(v(t), v_k(t)) + \psi_k(t, v_k(t)) \leq d(v(t), x_k) + \psi_k(t, x_k) ,$$

we obtain that $\lim_k d(v(t), v_k(t)) = 0$ and $\lim_k \psi_k(t, v_k(t)) = 0$. On the other hand, since $\phi_k(t, 0) = 0$, we have $\psi_k(t, 0) = 0$ and thus

$$d(v(t), v_k(t)) + \psi_k(t, v_k(t)) \leq d(v(t), 0) + \psi_k(t, 0) \leq M ,$$

where $M = \sup\{d(x,y) : (x,y) \in B' \times B'\}$. Since $|m|$ is bounded, by Lebesgue's dominated convergence theorem, we deduce that the sequence $(\psi_k(\cdot, v_k(\cdot)))_k$ converges to 0 in $L^1_{\mathbb{R}}(T, \mathcal{B}(T), d|m|)$. So we have

$$\lim_k \int_T [\phi_k(t, v_k(t)) - \phi_\infty(t, v(t))]^+ \, d|m|(t) = 0 \, ,$$

and hence

$$\limsup_k \int_T \phi_k(t, v_k(t)) \, d|m|(t) \leq \int_T \phi_\infty(t, v(t)) \, d|m|(t) \, . \tag{3.5.3}$$

We set now $m_k = v_k|m|$. It is clear that $(m_k)_k$ is a bounded tight sequence in $\mathcal{M}^b(T, E')$. We prove that $(m_k)_k$ converges weakly to m: let $f \in \mathcal{C}^b(T, E)$. Since $(v_k(t))_k$ converges to $v(t)$ in B', we have $\lim_k \langle f(t), v_k(t) \rangle = \langle f(t), v(t) \rangle$. Moreover, since $v(t) \in B'$, it follows that $|\langle f(t), v(t) \rangle| \leq \|f(t)\|$. Since $|m|$ is bounded, $f \in L^1_E(T, \mathcal{B}(T), d|m|)$ and by Lebesgue's dominated convergence theorem, we deduce that

$$\lim_k \int_T \langle f(t), v_k(t) \rangle \, d|m|(t) = \int_T \langle f(t), v(t) \rangle \, d|m|(t) \, .$$

and hence $\lim_k \langle f, m_k \rangle = \langle f, m \rangle$. That proves that $(m_k)_k$ converges weakly to m in $\mathcal{M}^b(T, E')$. Moreover, we obviously have

$$\frac{dm_k}{d|m_k|}(t) = \frac{d(v_k|m|)}{d(\|v_k\| \, |m|)}(t) = \begin{cases} \frac{v_k(t)}{\|v_k(t)\|}, & \text{if } \|v_k(t)\| \neq 0, \\ 0, & \text{if } \|v_k(t)\| = 0. \end{cases}$$

Then we obtain

$$\begin{aligned} \int_T \phi_k(t, v_k(t)) \, d|m|(t) &= \int_T \phi_k(t, \frac{v_k(t)}{\|v_k(t)\|}) \|v_k\| \, d|m|(t) \\ &= \int_T \phi_k(t, \frac{dm_k}{d|m_k|}(t)) \, d|m_k|(t) \, . \end{aligned}$$

Finally, by (3.5.3), we find

$$\limsup_k \int_T \phi_k(t, \frac{dm_k}{d|m_k|}(t)) \, d|m_k|(t) \leq \int_T \phi_\infty(t, \frac{dm}{d|m|}(t)) \, d|m|(t) \, .$$

That is $\limsup_k I_{\phi_k}(m_k) \leq I_{\phi_\infty}(m)$ and the proof is complete. □

4. Lower semi-continuity of integral functionals on the space of measures

In this section, we give several lower semi-continuity results for integral functionals defined on the space of vector measures with bounded variation.

The following result is the generalization to Banach valued measures defined on a Polish space of a well-known result of Y. Reshetnyak ([20, Theorem 2]).

Theorem 4.1 *Let $\phi : T \times B' \longrightarrow [0, +\infty]$ be a lower semi-continuous function on $T \times B'$ such that for all $t \in T$, $\phi(t, \cdot)$ is convex and positively 1-homogeneous on B'. Let $(m_k)_k$ be a bounded tight sequence in $\mathcal{M}^b(T, E')$ which converges weakly to $m \in \mathcal{M}^b(T, E')$. Then we have*

$$\liminf_k \int_T \phi(t, \frac{dm_k}{d|m_k|}(t))\, d|m_k|(t) \geq \int_T \phi(t, \frac{dm}{d|m|}(t))\, d|m|(t) .$$

Proof. This proof follows some steps developed by Y. Reshetnyak ([20], [8]) in the case when T is locally compact, with necessary modifications.

Extracting a subsequence, we can suppose that

$$a = \liminf_k \int_T \phi(t, \frac{dm_k}{d|m_k|}(t))\, d|m_k|(t) = \lim_k \int_T \phi(t, \frac{dm_k}{d|m_k|}(t))\, d|m_k|(t) .$$

For each $k \in \mathbb{N}$, we consider the measure $\nu_k \in \mathcal{M}^b_+(T \times B', \mathbb{R})$, image of $|m_k|$ by the map $t \longmapsto (t, \frac{dm_k}{d|m_k|}(t)), T \longrightarrow T \times B'$. We have ([11, III(73)]),

$$\nu_k = \int_T \delta_t \otimes \delta_{\left(\frac{dm_k}{d|m_k|}(t)\right)}\, d|m_k|(t) ,$$

where δ_x denotes the Dirac measure at x. Since $(m_k)_k$ is bounded and since $\|\nu_k\| = \int d|m_k| = \|m_k\|$, the sequence $(\nu_k)_k$ is bounded too. Moreover, since the sequence $(m_k)_k$ is tight, for every $\varepsilon > 0$, there exists a compact subset K of T such that $\sup_k |m_k|(T \setminus K) \leq \varepsilon$. It follows that

$$\nu_k([T \times B'] \setminus [K \times B']) = \nu_k([T \setminus K] \times B') = |m_k|(T \setminus K) .$$

And so $\sup_k \nu_k([T \times B'] \setminus [K \times B']) \leq \varepsilon$. Since $K \times B'$ is compact, it proves that $(\nu_k)_k$ is tight. Thus, by Prokhorov's theorem (Theorem 2.1), it is relatively weakly compact in the space $\mathcal{M}^b_+(T \times B', \mathbb{R})$ of non-negative bounded Radon measures on $T \times B'$; observe that $\mathcal{M}^b_+(T \times B', \mathbb{R})$ equipped with the weak topology is a Polish space – see [5, Proposition 5.10]. Therefore, there exist a subsequence $(\nu_{k_p})_p$ of $(\nu_{n_k})_k$ and a measure $\nu \in \mathcal{M}^b_+(T \times B', \mathbb{R})$ such that $(\nu_{k_p})_p$ converges weakly to ν. Since ϕ is non-negative and lower semi-continuous on $T \times B'$, the map $\tau \longmapsto \int \phi\, d\tau$ defined on $\mathcal{M}^b_+(T \times B', \mathbb{R})$ is weakly lower semi-continuous ([11, Théorème III.55]). Then we obtain

$$\begin{aligned} a &= \lim_k \int_T \phi(t, \frac{dm_k}{d|m_k|}(t))\, d|m_k|(t) \\ &= \lim_p \int_{T \times B'} \phi(t, x)\, d\nu_{k_p}(t, x) \\ &\geq \int_{T \times B'} \phi(t, x)\, d\nu(t, x) . \end{aligned} \tag{4.1.1}$$

Let μ be the projection of ν onto T defined by

$$\mu = \int_{T \times B'} \delta_t\, d\nu(t, x) .$$

Thanks to a result on disintegration of measures (see [25, Théorème 2], [7], [23], [10], [15]), there is a μ-measurable function $\lambda: t \longmapsto \lambda_t$ from T into the space $\mathcal{M}_+^1(B')$ of Radon probabilities on B' equipped with the weak topology, such that

$$\nu = \int_T \delta_t \otimes \lambda_t \, d\mu(t) .$$

Coming back to (4.1.1), we find

$$a \geq \int_{T\times B'} \phi(t,x)\, d\nu(t,x) = \int_T \int_{B'} \phi(t,x)\, d\lambda_t(x)\, d\mu(t) . \tag{4.1.2}$$

Let now $b(\lambda_t) \in B'$ be the barycenter of λ_t. Since $b(\lambda): T \longrightarrow B', t \longmapsto b(\lambda_t)$ is measurable and for every $t \in T$, the function $\phi(t,\cdot)$ is convex and lower semi-continuous on B', we have

$$\int_{B'} \phi(t,x)\, d\lambda_t(x) \geq \phi(t, b(\lambda_t)), \qquad \forall t \in T .$$

It follows from (4.1.2) that

$$a \geq \int_T \phi(t, b(\lambda_t))\, d\mu(t) . \tag{4.1.3}$$

Let us show now that $m = b(\lambda)\mu$. Let $f \in \mathcal{C}^b(T,E)$. For each $k \in \mathbb{N}$, we have

$$\int_T f(t)\, dm_k(t) = \int_T \langle f(t), \frac{dm_k}{d|m_k|}(t)\rangle\, d|m_k|(t) = \int_{T\times B'} \langle f(t), x\rangle\, d\nu_k(t,x) .$$

On the other hand, since $(m_{k_p})_p$ converges weakly to m, we have

$$\int_T f(t)\, dm(t) = \lim_p \int_T f(t)\, dm_{k_p}(t) .$$

Moreover, since $\langle f(\cdot), \cdot\rangle \in \mathcal{C}^b(T \times B', \mathbb{R})$ and the sequence $(\nu_{k_p})_p$ converges weakly to ν, we have

$$\int_{T\times B'} \langle f(t), x\rangle\, d\nu(t,x) = \lim_p \int_{T\times B'} \langle f(t), x\rangle\, d\nu_{k_p}(t,x) .$$

Then, we put together the three previous equalities to obtain

$$\begin{aligned} \int_T f(t)\, dm(t) &= \int_{T\times B'} \langle f(t), x\rangle\, d\nu(t,x) = \int_T \int_{B'} \langle f(t), x\rangle\, d\lambda_t(x)\, d\mu(t) \\ &= \int_T \langle f(t), \int_{B'} x\, d\lambda_t(x)\rangle\, d\mu(t) = \int_T \langle f(t), b(\lambda_t)\rangle\, d\mu(t) \\ &= \int_T f(t)\, d(b(\lambda)\mu)(t) . \end{aligned}$$

This proves that $m = b(\lambda)\mu$ and thus $\frac{dm}{d\mu} = b(\lambda)$. We deduce now from (4.1.3) that

$$a \geq \int_T \phi(t, b(\lambda_t))\, d\mu(t) = \int_T \phi(t, \frac{dm}{d\mu}(t))\, d\mu(t) . \tag{4.1.4}$$

Let $|b(\lambda)|: T \longrightarrow [0,1], t \longmapsto \|b(\lambda_t)\|$. Since μ is a non-negative measure on T, we have $d|m| = |b(\lambda)|\, d\mu$ and so

$$dm = \frac{dm}{d|m|}\frac{d|m|}{d\mu}\, d\mu = \frac{dm}{d|m|}|b(\lambda)|\, d\mu\,.$$

We obtain for almost every $t \in T$,

$$\phi(t, \frac{dm}{d\mu}(t)) = \phi(t, \|b(\lambda_t)\|\frac{dm}{d|m|}(t)) = \|b(\lambda_t)\|\phi(t, \frac{dm}{d|m|}(t))\,.$$

Finally, with (4.1.4), we find

$$\begin{aligned} a &\geq \int_T \phi(t, \frac{dm}{d\mu}(t))\, d\mu(t) = \int_T \phi(t, \frac{dm}{d|m|}(t))\|b(\lambda_t)\|\, d\mu(t) \\ &= \int_T \phi(t, \frac{dm}{d|m|}(t))\, d(|b(\lambda)|\mu)(t) = \int_T \phi(t, \frac{dm}{d|m|}(t))\, d|m|(t)\,. \end{aligned}$$

This proves the result. □

Remarks 4.2 1. When T is a (non necessary metrizable) locally compact topological space, the first author proves in [8] the same result without tightness assumptions.

2. One can find a very different proof of Reshetnyak's theorem (with $E = \mathbb{R}^n$) in Buttazzo's book ([6, Theorem 3.4.3]).

We give now a result concerning integral functionals defined on the space of vector measures with bounded variation associated to *right* lower semicontinuous multifunctions. This result is inspired by the previous ones of C. Castaing ([9]), R. Rockafellar ([22]) and M. Valadier ([26]).

Let $T > 0$. We consider on $[0,T]$ the *natural* topology τ and the *right* (or *half-open interval*) topology τ_d. (See Kelley [17]). Obviously $\tau \subset \tau_d$. Moreover, it is well known ([17, p. 172]) that $([0,T], \tau_d)$ is a paracompact topological space.

Let $\mathrm{BRC}([0,T], E)$ be the space of all bounded right continuous (that is continuous on $[0,T]$ with respect to the right topology τ_d) functions from $[0,T]$ to E. Note that $\mathrm{BRC}([0,T], E)$ is a subspace of $\mathcal{L}^\infty_E([0,T])$. So the integral of $u \in \mathrm{BRC}([0,T], E)$ with respect to $m \in \mathcal{M}^b([0,T], E')$ is defined by

$$\int u\, dm = \int_{[0,T]} \langle u(t), \frac{dm}{d|m|}(t)\rangle\, d|m|(t)\,.$$

We equip the space $\mathrm{BRC}([0,T], E)$ with the weak topology induced by the separated duality with $\mathcal{M}^b([0,T], E')$.

A multifunction $\Gamma\colon [0,T] \rightrightarrows E$ with non-empty weakly compact convex values is said *bounded* if $\cup_{t\in[0,T]}\Gamma(t)$ is bounded. It is said to be *lower semi-continuous* at $t_0 \in [0,T]$ if for every open subset V of E such that $\Gamma(t_0) \cap V \neq \emptyset$, there exists a neighborhood U of t_0 in $[0,T]$ such that for each $t \in U$, $\Gamma(t) \cap V \neq \emptyset$.

We denote by $\mathcal{S}_\Gamma^{\mathrm{BRC}}$ the set of all right continuous selections of Γ and by $\mathcal{S}_\Gamma^\infty(|m|)$ the set of all selections of Γ in $\mathcal{L}_E^\infty([0,T],|m|)$.

The *normal cone* to $\mathcal{S}_\Gamma^{\mathrm{BRC}}$ at $u \in \mathrm{BRC}([0,T],E)$ with respect to the duality with $\mathcal{M}^b([0,T],E')$ is defined by

$$N_{\mathcal{S}_\Gamma^{\mathrm{BRC}}}(u) := \left\{m \in \mathcal{M}^b([0,T],E') : \langle u,m\rangle = \delta^*(m,\mathcal{S}_\Gamma^{\mathrm{BRC}})\right\},$$

where $\delta^*(\cdot,\mathcal{S}_\Gamma^{\mathrm{BRC}})$ denotes the support function of the set $\mathcal{S}_\Gamma^{\mathrm{BRC}}$ with respect to the duality with $\mathcal{M}^b([0,T],E')$.

Theorem 4.3 *Let $\Gamma\colon [0,T] \rightrightarrows E$ be a bounded right lower semi-continuous multifunction on $[0,T]$ with non-empty weakly compact convex values.*

Then $\mathcal{S}_\Gamma^{\mathrm{BRC}}$ is non-empty convex and weakly closed in $\mathrm{BRC}([0,T],E)$ *and the support function of $\mathcal{S}_\Gamma^{\mathrm{BRC}}$ is given by*

$$\delta^*(m,\mathcal{S}_\Gamma^{\mathrm{BRC}}) = \int_{[0,T]} \delta^*\!\left(\frac{dm}{d|m|}(t),\Gamma(t)\right) d|m|(t)$$

for all $m \in \mathcal{M}^b([0,T],E')$.

Moreover, the normal cone to $\mathcal{S}_\Gamma^{\mathrm{BRC}}$ at $u \in \mathrm{BRC}([0,T],E)$ is given by

$$N_{\mathcal{S}_\Gamma^{\mathrm{BRC}}}(u) = \left\{m \in \mathcal{M}^b([0,T],E') : \frac{dm}{d|m|}(t) \in N_{\Gamma(t)}(u(t)) \;\; |m|\text{-a.e.}\right\}$$

where $N_{\Gamma(t)}(u(t))$ is the normal cone to $\Gamma(t)$ at the point $u(t)$.

Proof. The non-emptiness of $\mathcal{S}_\Gamma^{\mathrm{BRC}}$ follows from Michael's continuous selection theorem ([18]): since $([0,T],\tau_d)$ is a paracompact topological space and $\Gamma\colon [0,T]_{\tau_d} \rightrightarrows E$ is lower semi-continuous and bounded, there exists a bounded τ_d-continuous selection u of Γ.

The convexity of $\Gamma(t)$ implies the one of $\mathcal{S}_\Gamma^{\mathrm{BRC}}$. Let us show that $\mathcal{S}_\Gamma^{\mathrm{BRC}}$ is closed for the weak topology:

Let $(u_\alpha)_\alpha$ be a net in $\mathcal{S}_\Gamma^{\mathrm{BRC}}$ which converges to an $u \in \mathrm{BRC}([0,T],E)$. For each $x' \in E'$ and $t \in [0,T]$, $x'\delta_t$ is in $\mathcal{M}^b([0,T],E')$. Therefore, we have

$$\langle u_\alpha(t),x'\rangle = \langle u_\alpha,x'\delta_t\rangle \xrightarrow{\alpha} \langle u,x'\delta_t\rangle = \langle u(t),x'\rangle .$$

That means that $(u_\alpha(t))_\alpha$ converges weakly to $u(t)$. Since $\Gamma(t)$ is closed, $u(t) \in \Gamma(t)$. Thus $u \in \mathcal{S}_\Gamma^{\mathrm{BRC}}$.

Let $\psi\colon [0,T]_{\tau_d} \times E \longrightarrow [0,+\infty]$ and $\phi\colon [0,T]_{\tau_d} \times E' \longrightarrow]-\infty,+\infty]$ be the two lower semi-continuous functions defined by

$$\psi(t,x) = \delta\,(x,\Gamma(t))\,, \qquad \forall (t,x) \in [0,T] \times E\,,$$

that is the indicator function of $\Gamma(t)$ at x and

$$\phi(t,x') = \delta^*\,(x',\Gamma(t))\,, \qquad \forall (t,x') \in [0,T] \times E'\,,$$

that is the support function of $\Gamma(t)$ at x'.

We fix any $\overline{m} \in \mathcal{M}^b([0,T],E')$. Let $I_\psi\colon L^\infty_E([0,T],|\overline{m}|) \longrightarrow]-\infty,+\infty]$ and $I_\phi\colon L^1_{E'}([0,T],|\overline{m}|) \longrightarrow]-\infty,+\infty]$ be the integral functionals associated to ψ and ϕ.

Thanks to a standard duality result ([10, Corollary VII.15]), we have, for each $v \in L^1_{E'}([0,T],|\overline{m}|)$,

$$I_\phi(v) = (I_\psi)^*\,(v) = \sup\{\langle v,u\rangle - I_\psi(u) : u \in L^\infty_E([0,T],|\overline{m}|)\}\,.$$

Since $\frac{d\overline{m}}{d|\overline{m}|} \in L^1_{E'}([0,T],|\overline{m}|)$, we obtain

$$\begin{aligned}
&\int_{[0,T]} \delta^*(\frac{d\overline{m}}{d|\overline{m}|}(t),\Gamma(t))\,d|\overline{m}|(t)\\
&\quad= \sup\left\{\int_{[0,T]}\langle u,\frac{d\overline{m}}{d|\overline{m}|}(t)\rangle\,d|\overline{m}|(t) : u \in \mathcal{S}^\infty_\Gamma(|\overline{m}|)\right\}\\
&\quad\geq \sup\left\{\int_{[0,T]}\langle u,\frac{d\overline{m}}{d|\overline{m}|}(t)\rangle\,d|\overline{m}|(t) : u \in \mathcal{S}^{\mathrm{BRC}}_\Gamma\right\}\\
&\quad= \delta^*(m,\mathcal{S}^{\mathrm{BRC}}_\Gamma)\,.
\end{aligned}$$

We have now to prove the reverse inequality. For that, let $\overline{u} \in \mathcal{S}^\infty_\Gamma(|\overline{m}|)$ and $\alpha \in \mathbb{R}$ such that

$$\alpha < \int_{[0,T]}\langle \overline{u}(t),\frac{d\overline{m}}{d|\overline{m}|}(t)\rangle\,d|\overline{m}|(t)\,. \tag{4.3.1}$$

We only have to show that there exists an $u \in \mathcal{S}^{\mathrm{BRC}}_\Gamma$ such that

$$\alpha < \int_{[0,T]}\langle u(t),\frac{d\overline{m}}{d|\overline{m}|}(t)\rangle\,d|\overline{m}|(t)\,.$$

Let $\beta > 0$ large enough such that $\|\overline{u}\|_\infty < \beta$ and let $\varepsilon > 0$ such that

$$\alpha < \int_{[0,T]}\langle \overline{u}(t),\frac{d\overline{m}}{d|\overline{m}|}(t)\rangle\,d|\overline{m}|(t) - 2\beta\varepsilon\,. \tag{4.3.2}$$

By Lusin's Theorem, there exists a τ-compact (and so τ_d-closed) set K such that $|\overline{m}|([0,T]\setminus K) < \varepsilon$ and such that $\overline{u}$ is τ-continuous (and therefore τ_d-continuous) on K. We consider the multifunction $\Lambda\colon [0,T] \rightrightarrows E$ defined for each $t \in [0,T]$ by

$$\Lambda(t) = \begin{cases} \overline{u}(t), & \text{if } t \in K,\\ \Gamma(t)\cap B(0,\beta), & \text{otherwise.}\end{cases}$$

It is clear that Λ has non-empty convex values. Let us show that $\operatorname{cl}\Lambda$ is τ_d-lower semi-continuous. (Here, $\operatorname{cl}\Lambda(t) = \operatorname{cl}(\Lambda(t))$.) Let U be an open subset of E. We have

$$\begin{aligned} \operatorname{cl}\Lambda^-(U) &= \{t \in [0,T] : \operatorname{cl}\Lambda(t) \cap U \neq \emptyset\} \\ &= \{t \in [0,T] : \Lambda(t) \cap U \neq \emptyset\} \\ &= \{t \in [0,T] : \Gamma(t) \cap (U \cap B(0,\beta)) \neq \emptyset\} \setminus \{t \in K : \overline{u}(t) \notin U\} \,. \end{aligned}$$

Since U is open, thanks to the τ_d-lower semi-continuity of Γ and the τ_d-continuity of $\overline{u}$, the set $\operatorname{cl}\Lambda^-(U)$ is open in $[0,T]_{\tau_d}$. That proves the τ_d-lower semi-continuity of $\operatorname{cl}\Lambda$.

Thus, by Michael's theorem ([18]), there exists a τ_d-continuous selection u of $\operatorname{cl}\Lambda$. Since u is obviously bounded, $u \in \mathcal{S}_\Gamma^{\mathrm{BRC}}$. Since $u = \overline{u}$ on K, using (4.3.1) and (4.3.2), we get

$$\begin{aligned} \int_{[0,T]} \langle u(t), \frac{d\overline{m}}{d|\overline{m}|}(t)\rangle \, d|\overline{m}|(t) &= \int_{[0,T]} \langle \overline{u}(t), \frac{d\overline{m}}{d|\overline{m}|}(t)\rangle \, d|\overline{m}|(t) \\ &\quad + \int_{[0,T]\setminus K} \langle u(t) - \overline{u}(t), \frac{d\overline{m}}{d|\overline{m}|}(t)\rangle \, d|\overline{m}|(t) \\ &\geq \int_{[0,T]} \langle \overline{u}(t), \frac{d\overline{m}}{d|\overline{m}|}(t)\rangle \, d|\overline{m}|(t) - 2\beta\, |\overline{m}|([0,T] \setminus K) \\ &> \alpha \,. \end{aligned}$$

That finishes to prove the first part of the theorem. Let $u \in \mathcal{S}_\Gamma^{\mathrm{BRC}}$. By the previous result, we find

$$\begin{aligned} & N_{\mathcal{S}_\Gamma^{\mathrm{BRC}}}(u) \\ &= \left\{ m \in \mathcal{M}^b([0,T],E') : \langle u, m\rangle = \delta^*(m, \mathcal{S}_\Gamma^{\mathrm{BRC}}) \right\} \\ &= \Big\{ m \in \mathcal{M}^b([0,T],E') : \\ &\qquad \int_{[0,T]} \left[\delta^*(\frac{dm}{d|m|}(t), \Gamma(t)) - \langle u(t), \frac{dm}{d|m|}(t)\rangle \right] d|m|(t) = 0 \Big\} \\ &= \left\{ m \in \mathcal{M}^b([0,T],E') : \delta^*(\frac{dm}{d|m|}(t), \Gamma(t)) - \langle u(t), \frac{dm}{d|m|}(t)\rangle = 0 \ |m|\text{-a.e.} \right\} \\ &= \left\{ m \in \mathcal{M}^b([0,T],E') : \frac{dm}{d|m|}(t) \in \partial\,[\delta(\cdot, \Gamma(t))]\,(u(t)) \ |m|\text{-a.e.} \right\} , \end{aligned}$$

where $\partial\,[\delta(\cdot,\Gamma(t))]\,(u(t))$ denotes the subdifferential of $\delta(\cdot,\Gamma(t))$ at $u(t)$. Since this subdifferential is actually the normal cone to $\Gamma(t)$ at $u(t)$ ([21]), that proves the second part of the theorem. □

References

[1] H. Attouch, *Variational Convergence for Functions and Operators*, Pitman, London,1984.

[2] E.J. Balder, *An extension of Prohorov's theorem for transition probabilities with applications to infinite-dimensional lower closure problems*, Rend. Circ. Mat. Palermo (2) **XXXIV** (1985), 427–447.

[3] G. Bouchitté & M. Valadier, *Multifonctions s.c.i. et régularisée s.c.i. essentielle*, Congrés franco-québécois d'analyse non-linéaire appliquée, Perpignan, 22-26 juin 1987, Ann. Inst. H. Poincaré. Anal. Non Linéaire **6-suppl.** (1989), 123–149.

[4] N. Bourbaki, *Eléments de Mathématique. Intégration*, Chapitres I-IV, Second edition, Hermann, Paris, 1965.

[5] N. Bourbaki, *Eléments de Mathématique. Intégration*, Chapitre IX, Hermann, Paris, 1969.

[6] G. Buttazzo, *Semicontinuity, Relaxation and Integral Representation in the Calculus of Variation*, Longman Scientific & Technical, Harlow, 1989.

[7] C. Castaing, *Application d'un théorème de compacité à la désintégration des mesures*, C.R. Acad. Sci. Paris. Ser. A, **273** (1971), 1056–1059.

[8] C. Castaing, *Validité du théorème de Reshetnyak dans les espaces hilbertiens*, Sém. Anal. Convexe **17** (1987), 8.1–8.9.

[9] C. Castaing, *Quelques résultats de convergence dans les inclusions différentielles*, Sém. Anal. Convexe **17** (1987), 12.1–12.37.

[10] C. Castaing & M. Valadier, *Convex Analysis and Measurable Multifunctions*, Lecture Notes in Math., vol. 580, Springer-Verlag, Berlin, 1977.

[11] C. Dellacherie & P.-A. Meyer, *Probabilities and Potential*, North-Holland, Amsterdam, 1978.

[12] J. Diestel & J.J. Uhl, *Vector Measures*, Mathematical Survey, vol. 15, Amer. Math. Soc., Providence, Rhode Island, 1977.

[13] N. Dinculeanu, *Vector Measures*, Pergamon Press, Oxford, 1967.

[14] C. Goffman & J. Serrin, *Sublinear functions of measures and variational integrals*, Duke Math. J. **31** (1964), 159–178.

[15] C. Ionescu Tulcea, *Two theorems concerning the disintegration of measures*, J. Math. Anal. Appl. **26** (1969), 376–380.

[16] G.W. Johnson, *The dual of* $C(S,F)$ Math. Ann. **187** (1970), 1–8.

[17] J.L. Kelley, *General Topology*, Graduate Texts in Math., vol. 27, Springer-Verlag, Berlin, 1955.

[18] E. Michael, *Continuous selections*, I, Ann. of Math **63** (1956), 361–382.

[19] M.-F. Nouguès (Sainte-Beuve), *Some topological properties of vector measures with bounded variation and its applications*, Ann. Mat. Pura Appl. (4) **CXVI** (1978), 317-379.

[20] Y.G. Reshetnyak, *Weak convergence of completely additive vector functions on a set*, Siberian Math. J. **9** (1968), 1039–1045.

[21] R.T. Rockafellar, *Convex Analysis*, Princeton University Press, New Jersey, 1970.

[22] R.T. Rockafellar, *Integrals which are convex functionals.* II, Pacific J. Math. **39** (1971), 439–469.

[23] J. Saint-Pierre, *Désintégration d'une mesure non bornée*, Ann. Inst. H. Poincaré Probab. Statist. **11** (1975), 275–286.

[24] A. Salvadori, *On the M-convergence for integral functionals on* L^p_X, Sém. Anal. Convexe **15** (1985), 5.1–5.25.

[25] M. Valadier, *Désintégration d'une mesure sur un produit*, C.R. Acad. Sci. Paris Ser. A **276** (1973), 33–35.

[26] M. Valadier, *Une propriété de l'ensemble des sélections à variation bornée d'une multi-application à rétraction bornée*, Sém. Anal. Convexe **7** (1977), 13.1–13.7.

[27] M. Valadier, *Young measures*, Methods of non convex analysis (A. Cellina, ed.), Lecture Notes in Math., vol. 1446, Springer-Verlag, Berlin, 1990, pp. 152–188.

Multiple–Phase Economic Dynamics

Richard H. Day*
Department of Economics
University of Southern California
Unversity Park, Los Angeles
CA 90089-253, U.S.A.

We observe in economics, as in other fields, that quite different forces or relationships govern behavior in differing situations of state. Multiple–phase dynamic models formalize this fact.

In 1951 Georgescu–Roegen attributed to Le Corbeiller the idea that the theory of relaxation oscillations discussed earlier by van der Pole could give multiple–regime explanations of business fluctuations. In the same year Goodwin published his famous "Nonlinear Acceleration and the Persistence of Business Cycles," telling us of his inspiration from the same source. Georgescu's contribution was an outgrowth of Leontief's dynamic multisector model which was also based on the acceleration principle and which, likewise, led to the problem of multiple–regimes. This literature, together with Simon's (1951) insightful analysis of "trigger effects" in mathematical programming, was the inspiration for my own use of multiple phase dynamics to model structural change in microeconomic processes (Day (1963) and Day and Cigno (1978)) and in macroeconomic growth (Day and Walter (1989)). Haavelmo (1958) had already emphasized the inevitable intrusion of multiple–regimes in the study of growth, and subsequently Malinvaud (1980) took the same approach in his macroeconomic reconsideration of unemployment. Despite these seminal contributions and despite the likelihood that many apparent anomalies can be explained in these terms, multiple–phase dynamics is conspicuous by its under–representation in the theoretical literature.

My purpose in this lecture is to review and illustrate the basic concepts of discrete time multiple–phase dynamics and to indicate why their study is fundamental to the development of dynamic economic theory. Multiple regime systems are defined first and their multiple–phase character described. The construction of alternative development scenarios is outlined and the basic concepts illustrated with a family of "mountainous" maps. I then develop an

*Dedicated to Nicholas Georgescu-Roegen

"abstract adaptive economy" as a way of showing why an economy of rational agents should exhibit multiple phase dynamics.

1. Multiple–Phase Dynamics

1.1 Multiple–Phase Dynamical Systems

For simplicity, I shall have in mind discrete time dynamics in a finite dimensional vector space $\mathbf{X}$. Consider a family of single–valued mappings $\theta_p : \mathbf{X} \to \mathbf{X}$, $p \in \mathcal{N}$ where $\mathcal{N}$ is a set of indexes. Each map $\theta_p(\cdot)$ is called a *phase structure* and is defined on a set $D^p \subset \mathbf{X}$ called the pth *phase domain.* A *regime* is a pair $\mathcal{R}_p := (\theta_p, D^p)$. The dynamics within a given phase domain is given by the *phase equation*

$$x_{t+1} = \theta_p(x_t), \; x_t \in D^p \tag{1}$$

where it is assumed that $D^p \cap D^q = \emptyset$ for all $p \neq q \in \mathcal{N}$. Defining the map

$$\theta(x) := \theta_p(x), \; x \in D^p \tag{2}$$

with domain $D := \cup_p D^p$, we have the usual dynamical system (θ, D) with dynamics

$$x_{t+1} = \theta(x_t), \; x_t \in D.$$

Consequently, trajectories and solutions have the usual meanings. In particular, for initial conditions $x \in D$ a trajectory is given by the iterated map $\theta^t : D \to \mathbf{X}$. That is, $x_t = \theta^t(x)$.

Let $\chi_S(x)$ be the indicator function. Another way to write (2) is

$$x_{t+1} = \theta(x_t) = \sum_{p \in \mathcal{N}} \chi_{D^p}(x_t) \cdot \theta_p(x_t). \tag{3}$$

This equation is the mathematical analog of Haavelmo's admonition that, "we must face the fact that the form of the equations of a realistic model may have to be regarded as a function of the variables involved."

Suppose that the domain of a given dynamical system can be partitioned into nonintersecting subsets D^p, $p \in \mathcal{N}$ within each of which behavior is distinct, for example, monotonic growth, monotonic decline, cycles, or erratic behavior. Then the restriction of $\theta(\cdot)$ to D^p can be denoted θ_p and the pair (θ_p, D^p) can be considered a regime.

1.2 The Null Regime

The *null domain* $D^0 := \backslash D$ is the set for which no structure generating behavior has been defined. Define the *null phase structure* θ_0 that has the property that $\theta_0(x) \in D^0$ for all $x \in D^0$. That is, once a trajectory enters D^0 it cannot escape. In most economic applications the null domain is naturally thought of as the set of nonpositive states. One can define the null phase structure by

$$\theta_0(x) \equiv 0 \quad \text{for all} \quad x \in D^0. \tag{4}$$

Assume that the set of phase indexes is $\mathcal{N} := \{0, 1, \ldots, n\}$. From now on we will consider every multiple–phase dynamical system to be extended to the entire space $\mathbf{X}$ in this way.

1.3 Escape and Regime Switching

Suppose that a given trajectory enters a given regime with a value $y \in D^p$ and that there is a least integer n such that $\theta_p^n(y) \notin D^p$, then the trajectory from y is said to *escape* regime p after n periods. The map $\theta_p^n(\cdot)$ is the nth iterated map of the map $\theta_p(\cdot)$. If the trajectory escapes regime p, it must *switch* to another regime, or it must enter the null domain, in which case it is said to *self–destruct*.

1.4 Epochal Evolution

Evidently, a trajectory of a system can be characterized by the sequence of regimes through which it passes. Define the regime index of a given state by

$$I(x) := \sum_p p \cdot \chi_{D^p}(x). \tag{5}$$

Then the sequence

$$I\left[\theta^t(x)\right], \; t = 0, 1, 2, 3, \ldots \tag{6}$$

gives the dynamics of the system as a sequence of regimes. A given history can be decomposed into a denumerable sequence of *episodes*, each one of which represents a *sojourn* within a given regime. Let

$$\begin{array}{rclcl} 0 = s_1(x) \le & t & < s_2(x) & , \; I(\theta^t(x)) & = \; p_1, \\ s_2(x) \le & t & < s_3(x) & , \; I(\theta^t(x)) & = \; p_2, \\ s_3(x) \le & t & < s_4(x) & , \; I(\theta^t(x)) & = \; p_3, \\ \vdots & \vdots & \vdots & \vdots & \vdots \end{array} \tag{7}$$

The triple

$$(\theta_{p_i}(\cdot),\ D^{p_i}, T_i),\ i = 1, 2, \ldots \tag{8}$$

is called the ith *episode*; the period $s_i(x)$ is called the ith *entry time*; the state $\theta^{s_i(x)}(x)$ is called the ith *entry state*; the sequence of integers $T_i :=$ $\{s_i(x),\ s_i(x)+1, \ldots,\ s_{i+1}(x)-1\}$ is the ith *sojourn*, and the number $s_{i+1}(x) -$ $s_i(x)$ the *duration* of the ith episode. The sequence of episodes (8) associated with a given trajectory is an *epochal evolution*.

Suppose an epochal evolution has n episodes, $n < \infty$. Then $s_{n+1}(x) = \infty$ and the trajectory is trapped in the phase domain D^{p_n}. This does not mean that the trajectory converges to a stationary, steady, or periodic state, however, but only that the phase structure governing change does not switch once the p_n^{th} phase domain is entered. If the p_n^{th} phase zone is the null domain, the system self–destructs. If the sequence of episodes in an epochal evolution is infinite, then the system does not self–destruct but switches regimes indefinitely.

If for every p_i (except the last if the sequence is finite) there is a $j > i$ such that $p_j > p_i$, the history of episodes forms a *progression*. If for some $k > 0$, $p_{i+k} = p_i, i = 1, 2, 3, \ldots$ the evolution is *phase cyclic*. Note that a phase cycle does not imply that trajectories are periodic.

1.5 The Inverse Map

The inverse map of θ is defined by $\theta^{-1}(y) := \{x \in \mathbf{X} | \theta(x) = y\}$. The inverse of θ_p is defined analogously. Note that in general the inverse of a map is set valued. Let $S \subset \mathbf{X}$. Then

$$\theta^{-1}(S) := \cup_{y \in S}\{x \in \mathbf{X} | \theta(x) = y\},\ S \subset \mathbf{X}.$$

Analogously,

$$\theta_p^{-1}(S) := \cup_{y \in S}\{x \in D^p | \theta(x) = y\},\ S \subset D^p.$$

The inverse maps $\theta_p^{-n}(S)$ are defined recursively. Thus,

$$\theta_p^{-n}(S) := \theta_p^{-1}\left(\theta^{-n+1}(S)\right),\ n = 1, \ldots,\ S \subset D^p.$$

Note that $\theta_p^{-n}(S) \subset \theta^{-n}(S)$. In particular it can happen that $\theta_p^{-n}(D^p) = \emptyset$ while $\theta^{-n}(D^p) \neq \emptyset$. Note also that every map in $\{\theta^i, i \in \mathbb{N}\}$ has an inverse so that the set of positive and negative iterates together with the power set $\mathcal{P}(D)$ constitute a dynamical system.

Let $y \in \theta^{-n}(x)$. Then $\theta^n(y) = x$. The implication is that

- in a discrete time dynamical system trajectories *from* a given initial condition are unique, but different trajectories could pass *through* a given state (*from* which point they would be identical).

2. Stable and Unstable Regimes

The qualitative analysis of a multiple–phase dynamical system is carried out via an analysis of the dynamics of its constituent regimes. For this purpose the ideas of stability and instability need to be defined appropriately.

2.1 Stable Regimes

If any trajectory that enters a given regime, say the $\mathcal{R}_p$, remains there, then the regime is called *stable*. Formally, a regime $\mathcal{R}^p$ is stable if and only if for any entry state $y \in D^p$, $\theta^s(y) \in D^p, s = 1,2,3,\ldots.$
Or, equivalently,

- A regime $\mathcal{R}_p$ is stable if and only if $\theta_p(D^p) \subset D^p$.

In particular, the null regime is stable.

2.2 Locally Stable, Locally Unstable and Unstable Regimes

It is possible that for some entry states the trajectories escape. We shall call the set of such points the *unstable set* in phase domain D^p and denote it U^p. For any entry state $y \in U^p$ there is a minimum integer $k \geq 1$, such that $\theta^s(y) \in D^p, s = 0,1,\ldots,k-1$ and $\theta^k(y) \notin D^p$. The set $S^p := D^p \setminus U^p$ consists of all the entry states whose trajectories remain in D^p. We call it the *stable set*. Of course, $\mathcal{R}_p$ is stable if and only if $S^p = D^p$, that is, if $U^p = \emptyset$. If $U^p \neq \emptyset$, then $\mathcal{R}_p$ is not stable. Suppose $U_p \neq \emptyset$ and $S_p \neq \emptyset$. Then $\mathcal{R}_p$ is called *locally stable* on S_p and *locally unstable* on U_p. In this case, $\theta(S^p) \subset S^p$ and $\theta(U^p) \supset\supset U^p$.

If $U^p \neq \emptyset$, then there must exist a nonempty *escape set* $E^p \subset U^p$,

$$E^p = \{x \in D^p,\ \theta_p(x) \in \setminus D^p\}.$$

Any trajectory that enters E^p escapes the pth regime in one period. If $E^p \neq \emptyset$ and $E^p \neq U^p$, there must be a set in D^p, say E^p_2, whose trajectories escape in two periods and so on, recursively, there could be a set that escapes in n periods:

$$E^p_n := \theta_p^{1-n}(E^p), n = 1,2,3,\ldots. \tag{9}$$

We call E^p_n the nth escape set in U^p, the first one of which is just called the escape set. All these sets could be empty for $n \geq 2$.

Evidently, the set of points that eventually escape a given regime, that is, the unstable set, is the union of all these escape sets E^p_n, that is,

$$U^p = \cup_{n=0}^{\infty} E_n^p. \tag{10}$$

Obviously, if the pth regime is ***unstable***, then $U^p = D^p$.

A given phase domain D^p may contain a set of periodic states, say P^p. Let

$$C^p := \cup_{n=0}^{\infty} \theta_p^{-n}(P^p). \tag{11}$$

This is the set of points in D^p that are ***eventually periodic***, so we will call it the ***eventually periodic set*** in D^p. What happens to trajectories whose entry states do not belong to the eventually periodic set? They could converge to a periodic orbit in P^p; or they could wander erratically in D^p, never converging to a cycle of any order. If either event were true for any entry state, the regime would be stable. But even points that come close to a point in C^p might eventually exit and this could be true for almost all points. Note that a periodic orbit may intersect several phase zones. Trajectories with such orbits, or that converge to them, will be phase cyclic.

2.3 Ergodic Multiple–Phase Systems

Assume that Σ is a σ–algebra of subsets of D and $\mu(\cdot)$ is a probability measure defined on Σ. Recall that a map θ is ***μ–preserving*** and the measure μ is called ***invariant with respect to θ*** if $\mu(\theta^{-1}(X)) = \mu(X)$ for all $X \in \Sigma$. Recall also, that a semi–dynamical system is ***μ–ergodic*** if the μ–measure of every invariant set is either zero or unity.

As a direct consequence of the Poincaré Recurrence Theorem and the Birkhoff Ergodic Theorem, we have

- *If (θ, D) is a multiple-phase dynamical system with invariant ergodic probability measure μ, and if B is the basin of attraction for the attractor supp μ, then* in the limit, *for almost all $x \in B$, the relative frequency of states that occur in regime i is $\mu(D^i)$. Alternatively, we can say that the proportion of time the system spends in a given regime is equal to its measure.*

If $0 \leq \mu(D^i) < 1$, we shall say that the system is ***μ–unstable***. If $\mu(D^i) = 1$, then the regime is μ–stable. A regime that is μ–unstable may have a stable subset of Lebesgue measure zero. In particular, this set would contain all the periodic orbits that are strictly contained in D^p. Each periodic orbit would have an atomic measure, say $p(\cdot)$, which is p–stable.

2.4 Decomposability, Indecomposability and Switching Regimes

If every regime $\mathcal{R}_p, p \in \mathcal{P}$ of a multiple–phase semi–dynamical system is stable, then it is a closed semi–dynamical system and the multiple–phase system is *decomposable.*

Suppose every regime is μ–unstable except the null regime $\mathcal{R}_0$, then every trajectory must involve switching regimes μ–almost surely. The number of episodes in any nondegenerate evolution is countable, either finite or infinite. If the number of episodes is infinite, then every evolution must involve phase cycling or it must exhibit a nonperiodic sequence of regimes. If the number of episodes is finite, then the trajectory must eventually be trapped in the "last" phase zone (which could be the null domain).

3. Possible Evolutions

In various applications it is of interest to consider the possibility of explaining empirically identifiable phases of the business cycle or "patterns of evolution" in economic development in terms of a formal multiple–regime model. The question is one of *qualitative econometrics*: given an observed sequence of empirically identified stages or phases, could the model generate corresponding sequences or patterns? In answering this question, one need not demand that the *quantitative* estimates generated by the model be accurate. That is in general an unreasonable demand for nonlinear dynamic models. But it is not unreasonable — indeed, it is an essential requisite of a reasonable model — that it be capable of generating qualitative patterns like those observed in the data. This constitutes an existence question for possible evolutions. We can formalize this question by defining conditional evolutions.

3.1 Conditional Evolutions

Let $\{\mathcal{R}_p, p \in \mathcal{P}\}$ be a multiple–phase dynamical system and consider a finite permutation of regimes

$$p_1, p_2, \ldots, p_k. \tag{12}$$

which we shall call a *conditional evolution.* Could there exist a trajectory with initial condition $x \in D$ and sojourns T_i, $i = 1, \ldots, k-1$, such that

$$I\big(\theta^t(x)\big) = p_i \quad \text{for} \quad t \in T_i,\ i = 1, \ldots, k? \tag{13}$$

3.2 Switching and Transition Sets

Obviously, for such a given sequence of episodes to occur, it must be possible to switch from one regime to another *in the specified sequence.* This means that each regime in the sequence up to the last must be locally unstable. Define a *switching set*

$$E^{i,j} := \{x \in E^i | \theta_i(x) \in D^j\}, \ i \neq j. \tag{14}$$

It is the set of points in the ith regime's escape set E^i that map outside D^i and into the jth phase domain. Obviously, if $E^{i,j} = \emptyset$, such a switch can't occur. Of course, $E^i = \cup_j E^{i,j}$.

Now define a *transition set* $U^{i,j}$

$$U^{i,j} := \bigcup_{n=0}^{\infty} \theta_i^{-n}(E^{i,j}). \tag{15}$$

This is a subset of the unstable set U^i containing all the points in D^i that eventually escape and enter D^j. Of course, $U^i = \cup_j U^{i,j}$.

3.3 Existence of Conditional Evolutions

Our question can now be framed in terms of these transition sets. Can there exist an initial condition $x \in D$ and entry times $s_i(x)$, $i = 1, \ldots, k$ such that $s_i(x) < s_{i+1}(x)$ and such that for $T_i = s_{i+1}(x) - s_i(x)$,

$$\theta^t(x) \in U^{p_i, p_{i+1}}, t \in T_i, i = 1, \ldots, k-1? \tag{16}$$

For this to happen it is necessary that each transition set in the sequence be nonempty. This will be true if each escape set in the sequence is nonempty. But this is not sufficient to insure that a trajectory can traverse a given conditional evolution because the trajectory must not only eventually switch from one regime to the next, it must enter each successive transition set. For this to happen, $\theta^{s_{i-1}(x)}(x)$ must enter the correct subset of U^{p_{i-1}, p_i}, and this must be true phase by phase so the initial condition x must belong to a special subset of U^{p_1, p_2} that will lead from one transition set to another in the proscribed sequence.

To determine what this special set is, we can proceed backwards from the last regime in the conditional evolution. Thus, we must have

$$\theta_{p_{k-1}}(D^{p_{k-1}}) \cap D^{p_k} \neq \emptyset.$$

This implies that the set of points that escape $D^{p_{k-1}}$ and enter D^{p_k} is nonempty, i.e.,

$$E^{p_{k-1}, p_n} := \theta_{p_{k-1}}^{-1}\left(\theta_{p_{k-1}}(D^{p_{k-1}}) \cap D^{p_k}\right) \neq \emptyset$$

so

$$U^{p_{k-1},p_k} := \cup_{n=0}^{\infty}\theta^{-n}(E^{p_{k-1},p_k}) \neq \emptyset.$$

Any trajectory that enters U^{p_{k-1},p_k} eventually arrives in D^{p_k}.

To enter U^{p_{k-1},p_n} from $D^{p_{k-2}}$ we must have

$$\theta(D^{p_{k-2}}) \cap U^{p_{k-1},p_n} \neq \emptyset.$$

This is a subset of $E^{p_{k-2},p_{k-1}}$ and we now have to get the corresponding subset of its transition set, which we shall denote

$$U^{p_{k-2},p_{k-1},p_k} := \cup_{n=1}^{\infty}\theta_{p_{k-2}}^{-n}\Big(\theta(D^{p_{k-2}}) \cap U^{p_{k-1},p_k}\Big).$$

Any entry state $\theta^{s_{k-2}(x)}(x) \in U^{p_{k-2},p_{k-1},p_k}$ will traverse $\mathcal{R}_{p_{k-2}}, \mathcal{R}_{p_{k-1}}, \mathcal{R}_{p_k}$ in succession.

In like manner we can define

$$U^{p_{k-i},\ldots,p_k} := \cup_{n=1}^{\infty}\theta_{p_{k-i}}^{-n}\Big(\theta_{p_{k-i}}(D^{p_{k-i}}) \cap U^{p_{k-i+1},\ldots,p_k}\Big)$$

and so on, recursively, $i = 1, \ldots, k-1$ until we get the condition $\theta_{p_1}(D^{p_1}) \cap U^{p_2,\ldots,p_k} \neq \emptyset$.

We thus arrive at the following existence condition.

- The necessary and sufficient condition for a trajectory $\tau(x)$ to follow the conditional evolution defined by $p_1, p_2, \ldots, p_k$ is that

$$x \in U^{p_1,\ldots,p_k} := \cup_{n=1}^{\infty}\theta_{p_1}^{-n}\Big(\theta_{p_1}(D^{p_1}) \cap U^{p_2,\ldots,p_k}\Big) \neq \emptyset. \tag{17}$$

It should be noted for a given system that many, perhaps even most, conditional evolutions among all those sequences that can be written down will not be possible. This implies a restriction on the possible empirical patterns that can be explained in terms of the given system.

3.4 Probable Conditional Evolutions

If initial conditions are not known but can be located with varying degrees of reasonable belief within certain sets, one could then give a precise meaning to the probability of occurrence of a given conditional evolution.

Suppose then that μ is a probability measure on D^1. The number $\mu(S)$, $S \subset D^1$ can be interpreted to be the degree of reasonable belief that the initial condition $x \in S$. Because trajectories from any initial condition are unique (within the class of dynamical systems discussed in this paper), they establish a one–to–one map $x \rightarrow \tau(x)$ where $\tau(x) := \{x, \theta(x), \theta^2(x), \ldots\}$. This in turn enables conditional evolutions to be associated with a degree of reasonable belief.

- *Let $\mu(\cdot)$ be a probability measure defined on D^1. The "chance" that a trajectory follows a conditional evolution defined by the permutation of phases $p_1, \ldots, p_k$ is just*

$$\mu(U^{p_1,\ldots,p_k}).$$

4. Examples

Most of the concepts outlined above can be illustrated with a family of tent maps defined by

$$\theta(x) = \begin{cases} \theta_1(x) := a(x-m) & , x \in D^1 \\ \theta_2(x) := b(n-x) & , x \in D^2 \end{cases}$$

where a, b, m, n satisfy $a, b > 1$ and $0 < m < n$,

$$D^1 := \left(m, \frac{am+bn}{a+b}\right], \quad D^2 := \left(\frac{am+bn}{a+b}, n\right].$$

Note that $D^0 := (\infty, m] \cup (n, \infty)$ and $D = (m, n]$.

4.1 Ergodic Behavior and Phase Cycling on the Unit Interval

Let $m = 0$, $n = 1$, $a = b = 2$. Then θ is Lebesgue measure preserving, ($\lambda(\cdot)$ is invariant w.r.t. θ), and $\lambda(D^1) = \lambda(D^2) = \frac{1}{2}$ where $\lambda(\cdot)$ is Lebesgue measure. This means that both regimes are λ–unstable and in the limit of the system spends half its time in phase zone D^1 where the value of x increases and half its time in phase zone D^2 where values of x exhibit expanding oscillations. (Actually, it is easy to see that both regimes are *unstable*). The system is phase cyclic with phase sequence $(1,2,1,2,\ldots)$ or $(2,1,2,1,\ldots)$, depending on whether or not the initial condition is in zone 1 or 2. But trajectories are chaotic almost surely. This means that the sojourns within each phase zone are of chaotically varying lengths of time even though in the limit the average fraction of time in each zone is $\frac{1}{2}$.

4.2 Self–Destructive Behavior on the Unit Interval

Let $m = 0$, $n = 1$ as before, but let $a = b = 2 + e$. Then there exists an escape interval $E = (\frac{1}{2+e}, \frac{1+e}{2+e}) \subset D$ with Lebesgue measure $\lambda(E) = \frac{e}{2+e}$ which approaches zero as e approaches zero. However, for any positive e, no matter how small, there exists a *conditionally invariant measure* μ such that $\mu(U) = 1$. That is, trajectories enter the null domain almost surely. See Day

(1994), §8.1 and Theorem 9.1. The regimes will still alternate in sequence until escape occurs. Moreover, cycles of all orders still exist, but the set of eventually periodic points has measure zero.

4.3 Mountainous Maps and Evolving Regimes

Now we consider a "mountainous" multiple regime system made up of a family of tent or mountain–like maps that are analogous to ones that arise in the theory of economic development as described in Day and Walter (1989). Define

$$\theta_i(x) := \min\Big\{a_i(x - m_i),\ b_i(n_i - x)\Big\}$$

for $x \in [m_i, n_i]$ with a_i, m_i, b_i, n_i all positive real numbers and with $m_1 = 0, m_i < m_{i+1}, n_i < n_{i+1}$ and $m_i < n_i$, $i = 1, \ldots, k$. Each $\theta_i(\cdot)$ has a tent–like appearance on the internal $[m_i, n_i]$. If a_i and b_i are greater than unity, each map is expansive. Let

$$\theta(x) = \max\Big\{\theta_i(x)\Big\},\ x \geqq \mathbf{R}.$$

If each $m_{i+1} < m_i$, $i = 1, \ldots, k-1$, then the maps overlap and $\theta(x) > 0$, $x \in (0, n_k)$. Let x^{**i} be the maximizer and let x^{Mi} be the maximum of θ_i on $[m_i, n_i]$. Then

$$x^{**i} = \frac{a_i m_i + b_i n_i}{a_i + b_i} \quad \text{and} \quad x^{Mi} = \frac{a_i b_i(m_i + n_i)}{a_i + b_i}.$$

Now consider where each map overlaps its predecessor and assume this occurs where the downward sloping segment of $\theta_i(x)$ intersects the upward sloping segment of $\theta_{i+1}(x)$. This is implied by and implies the condition

$$d_{i+1} := \frac{a_{i+1}m_{i+1} + b_i n_i}{a_{i+1} + b_i} > x^{**i}.$$

Given this condition, the map $\theta(x)$ has a local minimum at each d_i and a local maximum at each x^{**i}. Moreover, it is easy to see that

$$\theta(x) = \theta_i(x),\ x \in [d_i, d_{i+1}].$$

Let $d_1 = 0$, define

$$D_i = (d_i, d_{i+1}],\ i = 1, \ldots, k,$$

and let $D_0 = (-\infty, 0] \cup [m_k, \infty)$. Then the $\{\theta_i(x), D_i,\ i = 0, \ldots, k\}$ constitutes a multiple regime dynamical system.

It can be seen that there exist exactly two stationary states in each regime

$$\tilde{x}^{\ell_i} = a_i m_i/(a_i - 1) \quad \text{and} \quad \tilde{x}^{u_i} = b_i n_i/(b_i + 1).$$

Both are unstable. If

$$\tilde{x}^{\ell_i} < \theta_{i+1}(d_{i+1}),$$

$i = 1, \ldots, k-1$, then no trajectory that enters a given phase zone D^i can revert to a phase zone earlier in the sequence. Moreover, if

$$\theta_i(x^{**i}) > \tilde{x}^{\ell_{i+1}},$$

then every trajectory that enters a given zone must escape almost surely to a higher phase zone. Consequently, if we consider the sequence of phases $1, \ldots, k$ as a conditional evolution, then it is clear that for almost all $x \in D^1$ a monotonic progression of episodes occurs, i.e., for almost all $x \in D^i$ there exists entry times $s_i(x)$ $i = 1, \ldots, k$ such that

$$I\big(\theta^t(x)\big) = i, s_i(x) \leq t < s_{i+1}(x), \; i = 1, \ldots, k-1$$

and $I(\theta^t(x)) = k$ for $t = s_k(x)$. We cannot say what happens after time $t = s_k(x)$, but if

$$\theta_k(x^{Mk}) > \tilde{x}^{\ell k},$$

then all trajectories eventually become trapped in the set $[\theta_k(x^{Mk}), x^{Mk}]$. Then all epochal evolutions form monotonic progressions. But *trajectories* can be, but need not be, monotonic. Indeed, trajectories may grow, fluctuate, then grow again, then fluctuate, etc. Change will consist of spans of growth interspersed with nonperiodic oscillations. Nonetheless, given the assumptions of a finite number of regimes, all trajectories reach the last regime where fluctuations persist indefinitely.

By modifying the parameters, many different types of epochal evolutions can occur in a manner analogous to the possibilities described in the multiple–phase version of classical growth theory developed by Day and Walter.

5. Multiple–Phase Dynamics in an Abstract Adaptive Society

The multiple–phase theory of economic development outlined in Day and Walter that motivated the preceding example is a macrodynamic theory. To see how multiple–phases arise naturally in microeconomic processes, I consider an abstract adaptive society analogous to the abstract economy of Debreu, and Arrow and Debreu, and based on a simplified version of the adaptive, evolutionary framework outlined in Day (1975).

5.1 The Framework

Begin with an "external" state and with the actions that have just been taken by all the agents in the system. Agents observe this situation and update their information about it. On the basis of this information, plans are formed. Actions are not in general the same thing as plans but generally involve a more or less conscious attempt to control behavior according to plan, but contingent on the current state of the environment and on information not taken explicitly into account in the plan. The dependence of actions on states can cause deviation from plans despite the controlling function. As a result of the actions taken, a new state emerges. In this way, agents and environment coevolve.

Let $\mathcal{I} := \{1, \ldots, I\}$ be a finite set of agents. For each agent $i \in \mathcal{I}$, let

$\mathbf{A}^i$	be the action space,
$\mathbf{X}^i$	be the plans space,
$\mathbf{W}^i$	be the information space.

Also, let

$\mathbf{A} := \times_{i \in \mathcal{I}} \mathbf{A}^i$	be the joint action space,
$\mathbf{W} := \times_{i \in \mathcal{I}} \mathbf{W}^i$	be the joint information space,

and

$\mathbf{Z}$ be the "external" state space.

Define

δ^i	$: \mathbf{W}^i \times \mathbf{A} \times \mathbf{Z} \to \mathbf{W}^i,$	the information updating operator
ψ^i	$: \mathbf{W}^i \to \mathbf{X}^i,$	the planning operator
ζ^i	$: \mathbf{A}^i \times \mathbf{X}^i \times \mathbf{W}^i \times \mathbf{Z}^i \to \mathbf{A}^i,$	the control operator
ω	$: \mathbf{A} \times \mathbf{Z} \to \mathbf{Z},$	the state transition operator.

Begin with a state $z_t \in \mathbf{Z}$, a joint information $w_t := (w_t^i)_{i \in \mathcal{I}}$ and a joint action $a_t := (a_t^i)_{i \in \mathcal{I}}$. Each agent generates an updated information

$$w_{t+1}^i = \delta^i(a_t, w_t^i, z_t) \in \mathbf{W}^i.$$

While this information is being collected and processed, a new plan is drawn up on the basis of the most recent information; that is,

$$x_{t+1}^i = \psi^i(w_t^i).$$

Given this plan, given the new information that becomes available, and given the state, an action is taken

$$a_{t+1}^i = \zeta^i(a_t^i, x_t^i, w_{t+1}^i, z_t),$$

which we consider to be a modification or departure of the previous action a_t^i. By substitution,

$$a_{t+1}^i = \alpha^i(a_t, w_t, z_t) := \zeta^i\Big(a_t^i, \psi^i(w_t^i), \delta^i(a_t, w_t^i, z_t), z_t\Big).$$

In the meantime, a new state unfolds

$$z_{t+1} = \omega(a_t, z_t).$$

Let $u_t := (a_t, w_t, z_t)$ and define the *joint information map*

$$w_{t+1} = \delta(u_t) := \Big(\delta^i(w_t^i, a_t, z_t)\Big)_{i \in \mathcal{I}}.$$

Also define the *joint action map*

$$a_{t+1} = \alpha(u_t) := \Big(\alpha^i(u_t)\Big)_{i \in \mathcal{I}}.$$

The coevolution of agent and environment is then given by the discrete time dynamic system

$$\begin{aligned} w_{t+1} &= \delta(u_t) \\ a_{t+1} &= \alpha(u_t) \\ z_{t+1} &= \omega(a_t, z_t). \end{aligned}$$

Let

$$\mathbf{U} := \mathbf{A} \times \mathbf{W} \times \mathbf{Z} \quad \text{and} \quad \theta(u) := \Big(\alpha(u), \delta(u), \omega(a, z)\Big).$$

Then (24) can be re–expressed as

$$u_{t+1} = \theta(u_t). \tag{18}$$

and a trajectory from an initial condition $u^0 \in U$ can be represented by $u_t = \theta^t u_0$ where θ^t is the tth interated map of θ.

5.2 Viability

For a trajectory to exist, actions must be possible. Suppose each agent i has an admissibility *correspondence*

$$\Omega^i : \mathbf{A} \times \mathbf{Z} \to 2^{\mathbf{A}^i}$$

which describes the set of possible actions for agent i, given the current behavior of the society and the external state z. For the society as a whole, the map $\Omega(a,z) := (\Omega^i(a,z))_{i \in \mathcal{I}}$ defines the set of admissible joint actions given a joint action a and environmental state z.

Assume there exists a set $U \subset \mathbf{U}$ such that

$$\alpha(a,w,z) \in \Omega(a,z) \neq \emptyset \qquad \text{for all} \qquad (a,w,z) \in U. \tag{19}$$

A sufficient condition for the abstract adaptive society to be viable is that

$$\theta(u) \in U, \quad \text{all} \quad u \in U. \tag{20}$$

The set U is called the *viability domain.*

The condition (19) does not imply (20). It only says that each individual can do something using the strategies determined by the ζ^i, ψ^i and δ^i. There must be a further requirement that agents' actions are compatible. To formalize this notion, let $\mathcal{A} : u \to 2^{\mathbf{A} \times \mathbf{W}}$ be a *compatibility correspondence* defined by

$$\mathcal{A}(u) := \left\{(a',w') | \left(a',w',\omega(a,z)\right) \in U\right\}. \tag{21}$$

where the feasible action condition (19) holds for all $u \in U$. The agents will be *compatible* if

$$\left(\alpha(u), \delta(u)\right) \in \mathcal{A}(u) \neq \emptyset \tag{22}$$

for all $u \in U$. Suppose (19) and (22) *do* hold and suppose $u_0 = (a_0, \omega_0, z_0) \in U$. Then $\alpha(u_0) \in \Omega(a_0, z_0) \neq \emptyset$ because agents are admissible. Moreover, $(\alpha(u_0), \delta(u_0)) \in \mathcal{A}(u_0) \neq \emptyset$ because agents are compatible. This implies that $(a_1, w_1, z_1) = (\alpha(u_0), \delta(u_0), \omega(a_0, z_0)) \in U$. By recursion, $\theta^t(u) \in U$ for all $u = u_0 \in U$. We therefore have

- *If there exists a set U such that for all $u \in U$*

  (i) $\alpha(u) \in \Omega(u) \neq \emptyset$ (agents are viable); *(A)*

  (ii) $(\alpha(u), \delta(u)) \in \mathcal{A}(u) \neq \emptyset$ (agents are compatible).

  Then trajectories exist for all initial conditions $u \in U$.

Notice that the feasibility and compatibility conditions depend on agents' information as well as on their actions and that the compatibility correspondence (ii) implies that in general agents' actions and information cannot be independent.

5.3 L^* Optimizing Plans

Now assume that *plans* can be represented as the solution to multi–criterion mathematical programming problems of the L^* variety. Thus, agents estimate the constraints that limit or potentially limit their actions and choose plans that satisfy these constraints according to a priority ordering amongst multiple criteria. At each stage in the lexicographic choice sequence, choices are constrained, not only to be feasible, but also by the requirement that they satisfice all higher order utility indexes. Consequently, the more criteria are satisficed, the more constrained the choice. Obviously, if there is a single decision criterion, then the choice of plan is representable by a conventional constrained maximizing problems.

Formally, let us introduce for each agent, $i \in \mathcal{I}$, a family of *constraint functions* $\beta^i_j : \mathbf{X} \times \mathbf{W} \to \mathbb{R}$, $j \in \mathcal{J}^i$, and a family of *limitation functions* $\gamma^i_j : \mathbf{W}^i \to \mathbb{R}$, $j \in \mathcal{J}^i$, and assume that $\mathbf{X}$ is a vector space with the feasibility correspondence $\Gamma^i(w)$ defined by

$$\Psi^i_0(w^i_t) \equiv \Gamma^i(w_t) := \left\{x | \beta^i_j(x, w_t) \leq \gamma_j(w^i_t),\ j \in \mathcal{J}^i\right\}.$$

Introduce also a family of decision criteria (utility or objective functions) $\{\phi^i_k : \mathbf{X} \times \mathbf{W} \to \mathbb{R},\ k \in \mathcal{K}\}$ and a family of *satisficing functions* $\{\sigma_j : \mathbf{W} \to \mathbb{R},\ k \in \mathcal{K}\}$. Now let

$$u^i_k(x, w) := \min\left\{\phi^i_k(x^i, w^i),\ \sigma^i_k(w^i)\right\},\ k \in \mathcal{K}.$$

Define the correspondence

$$\Psi^i_k(w) := \text{Arg} \max_{x \in \Psi_{k-1}(w)} u^i_k(x^i, w^i),\ k \in \mathcal{K}. \tag{23}$$

This is the set of perceived feasible plans for the ith agent which maximizes the kth criterion given that the preceding $k-1$ criteria in the priority order are satisficed.

Now let

$$\Psi^i_\ell(w^i) := \bigcap_{k \in \mathcal{K}} \Psi^i_k(w^i). \tag{24}$$

This set is in general non–unique. Let $\psi^i_\ell : \mathbf{W}^i \to \mathbf{X}^i$ be an operator that for any w^i selects a plan $\psi^i_\ell(w_t) \in \Psi^i_\ell(w_t)$. This operator will depend on the procedure or algorithm used in finding a solution to the L^* programming problem (22). I shall call it the L^* planning operator.

The dependence of the decision criteria on the data allows for the adaptation of preferences to experience. The data themselves can only depend on past information actions and states, but may represent forecasts of the future so that the plan itself may represent a sequence of actions to be undertaken in the future. But these anticipated actions cannot depend on states that have not occurred but only projected states. Hence, by limiting the criteria to a single objective and defining Γ appropriately, the plan x_t selected at each point in time could represent a (provisionally) optimal trajectory of action which would be revised from time to time on the basis of the data operator δ. It is also possible within this framework to represent optimizing plans by a Bayesian dynamic programming procedure. Moreover, by appropriate definitions of data elements in w and the information operator δ, it is possible to represent plans as being modified at intervals longer than the unit time period, while actions may be adjusted to experience continually, in principle every unit period, by means of various behavioral rules incorporated into the control operator ζ. But these are details that will not concern us further here.

5.4 Existence

I now want to consider the implications of L^* optimizing plans when agents adapt out of equilibrium in response to experience and to changes in the environment.

L^* agents are rational in that their plans optimize an L^* preference ordering given their information strategies. In the present context, the feasibility correspondence $\Gamma(w)$ is an estimate of the viability correspondence $\Omega(a,z)$. That is, what the individual considers to be feasible may be distinct from what is actually the set of viable actions. It is essential that the control function make up for this possibility and that together the information, planning and control functions are *practical* in the sense that

$$\alpha^i(u) = \left(\zeta^i\left(a, \psi^i_\ell(w), \delta^i(a, w^i, z), z,\right)\right)_{i\in\mathcal{I}} \in \Omega^i(a,z) \tag{25}$$

at each stage of the process as it evolves through time. The mere requirement that $\psi^i(w^i) \in \Psi^i_\ell(w^i) \subset \Gamma^i(w)$ for all i does not guarantee this. The existence statement (A) needs to be modified slightly.

- *If there exists a U such that for all agents $i \in \mathcal{I}$, for all $u \in U$* *(B)*
    - *(i)* $\Gamma^i(\delta^i(u)) \neq \emptyset$ *(agents think they have feasible plans);*
    - *(ii)* $\psi^i(w^i) \in \Gamma^i(w^i)$ *(agents select a feasible plan);*
    - *(iii)* $\alpha^i(u) = \zeta^i(\psi^i(a^i, w^i),\ \delta^i(w^i, z), z) \in \Omega^i(a,z)$ *(agent's actions are admissible);*

(iv) $(\alpha(u), \delta(u)) \in \mathcal{A}(\omega(u)) \neq \emptyset$ *(agents' joint actions and joint information are compatible).*

Then trajectories exist for all initial conditions $u = u_0 \in U$.

(B) differs from (A) only by taking account explicitly of the need for rational agents to have feasible plans (otherwise they could not arrive at a decision). □

5.5 Multiple Phase Dynamics

Now take into account the structure of the L^* planning operator Ψ^i_ℓ. A plan $x^i \in \Psi^i_\ell(w)$ is characterized by a programming problem. This implies that each choice function, ψ^i_ℓ satisfies a dual system of equated constraints associated with that solution. This dual system depends on the subset of constraints that are equated, the subset of the criteria that are satisfied and on the gradient of the criterion maximized.

If the information on which plans are based changes from period to period, then this dual system of equated constraints will change also and the characterization of each ψ^i_ℓ in terms of this system will change correspondingly. There are a finite number of possible dual systems. Let us index these sets by a subscript $p \in \mathcal{P}^i$. Then associated with each w^i there is at least one system of equated constraints, e.g., the $p^i(w)^{\text{th}}$, a single-valued map $\psi^i_{p^i(w)}(\mathbf{W}^i \to \mathbf{X}^i)$ which I shall call a *planning rule* and there exists an *information zone* $\mathbf{W}^i_p \subset \mathbf{W}^i$ such that

$$\text{for each } w^i \in \mathbf{W}^i_p, \quad \text{there exists } p \text{ such that } \psi^i_p(w^i) \in \Psi^i_\ell(w^i).$$

Each situation will lead to a subsequent information in one or the other of these information zones. This induces a partition $\{U_{p_i}, p_i \in \mathcal{P}^i,\ i \in \mathcal{I}\}$ on $U \subset \mathbf{A} \times \mathbf{W} \times \mathbf{Z}$ such that for each $u \in U_{p_i}$

$$\delta^i(u) \in W^i_p.$$

Now define for each agent the p *action rules*

$$\alpha^i_p(u) := \zeta^i\Big(a^i, \psi^i_p(w^i), \delta^i(u), u\Big), \quad u \in U_{p_i} \in \mathcal{P}^i$$

and let $\theta_p(u) := (\alpha_p(u), \delta(u), \omega(u))$. Then we arrive at the multiple phase dynamical system

$$u_{t+1} = \theta(u_t) := \theta_{p_i}(u_t), \quad u_t \in U_{p_i}, \ p_i \in \mathcal{P}^i, \ i \in \mathcal{I}.$$

As the system evolves, from time to time u_t may cross the boundary from one phase zone to another, implying that the system of equated constraints for one or more agents will have changed. The phase structure switches endogenously. The implication is that for some agents different constraints (associated with different resources, for example) may be binding or a different set of needs and wants may be satisficed. The particular preferences governing choice at any time and the values imputed to the various constraining factors will therefore change. Moreover, the $\psi_p^i(w)$ may not be continuous. This means that trajectories can exhibit discrete jumps from time to time even when change may have been very gradual in between times.

5.6 Endogenous Priorities

In reality, people change their minds about priorities which are reordered from time to time. To account for this phenomenon within the adaptive abstract economy framework, assume that not only the criterion and satisficing functions depend on states, but their order does also. Let $N(w) := \{i_1(w), i_2(w), \ldots\} \subset N := \{1, 2, \ldots\} \subset \mathbb{N}^+$. We can then refer to the jth element in $N(w)$, $i_j(w)$ by the index j. Now consider the L^* family defined in (8) except replace N by $N(w)$. For any information datum w, the L^* decision operator is well defined and can be characterized by a Kuhn–Tucker dual optimization problem.

Given this interpretation, the number *and* order of criteria satisficed evolve. Extended to all agents in a adaptive abstract society, this implies an endogenous evolution in the distribution of satisficed wants and needs; a possible convergence — or divergence — of social values, a development of social accord or discord. These social developments could have an important feedback effect on action and on the potential viability of a given system.

6. Discussion

The abstract adaptive society as it is presented here goes a long way toward recognizing salient features of economic experience that have not been accounted for in formal theory, namely, the endogenous generation and switching of regimes which is induced by the way people decide and interact. The behavior that results is based on limited and evolving knowledge which implies that out of equilibrium, adaptation is more likely to characterize economic action than intertemporal or temporal equilibrium.

Once we have gotten this far, the essence of the theoretical existence problem boils down to the nature of the data operator — which defines how agents learn about the world — and the nature of the compatibility operator which defines how a society enables the actions of its individuals to be consistent, one with another when each has scope for independent behavior. The former is both a psychological and social problem because agents learn from each other and their values (decision criteria) depend on the evolving system as a whole. The latter is a sociopolitical problem because compatibility is determined by the rules of conduct and the various institutions that formalize them that appear as society evolves. This institutional nature of the existence problem involves organization which implies the existence of an evolving superstructure of coalitions among agents. I have not tried to formalize this very fundamental aspect of the existence problem in mathematical terms, but I have no doubt that the evolution of institutions can be explained, at least to some extent, in terms of the need to facilitate compatibility. I have written about this issue at some length elsewhere (Day, 1984 and 1993).

To go further than these very general remarks, we need to add more specific aspects of economic structure, in particular, a more explicit distinction between stocks and flows, (resources and capital, production and consumption), the incorporation of prices and a specific provision for exchange transactions out of equilibrium. We should then have a multiple–phase, evolutionary analog of the Arrow–Debreu abstract economy which would be a more realistic theoretical microfoundation for macroeconomics than is now available.

References

[1] Kenneth J. Arrow and Gerard Debreu, *Existence of an Equilibrium for a Competitive Economy*, Econometrica (1954), **22**, 265–290.

[2] Richard H. Day, *Recursive Programming and Production Response*, North-Holland Publishing Company, Amsterdam, 1963.

[3] Richard H. Day, *Adaptive Processes and Economic Theory,* in R. H. Day and T. Groves (eds.), *Adaptive Economic Models*, Academic Press, New York, 1975.

[4] Richard H. Day, *Complex Economic Dynamics, Volume 1, An Introduction to Dynamical Systems and Market Mechanisms*, MIT Press, Cambridge, 1994.

[5] Richard H. Day and Alessandro Cigno, *Modelling Economic Change*, Elsevier North-Holland, Inc., Amsterdam 1978.

[6] Richard H. Day, and Jean-Luc Walter, *Economic Growth in the Very Long Run: On the Multiple-Phase Interaction of Population, Technology, and Social Infrastructure,* Chapter 11 in W. Barnett, J. Geweke, K. Shell (eds.), *Economic Complexity: Chaos, Sunspots, Bubbles and Nonlinearity*, Cambridge University Press, Cambridge, 1989.

[7] Gerard Debreu, *A Social Equilibrium Existence Theorem,*, Proceedings of the National Academy of Sciences, **28** (1952), 886–893.

[8] Nicholas Georgescu-Roegen, *Relaxation Phenomena in Linear Dynamic Models*, Chapter 1 in T.C. Koopmans (ed.), *Activity Analysis of Production and Allocation*, John Wiley and Sons, New York, 1951.

[9] Richard Goodwin, *The Nonlinear Accelerator and the Persistence of Business Cycles,* Econometrica, 1951.

[10] Wassily W. Leontief, *Studies in the Structure of the American Economy,* Oxford University Press, London, 1958.

[11] Edmond Malinvaud, *Profitability and Unemployment*, Cambridge University Press, Cambridge, 1980.

[12] Herbert A. Simon, *Effects of Technological Change on a Linear Model,* in T.C. Koopmans (ed.), *Activity Analysis of Production and Allocation,* John Wiley and Sons, New York, 1951.

Stabilizing Unstable Systems

Takao Fujimoto
Department of Economics, University of Okayama
Tsushima, Okayama, 700, JAPAN

Abstract

A nonlinear difference equation which diverges cyclically may be stabilized when time-lag is introduced in a certain way. It is shown by using a numerical example that while stabilization is impossible when the resulting difference equation is of order two, it can be made stable when the order is not less than three. One application is made to stabilize the examples of global instability of competitive equilibrium by Scarf. Some suggestions are made concerning the computation of a fixed point, and the shrinkage of chaotic areas.

1. Introduction

Let us consider a linear differential equation in the n-dimensional real Euclidean space, $dx/dt = Ax$, where x is an n-column vector and A is a given n by n real matrix. We suppose that each eigenvalue of A has a negative or zero real part, and at least one pair of eigenvalues are pure imaginary numbers, thus in general exhibiting cyclical movements. When this equation is discretized, we may have

$$x(t+1) = x(t) + aAx(t), \tag{1.1}$$

where a is a positive scalar, and may be called the *speed of adjustment*. However small a may be, (1.1) remains unstable.

Now we introduce time-lags into (1.1) as

$$x(t+1) = (kx(t) + (1-k)x(t-1)) + aAx(t), \tag{1.2}$$

where k is a positive scalar such that $0 \leq k \leq 1$. Unfortunately, this device does not work in most cases. So, we generalize (1.2) as

$$x(t+1) = \left(\sum_{i=0}^{p}(k_i x(t-i))\right) + aAx(t), \tag{1.3}$$

where k_i's are positive scalars such that $\sum_{i=0}^{p} k_i = 1$, and $r = p+1$ shows the order of eq.(1.3). When r is not less than three, (1.3) may be stable with a suitable choice of k_i.

In section 2, we present a simple numerical example which cannot be stabilized in the form (1.2),i.e.,of order two, but can be stable in the form (1.3) with $r = 3$. A proof is given that the example cannot be stable when the order is two. In section 3, our method is applied to nonlinear difference equations. Specifically, we take up the discrete version of unstable examples by Scarf[5]. In the final section other possible applications are suggested.

2. An Example

Consider the Euclidean space of dimension two, R^2, and let A be

$$A \equiv \begin{bmatrix} 0 & -1 \\ 1 & 0 \end{bmatrix}.$$

Then,

$$x(t+1) = (kx(t) + lx(t-1) + mx(t-2)) + aAx(t), \qquad (2.1)$$

can be stable with a suitable choice of k,l,m, and a. This is to be confirmed by way of a simple program on a computer. We can prove that (1.2) or

$$x(t+1) = (kx(t) + lx(t-1)) + aAx(t), \qquad (2.2)$$

where $k + l = 1$, cannot be stable for any choice of k and l. That is, the matrix

$$B = \begin{bmatrix} 0 & 0 & 1 & 0 \\ 0 & 0 & 0 & 1 \\ l & 0 & k & -a \\ 0 & l & a & k \end{bmatrix}$$

has at least one pair of eigenvalues on the unit circle or outside. To prove this we use a theorem due to Cohn[2,p.115]. (See also Schur[6].)

Cohn's Theorem: Suppose the equation $f(z) = a_0 z^n + a_1 z^{n-1} + \cdots + a_n = 0$ $(a_0 = 0)$ has all its roots within the unit circle, then the equation $f(z) + \lambda f^*(z) = 0$ with $|\lambda| < 1$ also has all its roots within the unit circle, where

$$f^*(z) = \overline{a}_n z^n + \overline{a}_{n-1} z^{n-1} + \cdots + \overline{a}_0.$$

($\overline{a}$ is the conjugate of a.)

The eigenequation of B is

$$f(z) = z^4 - 2kz^3 + (k^2 - 2(1-k) + a^2)z^2 + 2k(1-k)z + (1-k)^2 = 0.$$

Putting $\lambda = -(1-k)^2$, we form $f(z)+\lambda f^*(z)$ and divide this by x to obtain

$$g(z) = (1+(1-k)^2)z^3 - 2(k^2-k+1)z^2 + (k^2+2k-2+a^2)z + 2(1-k) = 0.$$

Then, divide $g(z)$ by $(1+(1-k)^2)$, and putting $\lambda = 2(1-k)/(1+(1-k)^2)$, we form $g(z)+\lambda g^*(z)$ and arrange it to have

$$h(z) = k^2(k-2)^2z^2 - 2[\ldots]z + \{k^2(k-2)^2 + [(k-1)^2+1]a^2\} = 0.$$

Since the coefficient of z^2 is smaller than the constant term, this quadratic equation $h(z)$ cannot have its two roots within the unit circle. Thus, at least one pair of eigenvalues of B cannot lie within the unit circle.

When we consider the case of order three, an interesting phenomenon takes place. Eq.(2.2) is now changed to

$$x(t+1) = (kx(t) + lx(t-1) + mx(t-2)) + aAx(t), \tag{2.3}$$

where the coefficients k,l, and m are all nonnegative, and their sum is unity. The associated matrix is

$$B = \begin{bmatrix} 0 & 0 & 1 & 0 & 0 & 0 \\ 0 & 0 & 0 & 1 & 0 & 0 \\ 0 & 0 & 0 & 0 & 1 & 0 \\ 0 & 0 & 0 & 0 & 0 & 1 \\ m & 0 & l & 0 & k & -a \\ 0 & m & 0 & l & a & k \end{bmatrix}.$$

Let us fix $m = 0.3$, and we have the following eigenequation.

$$\begin{aligned} x^6 - 2kx^5 &+ (k^2+2k-1.4+a^2)x^4 + 2(k(0.7-k)-0.3)x^3 \\ &+(k^2-0.8k+0.49)x^2 + 0.6(0.7-k)x + 0.09 = 0. \quad (2.4) \end{aligned}$$

When $a < 1$, the modulus of each solution of (2.4) is less than unity whatever value k may take between 0 and 0.7, implying the asymptotic stability of the difference equation (2.3) of order 3. Once the speed of adjustment a becomes greater than 1.4 or so, at least one pair of solution is outside(or on) the unit circle regardless of the value of k. The 'quickest'(in a loose sense) convergence seems to be realized when a is around 1.26 and k is roughly between 0.5 and 0.7.

3. Stabilization of Scarf's Examples

In this section, we again consider R^2. One of the unstable examples in Scarf[5] is, when discretized and normalized(by choosing commodity 3 as the numeraire), is described as

$$x_1(t+1) = x_1(t) + aE_1(x_1(t), x_2(t)),$$

$$x_2(t+1) = x_2(t) + aE_2(x_1(t), x_2(t)),$$

where $E_1 = (-x_2)/(x_1 + x_2) + x_3/(x_3 + x_1)$, and $E_2 = x_1/(x_1 + x_2) + (-x_3)/(x_2 + x_3)$, and a is the speed of adjustment. Several authors have devised out more elaborate methods to calculate equilibrium prices because a simple tatonnement process fails to converge in this Scarf's example.(See Smale[7]). Using our 'conservative' mechanism to employ the past prices and average them, the above system can be stabilized in a simple and 'natural' way. That is,

$$x_i(t+1) = (kx_i(t) + lx_i(t-1) + mx_i(t-2)) + aE_i(x_1(t), x_2(t)) \quad \text{for } i = 1, 2.$$

By selecting k, l, m, a, and an initial vector properly, a solution path converges. Through a suitable choice, convergence is quite rapid, and the region of initial vectors which guarantees stability is large enough.

Let us define $b(x;3) \equiv (kx_1(t) + lx_1(t-1) + mx_1(t-2), kx_2(t) + lx_2(t-1) + mx_2(t-2))'$, and call $b(x;3)$ the benchmark vector of order 3. In the framework of competitive adjustment of tatonnement process, the benchmark vector provide a basis to conduct a conservative revision of prices on the side of auctioneer.

The second group of unstable examples in Scarf[5] can be handled in a similar way. In these examples, however, it seems that we need a system of order four.

4. Mean Value Methods

Some other simple methods have been proposed in Mann[4] and Zarantonello[8]. Mann's iteration can be represented in a special case as

$$x(t+1) = w(t)f(x(t)) + (1 - w(t))x(t), \quad \text{where } w(t) = 1/(1+t),$$

while Zarantonello's keeps $w(t)$ constant assuming the weak antitonality on $f(x)$. Mean value methods are useful also to tame down chaotic movements. A famous example, $x(t+1) = f(x(t)) = ax(t)(1 - x(t))$, can be handled. Since $f(x)$ is Lipschitz-continuous on the interval $[0,1]$, $w(t)$ is to be set constant. The reader may regard this case uninteresting because $w(t)a$ instantly decreases the critical value of the original coefficient a, thus avoiding chaos. Seemingly trivial, but this gives a straightforward explanation why averaging iteration can be useful.

When our method is applied to the above example by R.May, the iterates are

$$x(t+1) = s(ax(t)(1 - x(t)) + (k - s)x(t) + lx(t-1) + mx(t-2).$$

Provided $s < 0.33$, order two equations(i.e., $m = 0$) can stabilize this even when $a = 4$. (The reader is referred also to Bear and Conlisk[1].)

5. Concluding Remarks

What we have shown is that when the Jacobian matrix is of some types, the maximum modulus of its eigenvalues can be reduced by introducing time-lags into the benchmark variable. In this manner, wild movements of the orbit may be damped down, or lucky enough may converge to an equilibrium point with a suitable choice of an initial vector. We have also given an example which time-lags of order three can stabilize, while those of order two never. Some more remarks in brief.

(1) Our simple method can be applied to compute a fixed point in an easy way.

(2) When the above lagging approach is applied to a system which yields chaotic movements in a certain region of parameters, the region producing chaos shrinks.

(3) Nonlinear difference equations of the type $x(t+1) = x(t) + aF(x(t))$, can be classified according to the minimum order r of the benchmark vector by introducing which they can be made stable. If any orbit which starts within a compact region is bounded, can we say the minimum order is finite ?

(4) In the nonlinear case, even when stabilized, we have to ask how large the area is within which the initial vector is designated and the solution orbit converges to an equilibrium.

(5) Professor H. Matano suggested in the Conference an alternative proof to that in section 2 avoiding the use of Cohn's theorem.

(6) Professor R. Nussbaum has worked out more general theorems in the linear case in preliminary notes.

References

[1] D.V.T. Bear , J. Conlisk, *The effects of unlagging and speed of adjustment*, J. Econ. Theory **21** (1979).

[2] A. Cohn, *Über die Anzahl der Wurzeln einer algebraischen Gleichung in einem Kreise.*, Math. Zeit. **14** (1922).

[3] T. Fujimoto, *Order three may imply 'order', presented at Seminar on Nonlinear Analysis*, Kyoto University (1992).

[4] R. Mann, *Mean value methods in iteration*, Proc. Amer. Math. Soc. **4** (1953).

[5] H. Scarf, *Some examples of global instability of the competirive equilibrium*, Int. Econ. Rev. **1** (1960).

[6] I. Schur, *Über Potenzreihen, die im Innern des Einheitkreises beschränkt sind*, J. f. Math. **148** (1918).

[7] S. Smale, *A convergent process of price adjustment and global Newton method*, J. Math. Econ. **3** (1976).

[8] E. Zarantonello, *Solving functional equations by contractive averaging*, Math. Res. Center Rept. **160** (1960).

On First Order Sufficient Conditions for Constrained Optima.

Giorgio Giorgi*

Dipartimento di Ricerche Aziendali-Sezione
di Matematica Generale ed Applicata
University of Pavia, Via S. Felice, 5 PAVIA, ITALY.

Abstract

Some remarks are made on the results obtained by A. Mukherji[15] for mathematical programming problems involving quasi-concave functions and some extensions of the same results are established. Moreover, for the same problems and under no generalized concavity assumptions, some first order local optimality conditions are obtained.

1. It is well known that generalized concavity plays a key role in mathematical programming and economic analysis. After the pioneering paper of Arrow and Enthoven[1] many contributions were focused on the study of various generalizations of the concave functions, owing to different motivations arising from economic theory, management science, optimization theory, etc. (see, e.g., [2, 3, 4, 6, 11, 14, 16, 17, 18]). In spite of the large number of researches on generalized concavity, there are again, especially among theoretical economists, some misunderstandings concerning this topic. It is, e.g., the case of a quite recent article of A. Mukherji, appeared on Economic Letters[15].

The present paper is organized as follows: in Section 2 we make some remarks on the results obtained by A. Mukherji in [15]. In Section 3 we consider a mathematical programming problem with both inequality and equality constraints and give (sufficient) global optimality conditions under weaker assumptions than quasi-concavity. In Section 4 we present some first order local optimality conditions for a constrained optimization problem, with no (generalized) concavity assumption on the functions involved.

*The author is thankful to M. U. R. S. T. and CARIPLO for financial support of the present work.

2. Following A. Mukherji[15] consider the mathematical programming problems

$$\max_{x \in K_1} f(x) \tag{P$_1$}$$

and

$$\max_{x \in K_2} f(x), \tag{P$_2$}$$

where

$$K_1 = \{x | x \in X, h_i(x) = 0, i = 1, 2, \ldots, m\},$$
$$K_2 = \{x | x \in X, g_i(x) \geqq 0, i = 1, 2, \ldots, m\},$$

$X \subset R^n$ is an open set, $f, h_i, g_i : X \to R$ are differentiable functions on X.

Now consider (P_1) and assume further that every $h_i(x)$ is continuously differentiable at $\hat{x} \in X$ (the differentiability alone, as assumed by Mukherji, is not enough) and that the Jacobian matrix $\nabla h(\hat{x})$ is of full rank; it is then well known that if $\hat{x}$ is a point of local maximum for (P_1) there exists a vector $\hat{\lambda} \in R^m$ such that

$$\nabla f(\hat{x}) + \hat{\lambda} \nabla h(\hat{x}) = 0, \tag{1}$$

i.e. the Lagrangian function $L(x, \lambda) = f(x) + \lambda h(x)$ has a critical point at $(\hat{x}, \hat{\lambda})$.

As for the sufficient optimality conditions for problem (P_1), Mukherji presents the following

Theorem 1 *If $f(x)$ and $h_i(x)$, $i = 1, 2, \ldots, m$, are differentiable quasi-concave functions on X and if for some $\hat{x} \in K_1$ there is a vector $\hat{\lambda} \in R^m$ such that (1) holds, then $\hat{x}$ solves (P_1) provided $\nabla f(\hat{x}) \neq 0$ and $\hat{\lambda} \geqq 0$.*

About the previous assertion we make the following remarks:

Remark 1 It can be proved (see, e.g., [7, 8]) that a differentiable quasi-concave function $f(x)$, $x \in X \subset R^n$, is pseudo-concave at each point $x^0 \in X$ where $\nabla f(x^0) \neq 0$. We recall that $f(x)$ is said to be pseudo-concave at a point $x^0 \in X$ (with respect to X) if

$$(x - x^0)\nabla f(x^0) \leqq 0 \Longrightarrow f(x) \leqq f(x^0), \quad \forall x \in X$$

(see [14]).

In consequence of this fact, theorem 1 is contained in a result obtained by C. Singh[17] for a more general problem than (P_1).

Remark 2 On the ground of the previous remark it is therefore convenient to assume that the objective function is pseudo-concave at $\hat{x} \in K_1$ (with respect to X), so that also the case $\nabla f(\hat{x}) = 0$ is allowed. Moreover condition (1) is sufficient for $\hat{x}$ to solve problem (P_1) even if the quasi-concavity of the functions $h_i(x)$, $i = 1, 2, \ldots, m$, is substituted with the local quasi-concavity at $\hat{x}$ (with respect to X).

Remark 3 Condition (1) is sufficient for $\hat{x} \in K_1$ to solve problem (P_1) also if we assume that the Lagrangian function $L(x, \hat{\lambda})$ is pseudo-concave at $\hat{x}$ (with respect to X). We note that the linear combination of pseudo-concave or quasi-concave functions may not be respectively pseudo-concave or quasi-concave but also the reverse case does not hold: a function, given by the linear combination of some other functions can be pseudo-concave or quasi-concave even if the related components do not satisfy any (generalized) concavity conditions. Now if $L(x, \hat{\lambda})$ is pseudo-concave at $\hat{x} \in K_1$ and (1) holds, $L(x, \hat{\lambda})$ has a maximum at $\hat{x}$ on K_1 and it results, $\forall x \in K_1$, $L(x, \hat{\lambda}) = f(x)$.

As for what regards (P_2) Mukherji provides the following result, here presented with the variants due to our notation.

Theorem 2 *If $f(x), g_i(x), i = 1, 2, \ldots, m$, are differentiable quasi-concave functions on X and if there is $(\hat{x}, \hat{\lambda})$ satisfying*

$$\nabla f(\hat{x}) + \hat{\lambda} \nabla g(\hat{x}) = 0 \tag{2}$$

$$g(\hat{x}) \geqq 0, \quad \hat{\lambda} \geqq 0, \quad \hat{\lambda} g(\hat{x}) = 0, \tag{3}$$

then $\hat{x}$ solves (P_2) provided: i) $\nabla f(\hat{x}) \neq 0$ and ii) there exists some $x' \in K_2$ such that $g_i(x') > 0$ for all i.

Also on theorem 2 some remarks are appropriate.

Remark 4 What observed in remarks 1 and 2 can be applied also with regard to condition i) of theorem 2, in order to obtain, through (2) and (3) (see, e.g. [14], sufficient optimality conditions for (P_2). Assumption ii) is therefore superfluous.

Remark 5 It is well known that if each $g_i(x)$ is quasi-concave then K_2 is convex; obviously K_2 may be convex without the functions $g_i(x)$, $i = 1, 2, \ldots, m$, being quasi-concave. Even in this case it can be shown (see [6]) that conditions (2)–(3) are sufficient for $\hat{x}$ to solve (P_2), if $f(x)$ is pseudo-concave at $\hat{x}$.

3. Let us consider the following problem, embracing, as particular cases, problems (P_1) and (P_2).

$$\max_{x \in K_3} f(x), \tag{P_3}$$

where $K_3 = \{x \mid x \in X,\ g_i(x) \geqq 0\ \forall i \in M;\ h_j(x) = 0\ \forall j \in K\}$, X is an open subset of R^n, $M = \{1, 2, \ldots, m\}$, $K = \{1, 2, \ldots, k\}$, $f : X \to R$, $g : X \to R^m$, $h : X \to R^k$ are differentiable functions on X.

With respect to (P_3) we state some sufficient (local and global) optimality conditions which involve generalized concave functions and extend the results of Mukherji previously pointed out.

I) We first recall some definitions (see [11]). A differentiable function $f: X \to R$ is called *incave* or *η-concave* if

$$f(x^1) - f(x^2) \leqq \eta(x^1, x^2)\nabla f(x^2), \quad \forall x^1, x^2 \in X;$$

pseudo-incave or *η-pseudo-concave* if

$$\eta(x^1, x^2)\nabla f(x^2) \leqq 0 \Longrightarrow f(x^1) \leqq f(x^2), \quad \forall x^1, x^2 \in X;$$

quasi-incave or *η-quasi-concave* if

$$f(x^1) \geqq f(x^2) \Longrightarrow \eta(x^1, x^2)\nabla f(x^2) \geqq 0, \quad \forall x^1, x^2 \in X,$$

where $\eta(x^1, x^2)$ is some real vector-valued function defined on the Cartesian product $X \times X$.

Local definitions at a point $x^0 \in X$ can be given, likewise to what done for pseudo-concave and quasi-concave functions (see [13]). The following inclusion relationships hold for differentiable functions (see [3, 8, 13]):

a) A concave function is η-concave but the converse is not true; a pseudo-concave function is η-concave but the converse is not true.

b) A η-concave function is η-pseudo-concave and viceversa; i.e.(unlike pseudo-concave and concave functions) there is no distinction between η-concave and η-pseudo-concave functions.

c) A η-concave function is η-quasi-concave but the converse is not true; a quasi-concave function is η-quasi-concave but the converse is not true.

For $x^0 \in K_3$ let $I = \{i \mid g_i(x^0) = 0\}$, $J = \{i \mid g_i(x^0) > 0\}$. We can state now the following theorem.

Theorem 3 *Let $x^0 \in X$; suppose that $f(x)$ is η-concave at x^0 (with respect to X), $g_i(x), i \in I$, and $h_j(x), j \in K$, are η-quasi-concave at x^0 (with respect to X) and that exist vectors $u^0 \in R^m, v^0 \in R^k$ such that at x^0 are satisfied the conditions*

$$\nabla f(x^0) + u^0\nabla g(x^0) + v^0\nabla h(x^0) = 0 \tag{4}$$

$$g(x^0) \geqq 0, \quad h(x^0) = 0 \tag{5}$$

$$u^0 g(x^0) = 0, \quad u^0 \geqq 0, \; v^0 \geqq 0, \tag{6}$$

then x^0 solves (P_3).

Proof. consider the following problem

$$\max_{x \in K_4} f(x), \tag{P$_4$}$$

where $K_4 = \{x \mid x \in X, g_i(x) \geqq 0 \ \forall i \in M;\ h_j(x) \geqq 0 \ \forall j \in K\}$. Clearly $K_3 \subset K_4$ and if $S_4(x^0)$ is a sufficient condition for a point x^0 to be a solution of (P_4), then the condition

$$S_4(x^0),\ h(x^0) = 0 \tag{7}$$

will be a sufficient condition for x^0 to be a solution of (P_3). Now, under the assumption made on $f(x)$; $g_i(x)$, $i \in I$; $h_j(x)$, $j \in K$, $S_4(x^0)$ has been obtained for (P_4) by Kaul and Kaur in [13], theorem 3.3; on the ground of their results (7) are just conditions (4)–(6) and the proof is so complete.

Remark 6 As η-concave functions are those (differentiable) functions whose every stationarity point is a global maximum, the thesis of theorem 3 follows also if we assume that the Lagrangian function $L(x, u^0, v^0) = f(x) + u^0 g(x) + v^0 h(x)$ is η-concave and conditions (4)–(6) hold.

II) We obtain now for problem (P_3) a global optimality condition with no (generalized) concavity assumption on the functions $h_j(x)$, $j \in K$. Let $x^0 \in K_3$ and let:

$$C_0 = \{x \mid x \in R^n, x\nabla g_i(x^0) \geqq 0, \quad \forall i \in I\};$$

C_0 is the so-called *linearizing cone.*

$$B_0 = \left\{x \mid x \in R^n,\ x = \sum_{i \in I} \lambda_i \nabla g_i(x^0),\ \lambda_i \geqq 0\right\};$$

$$L_0 = \{x \mid x \in R^n,\ x = v\nabla h(x^o), v \in R^k\};$$

B_0 and L_0 can be called the *cone of gradients.* Consider then the *positive polar cone* of the set $A \subset R^n$, i.e. the set A^* defined through $x \in A^* \iff xy \geqq 0 \quad \forall y \in A$.

Lemma 1 *(see [10]). It is $B_0 = C_0^*$.*

Theorem 4 *Let $x^0 \in X$, $f(x)$ be pseudo-concave at $x^0, g_i(x), i \in I$, quasi-concave at $x^0, h_j(x), j \in K$, differentiable at x^0 and $L_0 \subset C_0^*$. Let conditions (4), (5) and $u^0 g(x^0) = 0, u^0 \geqq 0, v^0 \in R^k$, be verified. Then x^0 solves (P_3).*

Proof. As $L_0 \subset C_0^*$, thanks to lemma 1 it will be

$$v\nabla h(x^0) = \sum_{i \in I} \lambda_i \nabla g_i(x^0), \quad \lambda_i \geqq 0, \quad i \in I.$$

Let us now define $q \in R^m$ as:

$$q_i = \begin{cases} \lambda_i & \forall i \in I \\ 0 & \forall i \in J. \end{cases}$$

From (4) we obtain

$$\nabla f(x^0) + u^0 \nabla g(x^0) + q \nabla g(x^0) = 0.$$

Letting $w^0 = u^0 + q$ we obtain

$$\nabla f(x^0) + w^0 \nabla g(x^0) = 0$$

$$w^0 g(x^0) = u^0 g(x^0) + q g(x^0) = 0$$

$$g(x^0) \geqq 0, \quad h(x^0) = 0, \quad w^0 = u^0 + q \geqq 0.$$

From these relations and from the assumptions on $g_i(x)$, $i \in I$, the inequalities

$$g_i(x) \geqq g_i(x^0) = 0, \quad \forall i \in I, \quad \forall x \in K_3,$$

imply

$$(x - x^0) \nabla g_i(x^0) \geqq 0, \quad \forall i \in I, \quad \forall x \in K_3.$$

Thus, being $w^0 \geqq 0$ and being zero each multiplier associated to $g_i(x^0)$, $i \in J$, we can write

$$(x - x^0) w^0 \nabla g(x^0) \geqq 0, \quad \forall x \in K_3.$$

This last inequality implies therefore that it is

$$(x - x^0) \nabla f(x^0) \leqq 0, \quad \forall x \in K_3,$$

which, by the pseudo-concavity of $f(x)$ at x^0, gives

$$f(x^0) \geqq f(x), \quad \forall x \in K_3.$$

Remark 7 The thesis of theorem 4 follows even if we assume the more general conditions: $f(x)$ be η-concave at x^0 and $g_i(x)$, $i \in I$, η-quasi-concave at x^0.

In order to develop further our analysis we need some definitions.

Let $S \subset R^n$ and let $\overline{S}$ denote its closure. If $x^0 \in \overline{S}$, the *contingent cone to S at x^0* is the set (see, e.g., [9, 12]):

$$T(S, x^0) = \left\{ x \in R^n \;\middle|\; \exists \{t_k\} \subset R_+,\ \exists \{x^k\} \subset S : \lim_{k \to +\infty} t_k = 0, \lim_{k \to +\infty} \frac{x^k - x^0}{t_k} = x \right\}.$$

The set $T(S, x^0)$ is also called the *Bouligand-Abadie tangent cone to S at x^0* and can be defined in many equivalent manners, e.g.:

a) $T(S,x^0) = \left\{ x \in R^n \,\middle|\, \exists \{t_k\} \subset R_+,\ \exists \{x^k\} \subset S : \lim_{k\to+\infty} x^k = x^0,\ \lim_{k\to+\infty} t_k(x^k - x^0) = x \right\};$

b) $T(S,x^0) = \left\{ x \in R^n \,\middle|\, \exists \{t_k\} \subset R_+,\ \lim_{k\to+\infty} t_k = 0, \exists \{x^k\} \to x : x^0 + t_k x^k \in S \right\};$

c) $T(S,x^0) = \bigcap_{k=1}^{+\infty} \overline{\text{cone}}\,(S \cap N_{\frac{1}{k}}(x^0), x^0),$

where $N_\alpha(x^0)$ is an open ball about x^0, of radius α, and cone (A, x^0) is the *cone generated by A at x^0*: cone $(A,x^0) = \{y \in R^n \mid y = \lambda(x - x^0),\ \lambda \geqq 0,\ x \in A\}$.

From characterization c) it appears that the contingent cone to S at x^0 is a closed cone with apex at zero, but it is not necessarily convex. We note that if x^0 is an isolated point of S, then $T(S,x^0) = 0$ and if $x^0 \in \text{int}\, S$, then $T(S,x^0) = R^n$.

The contingent cone to S at x^0 genuinely represents a local conical approximation of S in a neighborhood of x^0 but generally there is no inclusion relation between S and $T(S,x^0)$. However, if S is convex, or even *convex at* x^0 (or *star-shaped at* x^0, i.e. $\lambda x + (1-\lambda)x^0 \in S,\ \forall x \in S,\ \forall \lambda \in [0,1]$), then $T(S,x^0)$ is a closed convex cone and in this case $S \subset T(S,x^0)$.

The closure of the convex hull of $T(S,x^0)$ is called the *pseudotangent cone* to S at x^0 and will be denoted by $P(S,x^0)$:

$$P(S,x^0) = \overline{\text{conv}}\, T(S,x^0).$$

Note that $T^*(S,x^0) = P^*(S,x^0)$.

Consider now problem (P_3), where X is any subset of R^n, *not necessarily open* (i.e. X is a *set constraint*) and the functions f, g_i, h_j are defined on an open set $D \subset R^n$ containing X. Denote by (P'_3) this problem; it is known that if $x^0 \in K_3$ is a local solution of (P'_3) and at x^0 a suitable constraint qualification (e.g. the Guignard-Gould-Tolle constraint qualification, but not *any* constraint qualification) is satisfied, then the Kuhn-Tucker conditions hold for (P'_3). In any case we have the following result, due to Gould and Tolle[10].

Denote by D_0 the linearizing cone formed by the constraints h_j:

$$D_0 = \{x \mid x \in R^n,\ x\nabla h_j(x^0) = 0,\ \forall j\}$$

and let $E_0 = C_0 \cap D_0$.

Theorem 5 *If x^0 is a local solution of (P'_3), with f, g_i $(i \in I)$, h_j differentiable at x^0, g_i $i \notin I$, are continuous at x^0, there exist scalars $\lambda_i \geqq 0$ $(i \in I)$, $\mu_j \in R$, such that*

$$-\left[\nabla f(x^0) + \sum_{i \in I} \lambda_i \nabla g_i(x^0) + \sum_{j=1}^{k} \mu_j \nabla h_j(x^0)\right] \in T^*(K_3, x^0) \setminus E_0^* \cup \{0\}.$$

The necessary condition given in Theorem 5 becomes sufficient for x^0 to be a solution of (P'_3), as it is a special case of the following more general result. Let us premise the following definition.

The set $D \subset R^n$ is said to be *η-invex* at $x^0 \in D$ when there exists a function $\eta : D \times D \to R^n$ such that $\eta(x, x^0) \in P(D, x^0)$, $\forall x \in D$.

This definition obviously generalizes the concept of convexity of a set and even of convexity of a set at a point x^0("star-shaped set").

An example of η-invex set is : $A = \{x \in R \mid x = 0 \text{ or } x = 1/n, n \in N\}$; this set is η-invex at $x^0 = 0$ for $\eta(x, x^0) = x - x^0$ and also for, e.g., $\eta(x, x^0) = (x - x^0)^2$.

Theorem 6 *Consider (P'_3); let $x^0 \in K_3$, let K_3 be η-invex at x^0 and f be η-pseudo-concave at x^0, with respect to the same function η. If g_i $(i \in I)$ and h_j are differentiable at x^0 and there exist scalars $\lambda_i \geqq 0$ $(i \in I)$ and $\mu_j \in R$ such that*

$$-\left[\nabla f(x^0) + \sum_{i \in I} \lambda_i \nabla g_i(x^0) + \sum_{j=1}^{k} \mu_j \nabla h_j(x^0)\right] \in P^*(K_3, x^0),$$

then x^0 solves (P'_3).

Proof. Let $\varphi = (g_i, h_j)$, $i \in I$ and $j = 1, \ldots, k$ and denote by $B = \varphi(K_3)$ the image space of the constraints. From the invexity of K_3 at x^0, it will exist a vector-valued function $\eta : D \times D \to R^n$ such that $\eta(x, x^0) \in P(K_3, x^0)$, $\forall x \in K_3$.

Given an arbitrary $y \in T(K_3, x^0)$, by a well-known property of tangent cones it is $\nabla \varphi(x^0) y \in T(B, \varphi(x^0))$. Being $\eta(x, x^0) \in P(K_3, x^0)$, it will be $\eta(x, x^0) = \lim_{n \to +\infty} y^n$, with $y^n \in \operatorname{conv} T(K_3, x^0)$, i.e. $\eta(x, x^0) = \lim_{n \to +\infty} (t_n y_1^n + (1 - t_n) y_2^n)$, with $y_1^n, y_2^n \in T(K_3, x^0)$, $0 \leqq t_n \leqq 1$. Then it results:

$$\begin{aligned} \nabla \varphi(x^0) \eta(x, x^0) &= \lim_{n \to +\infty} [t_n \nabla \varphi(x^0) y_1^n + (1 - t_n) \nabla \varphi(x^0) y_2^n] \\ &\in \lim_{n \to +\infty} [t_n T(B, \varphi(x^0)) + (1 - t_n) T(B, \varphi(x^0))] \\ &= \overline{\operatorname{conv}} T(B, \varphi(x^0)) = P(B, \varphi(x^0)). \end{aligned}$$

Now let us prove that $u = (\lambda_i, \mu_j)$, $i \in I, \lambda_i \geqq 0$, belongs to $P^*(B, \varphi(x^0))$. Let us consider the quantity ut, where $t \in T(\varphi(K_3), \varphi(x^0))$ i.e. $t =$

$\lim_{n\to+\infty} \lambda_n^+(\varphi_n(K_3)-\varphi(x^0))$. Then $ut = \lim_{n\to+\infty} \lambda_n^+(u\varphi_n(K_3)-u\varphi(x^0)) = \lim_{n\to+\infty} \lambda_n^+ (u\varphi_n(K_3)) \geqq 0$. Then if $p \in P(\varphi(K_3), \varphi(x^0))$, i.e., $p = \lim_{n\to+\infty}\{t_n\tau_1^n + (1-t_n)\tau_2^n)\}$, with $0 \leqq t_n \leqq 1$, $\tau_1^n, \tau_2^n \in T(\varphi(K_3), \varphi(x^0))$, it is $up = \lim_{n\to+\infty} t_n u\tau_1^n + (1-t_n)u\tau_2^n \geqq 0$, i.e. $u \in P^*(\varphi(K_3), \varphi(x^0))$.

We have therefore $[u(\nabla\varphi(x^0)\eta(x, x^0))] \geqq 0$.

From $-[\nabla f(x^0 + u\nabla\varphi(x^0)] \in P^*(K_3, x^0)$, it follows, $\forall x \in K_3$, $-\nabla f(x^0)\eta(x, x^0) \geqq u\nabla\varphi(x^0)\eta(x, x^0) \geqq 0$, i.e. $\nabla f(x^0)\eta(x, x^0) \leqq 0$, $\forall x \in K_3$ and by the η-pseudo-concavity of f at x^0 it is $f(x) \leqq f(x^0)$, $\forall x \in K_3$.

4. We obtain now some first order sufficient local optimality conditions for (P_3), under no (generalized) concavity assumption on the functions involved in the same problem.

Lemma 2 *(see[12]). Let x^0 be a point of the closed set $S \subset R^n$ and let $f : R^n \to R$ be differentiable at x^0. If*

$$y\nabla f(x^0) < 0, \ \forall y \in T(S, x^0) \quad y \neq 0, \tag{8}$$

then there exists a neighborhood $N(x^0)$ of x^0 and a positive number α such that

$$f(x^0) \geqq f(x) + \alpha\|x - x^0\|, \ \forall x \in S \cap N(x^0),$$

i.e. x^0 is an isolated local maximum point of $f(x)$ on S.

We must point out that condition (8) cannot be verified if $x^0 \in \operatorname{int} S$, as in this case $T(S, x^0) = R^n$. Moreover, even if S is a closed convex set, it is not allowed to substitute $T(S, x^0)$ with S, being in this case $S \subset T(S, x^0)$. All what can be obtained, when this substitution is made, is an isolated *radial* local maximum point.

Lemma 3 *Let x^0 be a point of the closed convex set $S \subset R^n$ and let $f : R^n \to R$ be differentiable at x^0. If*

$$(x - x^0)\nabla f(x^0) < 0, \ \forall x \in S \setminus \{x^0\}, \tag{9}$$

then x^0 is an isolated radial local maximum point of $f(x)$ on S.

Proof. Consider the function of one variable $\varphi : [0, 1] \to R$, defined as $\varphi(\lambda) = f(x^0 + \lambda(x - x^0))$, $x \in S$. Being $f(x)$ differentiable at x^0, it will be $\varphi'(0) = (x - x^0)\nabla f(x^0) < 0$, $\forall x \in S \setminus \{x^0\}$. Hence $\varphi'(0) < 0$, i.e. $\lambda = 0$ is an isolated local maximum for φ on $[0, 1]$. Therefore there exists a scalar λ_0, $0 < \lambda_0 \leqq 1$, such that $\varphi(0) > \varphi(\lambda)$, $\forall\lambda \in (0, \lambda_0]$, i.e. $f(x^0) > f(x^0 + \lambda(x - x^0))$, $\forall\lambda \in (0, \lambda_0]$.

As S is convex and $0 < \lambda_0 \leqq 1$, it will be $x^0 + \lambda(x - x^0) \in S$ and therefore the last inequality characterizes x^0 as an isolated radial local maximum point of f on S.

Remark 8 The convexity assumption on S can be substituted with one of the more general conditions:

a) the set S is star-shaped at x^0; or

b) there exists a sequence $\{\lambda_n\} \to 0^+$ such that $x^0 + \lambda_n(x - x^0) \in S$.

It is well known (see e.g. [18]) that a necessary and sufficient condition that x^0 be a *global* maximum point over a set S is that x^0 be a *global* radial maximum point on S. However a *local* radial maximum point on a set is not necessarily a local maximum point on the same set: the first property is only necessary for the second one as it is shown by the following example due to Peano. Consider the function $f(x, y) = (y^2 - x)(x - 2y^2)$ defined on R^2; the point $(0, 0)$ is a local radial maximum point of $f(x, y)$, but not a local maximum point for the same function. It is therefore interesting to look for conditions assuring that a local radial maximum is also a local maximum. Here we consider the following ones:

I) If $f(x)$ is quasi-concave, then a local radial maximum point is a local maximum point of $f(x)$, on the ground of a result of Thompson and Parke [18]. On the other hand, if $f(x)$ is quasi-concave at x^0, from relation (9) it follows $f(x) < f(x^0)$, $\forall x \in S \setminus \{x^0\}$, i.e. x^0 is an isolated global maximum point of $f(x)$ on S.

II) If S is a *closed cone* with vertex at x^0 and if we denote by D the set of the directions of S, i.e.

$$D = \{y \in R^n \mid y \neq 0,\ x^0 + y \in S\},$$

it is then $T(S, x^0) \setminus \{0\} = D$ and therefore the condition

$$y\nabla f(x^0) < 0,\ \forall y \in D$$

is sufficient for x^0 to be an isolated local maximum of $f(x)$ on S.

III) See [5] for other less operative conditions assuring that a radial maximum is also a local one.

Let us now return to problem (P_3).

Theorem 7 *Let $x^0 \in K_3$ and let conditions (4), (5), $u^0 g(x^0) = 0$, $u^0 \geqq 0$ be verified. Suppose further that there is no vector $y \neq 0$ for which it is $y\nabla f(x^0) = 0$ and such that the system*

$$\begin{cases} y\nabla g_i(x^0) \geqq 0, & \forall i \in I \\ y\nabla h_j(x^0) = 0, & \forall j \in K \end{cases} \tag{10}$$

admits solution. Then x^0 is an isolated local maximum point of $f(x)$ on K_3.

Proof. Under the assumptions of the theorem it is

$$y\nabla f(x^0) = -y\left(\sum_{i\in I} u_i^0 \nabla g_i(x^0) + v^0 \nabla h(x^0)\right)$$

and for all $y \neq 0$ solution of (10) it is $y\nabla f(x^0) \leqq 0$, but being $y\nabla f(x^0) = 0$ excluded by assumption, it will be $y\nabla f(x^0) < 0$, $\forall y \in E_0$, where E_0 is the solutions set of (10), i.e. the *linearizing cone* at x^0 for problem (P_3).

Now, as it is always $T(K_3, x^0) \subset E_0$ (see e.g. [12]), it will be $y\nabla f(x^0) < 0$, $\forall y \in T(K_3, x^0)$, $y \neq 0$, and hence the result follows from Lemma 2.

Theorem 8 *Let $x^0 \in K_3$ and let the rank of A be n, where*

$$A = \begin{bmatrix} \nabla f(x^0) \\ \nabla g_I(x^0) \\ \nabla h(x^0) \end{bmatrix},$$

$g_I(x^0)$ being the vector-valued function whose components are $g_i(x)$, $i \in I$. If there exist multipliers $u_i > 0$, $i \in I$, and $v \in R^k$ such that

$$\nabla f(x^0) + \sum_{i\in I} u_i \nabla g_i(x^0) + v\nabla h(x^0) = 0, \tag{11}$$

then x^0 is an isolated local maximum point of $f(x)$ on K_3.

Proof. Let y be a solution of the system

$$\begin{cases} y\nabla f(x^0) \geqq 0 & \\ y\nabla g_i(x^0) \geqq 0, & i \in I \\ y\nabla h_j(x^0) = 0, & j \in K. \end{cases} \tag{12}$$

Obviously the solutions set of system (12) is non void; then, thanks to (11), it is

$$0 \leqq y\nabla f(x^0) = -\left(\sum_{i\in I} u_i y \nabla g_i(x^0) + \sum_{j=1}^{k} v_j y \nabla h_j(x^0)\right) = -\sum_{i\in I} u_i y\nabla g_i(x^0) \leqq 0.$$

This implies (it is $u_i > 0$, $\forall i \in I$):

$$y\nabla f(x^0) = 0;\ y\nabla g_i(x^0) = 0,\ \forall i \in I;\ y\nabla h_j(x^0) = 0,\ \forall j \in K.$$

Since n among the vectors $\nabla f(x^0)$; $\nabla g_i(x^0)$, $i \in I$; $\nabla h_j(x^0), j \in K$, are linearly independent, then system (12) admits the unique solution $y = 0$.

Now let us assume absurdly that x^0 is not an isolated local maximum for problem (P_3). Then there exists a sequence $\{x^k\} \subset K_3$, $x^k \neq x^0$, $\lim_{k\to+\infty} x^k = x^0$, such that $f(x^k) \geqq f(x^0)$ for all k. The generic vector x^k can then be written as $x^k = x^0 + \delta_k y^k$, where $\|y^k\| = 1$ and δ_k are positive numbers with $\lim_{k \to +\infty} \delta_k = 0$.

By the Bolzano-Weierstrass theorem, the sequence $\{y^k\}$ contains a convergent subsequence we can again denote by $\{y^k\}$; so $\lim_{k \to +\infty} y^k = y$, with $\|y^k\| = 1$.

From $f(x^k) \geqq f(x^0)$; $g_i(x^k) - g_i(x^0) = g_i(x^k) \geqq 0$, $i \in I$; $h_j(x^k) - h_j(x^0) = 0$, $j \in K$, we have

$$\frac{f(x^0 + \delta_k y^k) - f(x^0)}{\delta_k} \geqq 0;$$

$$\frac{g_i(x^0 + \delta_k y^k) - g_i(x^0)}{\delta_k} \geqq 0, \quad i \in I;$$

$$\frac{h_j(x^0 + \delta_k y^k) - h_j(x^0)}{\delta_k} = 0, \quad j \in K.$$

Taking the limits of these quotients, for $k \to +\infty$, namely taking the directional derivatives in the direction y ($\|y\| = 1$), we have

$$y \nabla f(x^0) \geqq 0; \; y \nabla g_i(x^0) \geqq 0, \; i \in I; \; y \nabla h_j(x^0) = 0, \; j \in K.$$

Thus the direction y is a solution of system (12) and, under our assumptions, this leads to a contradiction.

We now apply the results of Lemma 3 to a constrained optimization problem. For simplicity we refer to problem (P_2).

Following [16] a differentiable function $f : X \to R$ is called *strictly pseudo-concave at* x^0, with respect to X, if

$$x \in X, \; x \neq x^0, \; f(x) \geqq f(x^0) \Rightarrow (x - x^0) \nabla f(x^0) > 0.$$

Theorem 9 *Let $x^0 \in K_2$; if $g_i(x)$, $i = 1, 2, \ldots m$, are quasi-concave at x^0, conditions (2)-(3) are satisfied at x^0 and there exists at least a positive multiplier $\hat{\lambda}_r > 0$ in correspondence of which $g_r(x)$ is strictly pseudo-concave at x^0, then x^0 is an isolated local radial maximum point of $f(x)$ on K_2.*

Proof. The proof is quite similar to the one of theorem 4; note first that on the ground of the assumptions it is K_2 star-shaped at x^0. Let $x \in K_2$; then for each $i \in I$ it will be $(x - x^0) \nabla g_i(x^0) \geqq 0$. Being $\hat{\lambda}_i \geqq 0$, $\forall i \in I$, and being strictly pseudo-concave at x^0, in correspondence to a positive multiplier, at least one function $g_r(x)$, $r \in I$, it will be

$$(x - x^0) \sum_{i \in I} \hat{\lambda}_i \nabla g_i(x^0) > 0, \; \forall x \in K_2 \setminus \{x^0\}.$$

Therefore, thanks to condition (2) it follows

$$(x - x^0) \nabla f(x^0) < 0, \; \forall x \in K_2 \setminus \{x^0\}$$

and, thanks to Lemma 3, the thesis follows.

Remark 9 In the previous theorem the quasi-concavity at x^0 of the functions $g_i(x)$, can be substituted by the more general η-quasi-concavity at x^0. The strict pseudo-concavity of at least one function $g_r(x)$, together with the positiveness of the associated multiplier, can be substituted by the assumption: the unique point of the hyperplane $(x - x^0)\nabla f(x^0) = 0$ which belongs to K_2 is just x^0. As already pointed out, if $f(x)$ is quasi-concave, theorem 7 provides an isolated global maximum point for (P_2), but, on the ground of what said in section 2 and contrary to what asserted in [4], this is only an apparent weakening of the existing results.

References

[1] K.J. Arrow and A.C. Enthoven, *Quasi-concave programming*, Econometrica, **29** (1961), 799–800.

[2] M. Avriel, W.E. Diewert, S. Schaible and I. Zang, *Generalized concavity*, Plenum Press, New York, 1988.

[3] A. Ben-Israel and B. Mond, *What is invexity?*, J. Austral. Math. Soc., Ser. B, **28** (1986), 1–9.

[4] C. R. Bector, S. Chandra and M. K. Bector, *Sufficient optimality conditions and duality for a quasiconvex programming problem*, J. O. T. A., **59** (1988), 209–221.

[5] A. Capozzi and C. Resina, *Caratterizzazione di condizioni di ottimalita' in ottimizzazione non lineare*, Bollettino U. M. I., **18**-A (1981), 259–265.

[6] G. Giorgi, *Quasi-convex programming revisited*, Calcolo, **21** (1984), 307–316.

[7] G. Giorgi, *A note on quasiconvex functions that are pseudoconvex*, Trabajos de investigacion operativa, **2** (1987), 111–116.

[8] G. Giorgi, *A note on the relationships between convexity and invexity*, J. Austral. Math. Soc., Ser. B, **32** (1990), 97–99.

[9] G. Giorgi and A. Guerraggio, *On the notion of tangent cone in mathematical programming*, Optimization, **25** (1992), 11–23.

[10] F. J. Gould and J. W. Tolle, *Geometry of optimality conditions and constraint qualifications*, Math. Programming, **2** (1972), 1–18.

[11] M. A. Hanson, *On sufficiency of the Kuhn-Tucker conditions*, J. Math. Anal. Appl., **80** (1981), 545–550.

[12] M. R. Hestenes, *Optimization theory: the finite dimensional case*, J. Wiley, New York, 1975.

[13] R. N. Kaul and S. Kaur, *Optimality criteria in nonlinear programming involving nonconvex functions*, J. Math. Anal. Appl., **105** (1985), 104–112.

[14] O. L. Mangasarian, *Nonlinear programming*, Mc GrawHill, New York, 1969.

[15] A. Mukherji, *Quasi-concave optimization. Sufficient conditions for a maximum*, Economic Letters, **30** (1989), 341–343.

[16] J. Ponstein, *Seven kinds of convexity*, SIAM Review, **9** (1967), 115–119.

[17] C. Singh, *Sufficient optimality criteria in nonlinear programming for generalized equality-inequality constraints*, J. O. T. A., **22** (1977), 631–635.

[18] W. A. Thompson and D. W. Parke, *Some properties of generalized concave functions*, Op. Res., **21** (1973), 305–313.

Expectations Driven Nonlinear Business Cycles

Jean-Michel Grandmont
CEPREMAP, 142, Rue du Chevaleret
75013 Paris, FRANCE *

There are two traditional conflicting views about the workings of a market economy. The socalled "Classical" school stresses the virtues of free markets and their intrinsic internal stability. According to that school, persistent, nonexplosive fluctuations should be essentially due to repeated external macroeconomic shocks to the "fundamental" characteristics of the system (technology, tastes, resources). Models of the "Classical" vintage typically assume that expectations are self-fulfilling, i.e. every individual's assessment of the future (a probability distribution) is correct at any moment given his information. According to that view, expectations cannot be an independent source of economic fluctuations. There is another view, often associated to socalled "Keynesian" thinking, that starts with the observation that individual prophecies about aggregate outcomes often tend to be self-fulfilling if shared by sufficiently many people, and that suggests this phenomenon as a potential source of significant internal instability of socioeconomic systems in the absence of any outside regulation. According to that view, aggregate changes of expectations are likely to occur rather frequently in an unpredictable way and are potentially major sources of endogenous business fluctuations (*market psychology, animal spirits*).

There has been a new spur of interest in the "Keynesian" approach in the past decade or so. The new class of models shows that given the "fundamental" characteristics of the system, a large multiplicity of such expectations-driven business cycles can occur even when expectations are required to be correct. Models of this type seem to succeed, under increasingly plausible assumptions, in producing complex trajectories that tend to replicate more and more closely the qualitative dynamic regularities exhibited by actual macroeconomic time series. The lecture is devoted to an introduction to the concepts and methods that are employed in the study of such expectations-driven business cycles and of the macroeconomic patterns they generate[1].

*A full version of the paper is published by Westdeutscher Verlag for the German Academy of Sciences (1992), and in the proceedings of a Stockholm Conference (FIEF studies on business cycles) by Oxford University Press (1993).

[1]The material of this lecture is intended only to be a simple introduction to the subject.

The Framework

In order to fix ideas, we shall focus throughout on a simple version of the socalled "overlapping generations" model. People are assumed to participate in the market for two periods. They work and save their wages in the form of money when "young" (in the first period of their lifes) and consume when "old" (one would get similar results if people were assumed to work and consume in both periods). Generations overlap so that old members of a generation coexist and trade in any given period with the young members of the following generation. Population is stationary and there are no bequests. On the other hand, one unit of output (a perishable consumption good) can be produced from one unit of output in the same period (productive capital will be considered at the end of the lecture). Finally, the total money stock is assumed to be constant over time.

The essential feature of such an overlapping generations model, which will enable us to generate expectations-driven endogenous fluctuations, is that it has an important built-in capital market imperfection : the assets of the members of a generation are constrained to be zero at the end of their lifes. Such capital market imperfections are in a sense necessary to ensure that traders do not arbitrage away endogenous fluctuations although they are correctly foreseen (another possibility, not considered here, is to assume a high time preference so that traders do not care arbitraging away the fluctuations ; on this point see Boldrin's surveys (1988, 1991)).

The particular "story" we use to illustrate the analysis is chosen for its simplicity but the argument is much more general. For instance, one would get the same *local* results (i.e. near a steady state), if we assumed that people cared about the welfare of their offsprings and allowed bequests, but made the (realistic) assumption that nobody can borrow against the income of their children. Or if we assumed that people are infinitely long-lived but face liquidity (cash-in-advance) constraints expressing the difficulty of workers to borrow against future labor income. In this latter case, the interpretation of the length of a period would be much shorter — say a few months — than in the case of the overlapping generations model (see Woodford (1986)).

1. Deterministic Fluctuations

The "fundamentals" of the simple economic system under consideration are constant. We wish to analyze first how deterministic endogenous fluctuations can nevertheless arise.

For more extensive (and often more technical) surveys, see Baumol and Benhabib (1989), Boldrin (1988, 1991), Boldrin and Woodford (1990), Brock (1988), Brock and Dechert (1991), Chiappori and Guesnerie (1991).

Let us assume for simplicity that the aggregate behavior of all members of a generation can be represented by a "representative" houdehold. A representative household born at date t supplies the quantity of labor $0 \leqq l_t \leqq l^*$ in the current period (where $l^* > 0$ is the labor endowment) saves the amount of money $m_t \geqq 0$, and consumes $c_{t+1} \geqq 0$ of the good at date $t+1$. Tastes are represented by the separable utility function $V_1(l^* - l_t) + V_2(c_{t+1})$, in which $l^* - l_t$ can be interpreted as "leisure".

Markets are competitive. In equilibrium, the real wage should accordingly be equal to the marginal productivity of labor, which is 1. So we can identify in each period the money wage $w_t > 0$ with the price of the good p_t. Under this convention, the behavior of a representative household born at t is to maximize his utility function under the budget constraints $p_t l_t = m_t = p_{t+1} c_{t+1}$, in which p_{t+1} is the price expected to prevail in the future. The first order condition of this problem is $l_t V_1'(l^* - l_t) = c_{t+1} V_2'(ct + 1)$, in which $V_1'(l^* - l_t)$ is the marginal utility of leisure and $V_2'(c_{t+1})$ is the marginal utility of future consumption. If we identify the left hand and right hand sides of this equality with $v_1(l_t)$ and $v_2(c_{t+1})$, this yields

$$v_1(l_t) = v_2(c_{t+1}). \tag{1.1}$$

Under the usual assumption that marginal utility is decreasing, the function v_1 is increasing and has therefore an inverse v_1^{-1}. Then (1.1) can be rewritten as

$$l_t = v_1^{-1}[v_2(c_{t+1})] \equiv \chi(c_{t+1}). \tag{1.2}$$

The graph of χ is the locus of all optimal pairs (l_t, c_{t+1}) chosen by a trader and represents accordingly his *offer curve*. In equilibrium at t, the young household's labor supply l_t is equal to output y_t , which is itself equal to the old's consumption c_t. Under perfect foresight, expected consumption c_{t+1} is equal to the future actual output level y_{t+1}. Thus a deterministic equilibrium with perfect foresight is characterized by a sequence of outputs $y_t > 0$ that satisfy $y_t = \chi(y_{t+1})$ for all $t \geqq 1$.

The occurrence of cyclic deterministic equilibria will depend therefore upon the shape of the offer curve. Increasing the price of future consumption induces as usual substitution and income effects. The substitution effect results from the fact that future consumption becomes more expensive than current leisure. It should thus generate less future consumption and more leisure, or equivalently a lower current labor supply. The income effect is associated to the loss of purchasing power resulting from a higher price. It leads to a lower consumption of both the good and leisure, hence more labor supplied. Both effects work in the same direction for future consumption but in opposite directions for the current labor supply. The function χ representing the offer curve is thus increasing when the substitution effect dominates, decreasing otherwise. If one looks back at the definition of the offer curve given in (1.1), one sees that the substitution effect dominates whenever the "concavity" of the utility of future consumption, as measured by $-cV_2''(c)/V_2'(c)$, is small (less than 1) and that the income effect dominates when this concavity is large.

If the substitution effect dominates everywhere, χ is monotonically increasing also everywhere and there can be no cycle of period $k \geqq 2$. This is a case where the Classical viewpoint prevails : no persistent endogenous deterministic fluctuations can be observed. But assume that there is a strong conflict between substitution and income effects, so that the function χ is single peaked with a maximum occurring at the output level y^*, and displays a big hump : $\chi(y^*)$ is large while the second and third iterates, i.e. $\chi^2(y^*) = \chi(\chi(y*))$ and $\chi^3(y^*) = \chi(\chi^2(y^*))$, are smaller than y^*. This is a situation where the Keynesian viewpoint is valid, as it gives rise to multiple persistent nonexplosive endogenous fluctuations. In fact this configuration leads to the existence of a cycle of period 3. Such a cycle can be identified with a fixed point of χ^3 that differs from a stationary state. In the situation described here, one has $\chi^3(y) > y$ for $y > 0$ small while $\chi^3(y^*) < y^*$, and χ^3 must accordingly have a fixed point between 0 and y^*. Then from a theorem by Sarkovskii, there are in fact *infinitely many cycles with perfect foresight* : at least one cycle with period k for every positive integer k (Benhabib and Day (1982), Grandmont (1985a)).

2. Stochastic Endogenous Fluctuations

The foregoing deterministic cycles arise through a mechanism of self-fulfilling prophecies. Although the fundamentals are constant over time, traders collectively predict that prices and quantities will fluctuate and this prophecy turns out to be correct in equilibrium. There is no incentive for an individual household to deviate as he has no impact on the market as whole. There is no reason to stop there and one should also ask whether the same phenomenon can occur when traders predict that stochastic fluctuations will occur. The answer is indeed yes. There are also many nonexplosive stochastic equilibria generated by self-fulfilling expectations that vary randomly when the income effect is significant (Azariadis (1981), Azariadis and Guesnerie (1986), Chiappori and Guesnerie (1989), Farmer and Woodford (1987), Grandmont (1985b, 1986, 1989)).

Assume that a representative household believes at date t that the future price p_{t+1} of the good will be random. He seeks to maximize the expected value of his utility function, i.e. $V_1(l^* - l_t) + E_t V_2(c_{t+1})$, in which future consumption is now random since it has to satisfy the budget constraints $p_t l_t = m_t = p_{t+1} c_{t+1}$. The first order condition asociated to this decision problem is the analogue of (1.1)

$$v_1(l_t) = E_t v_2(c_{t+1}). \tag{2.1}$$

In equilibrium one has here again $y_t = l_t = c_t$. Under the assumption of self-fulfilling expectations, expected consumption c_{t+1} is equal to actual future output y_{t+1}. Thus a stochastic equilibrium with self-fulfilling expectations is characterized by a sequence of random outputs $y_t > 0$ such that

$$v_1(y_t) = E_t[v_2(y_{t+1})] \tag{2.2}$$

for all $t \geqq 1$. In the expression (2.2), the symbol E_t means that people condition their expectations at t on the information available at that date, which can be described by the values taken by an arbitrary exogenous stochastic process of signals s_t, that is

$$E_t = E[\ \cdot \mid (s_t, s_{t-1}, ...)].$$

The signals s_t are often referred to as "sunspots", as they describe random variables, the variations of which do not influence at all the "fundamentals" of the system. Yet if enough people base their expectations on such "sunspots", their variations may be translated into actual fluctuations of equilibrium prices and quantities, even under the constraint that expectations be self-fulfilling.

The general solution of (2.2) is

$$v_2(y_{t+1}) = v_1(y_t) + \varepsilon_{t+1} \tag{2.3}$$

in which ε_{t+1} is an arbitrary process of stochastic "shocks" satisfying $E_t e_{t+1} = 0$. It takes indeed the form generally encountered in empirical studies of economic time series, with an important difference in interpretation : *whereas the shocks ε_{t+1} are usually interpreted by econometricians finding an equation like (2.3) as shocks to the fundamentals, they should be interpreted here as shocks to expectations.*

A geometric characterization of stochastic equilibria is relatively easy in this simple framework. Let $[a, b]$ be the smallest interval containing all y_t with probability 1. Since the expectation operator E_t amounts essentially to taking averages, $v_1(y_t)$ belongs to $v_2([a, b])$. A continuity argument then shows that the minimal invariant interval $[a, b]$ must satisfy

$$v_1([a, b]) \subset v_2([a, b]) \text{ or } [a, b] \subset \chi([a, b]). \tag{2.4}$$

One can also show that if the inclusion in (2.4) is strict, i.e. if $[a, b]$ is contained in the interior of its image by χ, it is possible to construct a stochastic equilibrium satisfying (2.2) or (2.3) that stays in the interval at all times by choosing appropriately the disturbances ε_{t+1} (and the sunspot process). fact there are ***infinitely many*** of such stochastic equilibria for a ***given*** inv ant interval, see Grandmont (1985b, 1986, 1989) (one can impose stationa the existence of an invariant measure, etc ...).

This geometric characterization allows us to reduce the search for stoc tic equilibria to the much simpler search for invariant intervals. The follows immediately that no stochastic equilibrium sequence of outputs stay away from autarchy ($y = 0$) when the substitution effect dominates erywhere (when χ is everywhere increasing), except for the trivial seque $y_t = \overline{y} > 0$, in which $\overline{y}$ is the unique deterministic monetary stationary s defined by $\overline{y} = \overline{\chi}(y)$: the only invariant interval $[a, b]$ with $a > 0$ is inc $[\overline{y}, \overline{y}]$ in that case. The "Classical" viewpoint is thus valid when the s stitution effect dominates. By contrast, there are infinitely many invar intervals satisfying (2.4) when the income effect is strong enough near stationary state so that $\chi'(\overline{y}) < -1$. It suffices to take $a < \overline{y}$ arbitrary c to $\overline{y}$, and b in $(\chi^{-1}(a), \chi(a))$. The argument shows that such invariant in vals exist arbitrarily near the stationary state, i.e. in ***every neighborhoo*** $\overline{y}$. This is due to the fact that the (local) inverse χ^{-1} is contracting near stationary state. When the income effect is significant, the "Keynesian" v point is therefore correct : shifts in expectations can be a significant sourc economic fluctuations, independently of any variations of the fundament

3. Unit Roots (Local Bifurcations)

Many economic time series display a lot of "persistence" (the conseque of a shock tend to persist a long time). In other words, some "eigenvalu of the system tend to have a modulus close to 1. Some econometric have gone as far as to claim that quite a few economic time series car adequately represented as linear random walks. The recent literature nonlinear economic dynamics has shown that in such circumstances, ***a small amount of nonlinearity can dramatically affect the analysis*** (Grandm (1989)).

A convenient technique to study what happens when local eigenva have a modulus close to 1 is local bifurcation theory. To illustrate the p in the simple framework under consideration, consider a family of econon indexed by some parameter α, or equivalently of offer curves χ_α, and assu that the income effect is progressively increased so that the slope of the c curve at the stationary state $\overline{y}_\alpha$ (defined by $\overline{y}_\alpha = \chi_\alpha(\overline{y}_\alpha) > 0$) goes thro -1 when the index α goes up ($\chi'_\alpha(\overline{y}_\alpha)$ is greater than -1 when $\alpha < 0$, e to -1 when $\alpha = 0$ and less than -1 when $\alpha > 0$). The dynamic evolu of deterministic equilibria with perfect foresight near the stationary stat then given by inverting locally the equation $y_t = \chi_\alpha(y_{t+1})$. Thus they away from the steady state when $\alpha < 0$ and converge to it when $\alpha > 0$. As for stochastic equilibria, our previous arguments show that one can construct an infinity of them in every neighborhood of the stationary state if and only if $\chi'_\alpha(\overline{y}_\alpha) < -1$, i.e. if and only if $\alpha > 0$.

Were the map χ linear, these statements would be the only ones one could make and they would in fact hold *globally*. But the world is nonlinear and so is the offer curve. The generic picture one gets then is rather different. Indeed the family of maps χ_α undergoes a Flip bifurcation. If the bifurcation is supercritical, a period 2 cycle $(\overline{y}_{1\alpha}, \overline{y}_{2\alpha})$ appears after the bifurcation, i.e. for $\alpha > 0$, and deterministic equilibria with perfect foresight go away from it. For $\alpha > 0$, it is possible, as we already know, to construct invariant intervals $[a, b]$ and thus stochastic equilibria, but they cannot lie arbitrarily far away from the steady state as would have been predicted in a linear specification. Invariant intervals have here to be in the interior of the interval $(\overline{y}_{1\alpha}, \overline{y}_{2\alpha})$ determined by the cycle, and their union spans that interval.

If the family of maps χ_α undergoes a subcritical Flip bifurcation, the period 2 cycle appears for $\alpha < 0$, and it is attracting. When $\alpha > 0$, invariant intervals $[a, b]$, thus stochastic equilibria, can be constructed arbitrarily near the steady state. If $a < 0$, the map χ_a^{-1} is locally expanding and in a linear world we would be led to the conclusion that no nondegenerate invariant interval, hence no stochastic equilibrium, exists in such a case. This is not correct in the nonlinear case. True enough, there is no invariant interval $[a, b]$ with $a \neq b$ included in the open interval $(\overline{y}_{1\alpha}, \overline{y}_{2\alpha})$ determined by the cycle. But infinitely many such invariant intervals can nevertheless be constructed : each of them contains in its interior the period 2 cycle. Nonlinearity, however small, matters a lot in such a configuration.

4. Productive Investment and Endogenous Fluctuations

The model considered up to now owed its simplicity to the fact that it was "one-dimensional" : the state of the system at any date could be described by a single real number. Despite its pedagogical usefulness, this simplicity had a price. We had indeed to introduce a lot of nonlinearity in the system, i.e. to make the map χ representing the offer curve decreasing by assuming the presence of a strong income effect, in fact too strong to be empirically plausible. We wish to show now that if we take the (realistic) step of considering productive investment, most of the phenomena we have been talking about still occur (there are many endogenous expectations-driven business cycles) *even under the assumption that the substitution effect dominates*, i.e. even when the function χ is increasing everywhere.

The households' sector is the same as before. We assume now that output

y_t in period t is produced not only from the labor l_t supplied during the same period by the young household, but also from the stock of capital equipment k_{t-1} that is available at outset of the period, these two inputs having to be used in fixed proportions. Specifically, $y_t = \mathrm{Min}\{l_t, k_{t-1}/a\}$, where a is the capital-output ratio (or $1/a$ is the productivity of capital). Current output y_t is then partly directed toward consumption, partly invested. Thus in a market equilibrium at t, one has $y_t = c_t + i_t$, where c_t is consumption of the old and i_t is investment. The capital stock available for production at the outset of period $t+1$ is then

$$k_t = k_{t-1}(1-\delta) + i_t, \tag{4.1}$$

where $0 < \delta \leq 1$ is the (given) depreciation rate of capital equipment. To simplify matters, it is assumed that workers do not have access to direct or indirect ownership of capital, while the production sector is operated by entrepreneurs seeking to maximize profits [2].

We first look at deterministic equilibria with perfect foresight in this context. The households' optimal choices are still represented by the aggregate offer curve, i.e. by $v_1(l_t) = v_2(c_{t+1})$ or $l_t = \chi(c_{t+1})$. We focus on the standard case where the substitution effect dominates everywhere, i.e. $\chi(c)$ is increasing for all $c > 0$. The productive sector yields $y_t = l_t = k_{t-1}/a$. Under perfect foresight, expected consumption c_{t+1} is equal to actual future output y_{t+1} minus actual future investment i_{t+1}. But on account of (4.1)

$$i_{t+1} = k_{t+1} - k_t(1-\delta) = a[y_{t+2} - y_{t+1}(1-\delta)]. \tag{4.2}$$

Intertemporal equilibria with perfect foresight are therefore characterized by sequences of outputs $y_t > 0$ such that

$$y_t = \chi[(1 + a(1-\delta))y_{t+1} - ay_{t+2}], \tag{4.3}$$

for all $t \geq 1$.

We shall focus on the occurrence of endogenous expectations-driven business cycles near a stationary solution $y_t = \overline{y} > 0$ of (4.3), i.e. that satisfies $\overline{y} = \chi[(1 - a\delta)\overline{y}]$. It is easy to see that such a steady state exists (and is unique) whenever $a\delta < 1$, provided that the slope of the offer curve at the origin is large enough, i.e. $\lim_{c \to 0+} (\chi(c)/c) > 1/(1 - a\delta)$. The simplest technique to get cyclical behavior near the steady state is to generate a local bifurcation.

[2]Models of this type have been considered by Farmer (1986), Reichlin (1986), Woodford (1986). Here again one could assume that workers are in fact infinitely long lived but face liquidity (cash-in-advance) constraints stating that they cannot borrow against future labor income. One would get the same local results as in the analysis that follows (Woodford(1986)).

To this effect, one looks at the eigenvalues of the local dynamics obtained by linearizing (4.3) around the stationary solution $\overline{y}$ and one varies the "fundamental" characteristics of the system (or a parameter, say α, indexing these characteristics) so as to get a change of stability. The interesting point is that (4.3) is in effect two-dimensional since it involves two lagged state variables. It is then relatively easy to show that the only case which can occur when the map χ is increasing as here, is that a pair of complex conjugate eigenvalues crosses the unit circle — one gets then a socalled Hopf bifurcation. As argued earlier, analysis of a configuration of this sort is particularly relevant if one believes in the "unit root" hypothesis in macroeconomics, i.e. that economic time series display eigenvalues with a modulus close to 1.

It is possible to show that such a Hopf bifurcation will indeed occur if one holds the capital-output ratio constant and greater than $1/(1+\delta)$, and if one varies the "concavity" of an old trader's utility for consumption, as measured by the coefficient $-cV_2''(c)/V_2'(c)$, between 0 and 1. These assumptions are quite compatible with those which macroeconomists usually employ about utility functions and with actual data : the capital-output ratio is about 3 in industrialized countries. The steady state $\overline{y}$ is then first stable in the local dynamics determined by (4.3) and becomes unstable when the Hopf bifurcation occurs as the concavity of V_2 is increased. If the system was linear, that would be all one could say. But to be realistic, the offer curve must be nonlinear. Then an invariant closed curve must generically appear in the picture near the stationary solution in the state space (y_t, y_{t-1}), on which the motion is periodic (may be with a complicated long period) or quasi-periodic. If the Hopf bifurcation is supercritical, the invariant closed curve appears after the bifurcation and it is stable. If the bifurcation is subcritical, the invariant closed curve is present before the bifurcation but it is unstable.

The argument shows how endogenous deterministic fluctuations may occur near the steady state when the local characteristic roots of the system have close to unit modulus. As in the onedimensional case considered earlier, the story does not stop there and one should inquire about the occurrence of stochastic endogenous fluctuations as well. Stochastic equilibria with self-fulfilling expectations are here characterized by random sequences of output levels $y_t > 0$ that satisfy for all $t \geq 1$ a relation analogous to (4.3)

$$v_1(y_t) = E_t v_2[(1 + a(1-\delta))y_{t+1} - a y_{t+2}], \tag{4.4}$$

in which E_t means that traders condition their expectations on the observed values $(s_t, s_{t-1}, \ldots)$ of some exogenous stochastic process of "sunspots", the variations of which do not affect the "fundamentals".

A natural question is whether nonexplosive endogenous stochastic fluctuations of this sort can arise locally, i.e. near the steady state. As in the onedimensional case studied earlier, one can construct infinitely many such stochastic equilibria in every neighborhood of the stationary state if, and in general only if, it is stable in the local deterministic dynamics with perfect foresight defined implicitly by (4.3) — this will be the case if the capital-output ratio is held constant but the concavity of the utility of consumption, i.e. $-cV_2''(c)/V_2'(c)$, is small (see Woodford (1986)).

Consideration of what happens for systems displaying a high degree of persistence, or specifically when a local (Hopf) bifurcation occurs, yields also here interesting phenomena that are due to the presence of an invariant closed curve near the steady state. Assume that a Hopf bifurcation is generated, so that the steady state is first stable in the local deterministic equilibria defined by (4.3), and becomes unstable. If the bifurcation is subcritical, an invariant closed curve, which is unstable in the dynamics (4.3), exists before the bifurcation. In that case, stochastic equilibria appear only before the bifurcation, their supports have to lie in the interior of the region inside the invariant closed curve, and the union of these supports spans that region. If, on the other hand, the bifurcation is supercritical, the invariant closed curve appears after the bifurcation and it is stable in the dynamics (4.3). In that case, stochastic equilibria indeed exist before the bifurcation, in every neighborhood of the steady state. After the bifurcation there are no nondegenerate stochastic equilibria that would stay in an extremely small neighborhood of the steady state since it is then unstable. In fact, there are no stochastic equilibria, other than the steady state itself, that lie wholly in the interior of the region inside the invariant closed curve. But there are infinitely many stochastic equilibria, the support of which contains in its interior the invariant closed curve. Here is another case where a very small amount of nonlinearity changes significantly the conclusions of the analysis, by comparison with what we would have gotten had we (wrongly) chosen a linear specification.

I hope this last example will convince the reader that complex endogenous expectations-driven business cycles can arise under quite plausible assumptions and furthermore, that it is important to incorporate nonlinearities when studying such fluctuations, especially when the eigenvalues of the system have a modulus close to 1. The main "non standard" assumption that was responsible for the occurrence of endogenous fluctuations in the above model, apart from the presence of plausible capital markets imperfections, was that capital equipment and labor had to be combined in fixed proportions in production, or more generally that the elasticity of substitution between labor and capital was low. I would expect the same sort of results to hold with a more elastic technology if we incorporate in the model the realistic feature that there are significant costs to incur when adjusting the capital stock through investment, or when changing employment, so that the "effective" elasticity of substitution is in fact low. There is apparently still some way to go before models of this type are operational enough to change actual econometric practice. But I believe the studies reviewed here suggest strongly that endogenous business cycles models have become more and more credible alternatives to describe observed fluctuations in our economies.

References

[1] P. Anderson, K. Arrow and D. Pines (Eds.), *The economy as an evolving complex system*, vol. V, Santa Fé Institute Studies, Addison Wesley, 1988.

[2] C. Azariadis, *Self-Fulfilling Prophecies*, Journal of Economic Theory **25** (1981), 380–396.

[3] C. Azariadis and R. Guesnerie, *Sunspots and Cycles*, Review of Economic Studies **53** (1986), 725–737.

[4] W. Barnett, J. Geweke and K. Shell (Eds.), *Economic Complexity : Chaos, sunspots, bubbles and nonlinearity*, Cambridge University Press, 1989.

[5] W. Baumol and J. Benhabib, *Chaos : Significance, Mechanism and Economic Applications*, Journal of Economic Perspectives **3** (1989), 77–105.

[6] J. Benhabib and R. Day, *A Characterization of Erratic Dynamics in the Overlapping Generations Model*, Journal of Economic Dynamics and Control **4** (1982), 37–55.

[7] M. Boldrin, *Persistent Oscillations and Chaos in Dynamic Economic Models : Notes for a Survey*, in P. Anderson, K. Arrow and D. Pines (Eds.), 1988.

[8] M. Boldrin, *Perfectly Competitive Models of Endogenous Business Fluctuations*, European Economic Review **35** (1991), 300–305.

[9] M. Boldrin and M. Woodford, *Equilibrium Models Displaying Endogenous Fluctuations and Chaos*, Journal of Monetary Economics **25** (1990), 189–222.

[10] W. Brock, *Nonlinearity and Complex Dynamics in Economics and Finance*, in P. Anderson, K. Arrow and D. Pines (Eds.), 1988.

[11] W. Brock and W. Dechert, *Nonlinear Dynamical Systems : Instability and Chaos in Economics*, in W. Hildenbrand and H. Sonnenschein (Eds.) (1991).

[12] P.A. Chiappori and R. Guesnerie, *On Stationary Sunspot Equilibria of Order k*, in W. Barnett, J. Geweke and K. Shell (Eds.), 1989.

[13] P.A. Chiappori and R. Guesnerie, *Sunspot Equilibria in Sequential Markets Models*, in W. Hildenbrand and H. Sonnenschein (Eds.), 1991.

[14] R. Farmer, *Deficits and Cycles*, Journal of Economic Theory **40** (1986), 77–88.

[15] R. Farmer and M. Woodford, *Self-Fulfilling Prophecies and the Business Cycle*, Cuadernos Economicos de ICE, Mexico, 1987.

[16] J.M. Grandmont, *On Endogenous Competitive Business Cycles*, Econometrica **53**, (1985a) 995–1045.

[17] J.M. Grandmont, *Cycles Concurrentiels Endogènes*, Cahiers du Séminaire d'Econométrie, CNRS, Paris (1985b).

[18] J.M. Grandmont, *Stabilizing Competitive Business Cycles*, Journal of Economic Theory **40** (1986), 57–76.

[19] J.M. Grandmont, *Local Bifurcations and Stationary Sunspots*, in W. Barnett, J. Geweke and K. Shell (Eds.), 1989.

[20] R. Guesnerie, *Stationary Sunspot Equilibria in an N Commodity World*, Journal of Economic Theory **40** (1986), 103–127.

[21] W. Hildenbrand and H. Sonnenschein (Eds.), *Handbook of Mathematical Economics*, Vol. IV, North Holland, 1991.

[22] P. Reichlin, *Equilibrium Cycles in an Overlapping Generations Economy with Production*, Journal of Economic Theory **40** (1986), 89–102.

[23] M. Woodford, *Stationary Sunspot Equilibria in a Finance Constrained Economy*, Journal of Economic Theory **40** (1986), 128–137.

Cores, Almost Competitive Prices, and the Approximate Optimality of Walrasian Allocations in Discrete Spaces

Leonid Hurwicz[1,2]

Department of Economics, University of Minnesota
Minessota 55455, U.S.A

1. This paper is devoted to problems arising when all or some goods are indivisible. Part One deals with the properties of the core in such situations, Part Two with the optimality properties of the Walrasian (competitive) allocations. The purpose of Part One is to show how certain relationships between the core and prices can be extended (with modifications) to exchange economies in which all or some commodities are indivisible. The proposition[3] stating that a Walrasian allocation is in the core (provided preferences are selfish) remains valid without any additional assumptions when all goods are indivisible, as can be seen by inspecting the standard proofs. (see, e.g., Debreu and Scarf, Th-m 1, p.240, or Arrow and Hahn, Ch. 8, Th-m 1, p.187). [This is largely due to the fact that the blocking definition requires that strict preference hold for each member of a blocking coalition. Hence, in particular, a core allocation is only required to be *Weakly* Pareto Optimal (i.e., Pareto Efficient in the sense of Arrow and Hahn or Varian) rather than Pareto Optimal as defined, e.g., in Koopmans, p.46.[4] If a 'strict core' were defined as the

[1]The initial stimulus for investigating the relationship of core to Walrasian allocations in the context of discrete (hence denumerable) spaces was due to issues raised by the work of, and discussions with , Professor Alain A. Lewis, University of California at Irvine, concerning the 'effective decidability' of Edgeworth's Conjecture. See Lewis (1991). It seemed to me worthwhile to be able to study this problem in a situation free of complications due specifically to the commodity space being a continuum.

[2]I am indebted to my colleague Professor Marcel K. Richter, University of Minnesota, for numerous fruitful suggestions and encouragement. I am also grateful to Professor Robert M. Anderson, University of California at Berkeley, for a number of valuable suggestions (not all as yet implemented!), and for sharing with me his (September 1991) unpublished note "Relationship of Hurwicz (1991) to Approximate Cores." The sole responsibility for any errors is, unfortunately, the author's.

[3]Attributed to Shapley in Debreu and Scarf, p.240.

[4]A feasible allocation is Weakly Pareto Optimal if there is no feasible alternative strictly preferred by everybody. It is Pareto Optimal if there is no feasible alternative that is at least as good for everybody and better for some.

set of allocations 'strongly unblocked' in the sense of Def. 6, p.196, in Arrow and Hahn, then additional assumptions, such as local non-satiation, would be needed to guarantee that a Walrasian allocation belongs to the strict core.]

Our interest here is in the type of "partial converse" proposition asserting that, in an appropriate sense, a core allocation is 'approximately' Walrasian —whether asymptotically or in finite economies. Specifically, we shall study the problem of extending to discrete spaces the two results contained in Anderson's 1978 paper. His Theorem 1, called by Hildenbrand (1982, p.850) the fundamental Lemma[5] , applies to an economy of a given fixed size, while his Theorem 2 is asymptotic. Anderson's theorems assume preferences to be weakly monotonic and having the 'free disposal' property; they also assume that the commodity space is finite-dimensional Euclidean, i.e., that it is a continuum.

In section 5(ii) we consider 'mixed' commodity spaces, i.e., those with some goods divisible and possibly (but not necessarily) some indivisible, with continuum as a special case. It is shown that the assumption of weak monotonicity can be replaced by local non-satiation, and, in particular, by weak monotonicity with respect to a subset of divisible goods. This seems to yield a slight strengthening of Anderson's results.

Part Two is devoted to an examination of the extent to which, in discrete commodity spaces (i.e., with all goods indivisible), Walrasian allocations deviate from Pareto Optimality (as distinct from *Weak* Pareto Optimality, which is always attained). It is shown that for strictly monotone (complete) preorderings, subject to additional restrictions, this deviation does not exceed $n-1$ commodity units (where n is the number of traders).

Part One. The Core and Approximately Walrasian Prices in Discrete Commodity Spaces.

2. Our results for discrete commodity spaces can be roughly summarized as adding an extra unit or two in the bounds obtained by Anderson for certain average value discrepancies. In continuum economies, Anderson's bounds go to zero when the number of agents goes to infinity (while initial endowments remain bounded), and hence the average discrepancies also converge to zero. On the other hand, in discrete economies the extra unit or two in the bounds

[5]Anderson's Theorem 1 builds on earlier results, in particular those due to Vind (1965), Arrow and Hahn, 1971, Th-m 2, p.189, and subsequent extensions by Dierker (1975) and Keiding (1974); see Anderson, 1978, p.1483, and Hildenbrand, 1982, p.850. Versions of Anderson's Theorem 1 are found in Hildenbrand, 1982, pp.847–852, as well as in Hildenbrand and Kirman, 1988, pp.186–9. Closely related results are Proposition 7.4.1 in Mas-Colell, 1985, p.275, and Theorem 3.2, p.7, in Anderson, 1987. See also the proof of Theorem 3.4, p.8, in Anderson, 1987, where the existence of gap-minimizing prices' is established.

does not disappear asymptotically. This may be insignificant for core allocations involving large numbers of units, but very significant for allocations involving only very few units.

In any case, closeness in terms of the average value metric may conceal large individual deviations in the commodity space, as noted in Hildenbrand 1974, p.202, and illustrated by Mas-Colell, 1985, p.277, and Anderson, 1988, p.364. But this weakness is not peculiar to discrete commodity spaces.

For later reference, we state here the relations obtained by Anderson for an economy where all goods are divisible (i.e., the commodity space is R^k), and preferences have the two properties of (weak) monotonicity and free disposal.

In Anderson's Theorem 1:

$$\frac{1}{n}\sum_{a\in A}|p\cdot(f(a)-e(a))|\leq\frac{2M}{n}, \tag{1.i}$$

and

$$\frac{1}{n}\sum_{a\in A}|\, inf\{p\cdot(x-e(a)):x\succ_a f(a)\}|\leqq\frac{2M}{n}. \tag{1.ii}$$

In Anderson's Theorem 2:

$$\left[\frac{1}{n}\sum_{a\in A(n)}|p_n\cdot(f_n(a)-e_n(a))|\right]\to 0 \tag{2,i}$$

and

$$\left[\frac{1}{n}\sum_{a\in A(n)}|\, inf\{p_n\cdot(x-e_n(a)):x\succ_a f_n(a)\}|\right]\to 0. \tag{2.ii}$$

[For notation and terminology, see section 3 below.]

3. *Notation and assumptions.* Most of the notation follows that in Anderson, 1978, but we restate it here to make this note self-contained.

There are k goods and n agents. Denote by Ξ the commodity space. If Ξ is the k-dimensional Euclidean space we speak of (perfectly) divisible commodities or the continuum case. Anderson's results deal with that case. When Ξ is a lattice of equidistant points we speak of indivisible commodities. When the distance between points is normalized to equal one, we speak of the normalized indivisible case. This paper is primarily devoted to the latter case. (Of course, the normalization involves no loss of generality. See below, Section 5(i).) Let Z denote the set of all integers; Z_+ the set of all nonnegative integers. Z^k and Z^k_+ are the corresponding k-fold Cartesian products. In the normalized indivisible case the commodity space is Z^k and each agent's consumption set is assumed to be Z^k_+.

$|r|$ is the absolute value of the real number r. If x is a k-dimensional vector with real components x^i, the symbol $\|x\|$ denotes the norm of x. More explicitly, $\|x\|_\infty = max\{|x^i| : 1 \leqq i \leqq k\}$, while $\|x\|_1 = \sum_{i=1,\dots,k} |x^i|$. The vector in R^k with all components equal to 1 is denoted by $u = (1, \dots, 1)$. The inner ('dot') product of two vectors p and x is written $p \cdot x$.

Vector inequalities in R^k are written as follows; $x \geqq y$ means $x^i \geqq y^i$ for all i; $x \gg y$ means $x^i > y^i$ for all i; $x > y$ means $x \geqq y$ but not $x = y$. A strict preference relation is denoted by $\succ$. It is a binary relation on the nonnegative integer-valued orthant Z^k_+, i.e., on the set of all k-tuples of nonnegative integers, and it is assumed to have the following two properties: (i) *(weak) monotonicity*: $x \gg y$ implies $x \succ y$, and (ii) *free disposal*: $x \gg y$ and $y \succ z$ together imply $x \succ z$.It is to be noted that neither continuity nor convexity properties are assumed, and even transitivity is weakened. The set of all such preference relations is denoted by $\mathcal{P}$.

The (finite) *set of agents* is denoted by A, with the number of agents in A denoted by $\#A = n$. The *preference relation of agent* a in A is denoted by $\succ_a$. The *initial endowment of agent* a is denoted by $e(a)$. We assume $e(a) \in Z^k_+$. An *exchange economy* with indivisible commodities is a map $\mathcal{E} : A \to \mathcal{P} \times Z^k_+$. (In the preceding formula, the symbol $\times$ denotes the Cartesian product of sets.) An *allocation* is a map $f : A \to Z^k_+$ such that $\sum f(a) = \sum e(a)$, with both summations over all elements of A.

A *coalition* is a nonempty subset of A. A coalition S *improves on an allocation* f *via allocation* g if (1) g is *feasible* for S, i.e., if $\sum g(a) = \sum e(a)$ where both summations are over members of the coalition S, and (2) every member of S strictly prefers his/her component of g to that of f, i.e., if $g(a) \succ_a f(a)$ for all a in S. An *allocation* f *is blocked by the coalition* S if there is an allocation g such that S improves on f via g. The *core* of an exchange economy $\mathcal{E}$ is the set of all allocations in $\mathcal{E}$ that are not blocked by any coalition. A *price* (vector) p is an element of R^k_+ with elements normalized to add up to 1; i.e., $\sum_{h=1,\dots,k} p_h = 1$, or $\|p\|_1 = 1$. (Note that prices need not be rational numbers; they can be any reals.) The *set of all prices* is denoted by Δ. Hence $\Delta = \{p \in R^k_+ : \|p\|_1 = 1\}$.

For an economy $\mathcal{E}$, with k commodities and n agents, we define $M = M(\mathcal{E}) = sup\{\|e(a_1) + \cdots + e(a_k)\|_i nfty : a_1, \dots, a_k$ are elements of $A\} = sup_Q\{\|e_Q\|_\infty : A \supseteq Q, \#Q = k\}$ where e_Q is the sum of endowments held by agents belonging to the subset Q of A. [In most cases, the operation *sup* is applied to finite sets, in which case it becomes *max*.; similarly for *inf* and *min*.] For example, if (as in Hildenbrand and Kirman) all individual endowments are confined to a bounded interval $[0, b]$, with $b = sup\{e(a) : a \in A\}$, then $M = kb$. [6]

[6] More explicitly: Let there be $k < \infty$ goods. Denote by $e^j(a)$ the j-th good endowment of individual a, and write $e(a) = \langle e^1(a), \dots, e^k(a)\rangle$. Assume that there is a finite $b > 0$ such that $b = sup\{e^j(a) : a \in A, j \in \{1, \dots, k\}\}$, so that $0 \leqq e^j(a) \leqq b$ for all j in $\{1, \dots, k\}$ and all a in A. Since k is finite, such b necessarily exists when A is finite. In that case

4. THEOREM 1′. Let X be a finite exchange economy $\mathcal{E}$, with n agents and k normalized indivisible commodities. If f is an allocation in the core $C(\mathcal{E})$, then the exists a price p in Δ such that

$$\frac{1}{n}\sum_{a\in A}|p\cdot(f(a)-e(a))| \leqq (2M/n)+2 \tag{1.i'}$$

and

$$\frac{1}{n}\sum_{a\in A}|\, inf\{p\cdot(x-e(a)) : x \succ_a f(a)\}| \leqq (2M/n)+2. \tag{1.ii'}$$

REMARK. The primes distinguish the propositions (and sometimes entities) of this note from their counterparts in Anderson's 1978 paper. Our Theorem 1′ for indivisible commodities differs from Anderson's Theorem 1 for divisible commodities in that it has the added integer terms on the right-hand side of the inequalities.

PROOF. The proof follows Anderson in almost every detail. It involves defining the price p by the hyperplane separating two sets to be defined below. It is then shown that, with that price, the inequalities of the Theorem are satisfied.

Step 1. The two sets to be separated, denoted respectively by Π and W, are defined as follows. Let f be the core allocation specified in the Theorem, and define in the commodity space the set

$$\gamma(a) = \{x - e(a) \in Z^k : x \succ_a f(a)\}.$$

i.e., $\gamma(a)$ is the set of all *net* trades strictly preferred by agent a to his/her core allocation component $f(a)$. Next, enlarge the set $\gamma(a)$ by adjoining the origin; we thus define the commodity space set

$$\phi(a) = \{0\} \cup \gamma(a).$$

Then Π is defined by

$$\Pi = \sum_{a\in A}\phi(a).^7$$

We note for future reference the obvious fact that Π is a subset of Z^k.

In turn we define the set $W = \{w \in R^k : w \ll -z'\}$ where $z' = Mu = M(1,\ldots,1) = (M,\ldots,M)$.[8] Clearly, the set W is convex, but Π is not. To effect the separation we show that W is disjoint from the convex hull of Π, denoted by $con\,\Pi$; i.e., that

$$W \cap con\,\Pi = \emptyset. \tag{1}$$

there exists an individual a^* and a good j^* such that $b = e^{j^*}(a^*)$. Hence, in this case, $M = max\{\sum_{r\in\{1,\ldots,k\}} e^j(a_r) : \{a_1,\ldots,a_k\} \subset A\} = kb$.

[7]Anderson uses $\Phi = \frac{1}{n}\sum_{a\in A}\phi(a)$.

[8]Anderson uses $z = (M/n)u$.

PROOF OF (1).
Suppose (1) false so that there exists a vector $x \in W \cap con\,\Pi$. Since $x \in con\,\Pi$, it can be written [see Debreu(1959), p.24, 1.9(15)] as

$$x = \sum_{a \in A} r(a)$$

with $r(a) \in con\,\phi(a)$ for all a in A, but where (by the Shapley-Folkman Theorem) $r(a) \in \phi(a)$ for all but at most k agents $a_1, \ldots, a_q$, with $q \leqq k$. Write $\{a_1, \ldots, a_q\} = Q$, so that $\#Q = k$, and define $r^*(\cdot)$ by $r^*(a) = 0$ for a in Q while $r^*(a) = r(a)$ for a in $A \setminus Q$. Consider now

$$y' = \sum_{a \in A} r^*(a) = \sum_{a \in A \setminus Q} r(a) + (0 + \cdots + 0).$$

The last expression shows that y' belongs to Π since, by construction, $r(a) \in \phi(a)$ for all a, hence all a in $A \setminus Q$, and also $0 \in \phi(a)$ for all a in A, hence all a in Q.

On the other hand, since $x = \sum_{a \in A \setminus Q} r(a) + \sum_{a \in Q} r(a)$ while $y' = \sum_{a \in A \setminus Q} r(a)$, we have $y' = x - \sum_{a \in Q} r(a)$.

Now $\phi(a) \geqq -e(a)$, hence $con\,\phi(a) \geqq -e(a)$, and [since, for $a \in Q$, we have $r(a) \in con\,\phi(a)$] it follows that $-r(a) \leqq e(a)$ for all a in Q.

This yields the first inequality in

$$y' \leqq x + \sum_{a \in Q} e(a) \leqq x + z' \ll 0,$$

while the second inequality follows from the definitions of z' and M and the fact that $\#Q \leqq k$, and the third inequality is implied by the hypothesis $x \in W$, i.e., $x \ll -z'$.

We have now established that y' is both in R^k_- and in Π, hence in Z^k. It follows that y' is in Z^k_-. To obtain a contradiction it will suffice to show that

$$\Pi \text{ and } Z^k_- = \emptyset. \qquad [*]$$

(PROOF OF [*])
Suppose that, contrary to [*], there exists a vector $G' \ll 0$, $G' \in \Pi$, belonging to both sets. Since $G' \in \Pi$, it follows that, for all a in A, there is a selection g from ϕ, $g : A \to Z^k$, $g(a) \in \phi(a)$ for all a in A, with $G' = \sum_{a \in A} g(a)$.

We shall now show that the existence of the vector G' enables us to find a coalition B blocking the given core allocation f.Since a core allocation cannot be blocked, this contradiction will complete the proof that Π and Z^k_- are disjoint, and hence that $con\,\Pi$ and W are disjoint. The coalition B consists of all agents in A for whom $g(a) \neq 0$.[9]

[9]To see that B is non-empty, suppose to the contrary that $g(a) = 0$ for all a in A. Then $\sum_A g(a) = 0$, hence $G' = inf \sum_A g(a) = 0$, which is impossible since, by hypothesis, $G \ll 0$.

Let a_1 be any member of the coalition B. Then define the improving allocation h' by

$$h'(a_1) = g(a_1) + e(a_1) - G',$$

while

$$h'(a) = g(a) + e(a) \text{for all } a \text{ in } B \setminus \{a_1\}.$$[10]

Note that $h'(a)$ is in Z_+^k for all a in B because g is a selection from ϕ (so $g(a_1) \geqq -e(a_1)$) and $-G' \in Z_+^k$ because G' is in Z_-^k.

Since, on B, g is a selection from γ, we have (by definition of γ, ϕ, and g)

$$g(a) + e(a) \succ_a f(a) \text{for all } a \text{ in } B.$$

Hence $h'(a)$ is strictly preferred to $f(a)$ by all members of B other than a_1. But since $-G' \gg 0$, the free disposal assumption also yields $h'(a_1) \succ_{a_1} f(a_1)$. Hence h' improves on f provided it is feasible for the coalition B, i.e., provided that

$$\sum_{a \in B} h'(a) = \sum_{a \in B} e(a).$$

But this is the case because (denoting $B \setminus \{a_1\}$ by B^*) we have

$$\begin{aligned}
\sum_{a \in B} h'(a) &= h'(a_1) + \sum_{a \in B^*} h'(a) \\
&= g(a_1) + e(a_1) - G' + \sum_{a \in B^*} (g(a) + e(a)) \\
&= \sum_{a \in B} (g(a) + e(a)) - G' \\
&= \sum_{a \in B} (g(a) + e(a) - \sum_{a \in B} g(a) \\
&= \sum_{a \in B} e(a).
\end{aligned}$$

(The last equality follows from the definitions of G' and B.) So f is blocked by B, which contradicts the hypothesis that f is in the core. Hence Π and Z_-^k are disjoint; i.e., [*] is true.

But it was shown previously that if (1) were false, [*] would also be false. Since it has just been established that [*] is true, it follows that (1) is true; i.e., the sets *con* Π and W are disjoint.

[END PROOF (1).]

Since *con* Π and W are disjoint convex subsets of a finite dimensional Euclidean space, there is a hyperplane separating *con* Π from W, and hence separating Π from W. Because of the nature of W, this hyperplane can be

[10] In the corresponding part of Anderson's proof the counterpart of G' is distributed equally among members of the coalition. But this might result in fractional amounts of commodities and thus is inadmissible in the indivisible commodity space case.

taken as determined by a nonnegative k-vector whose components add up to 1, i.e., by a price vector p in Δ. Therefore there is a real number α such that $p \cdot z \geqq \alpha$ for all z in Π, and $p \cdot w \leqq \alpha$ for all w in W. Hence $p \cdot w \geqq p \cdot z$ for all z in Π and all w in W, and so

$$inf p \cdot \Pi \geqq sup p \cdot W. \tag{1*}$$

Step 2. In this section we derive inequalities to be used in step 2 of the proof.

As seen at the end of Step 1, $inf p \cdot \Pi \geqq sup p \cdot W$. But $sup p \cdot W = -p \cdot z' = -p \cdot Mu = -Mp \cdot u = -M$ (because $\|p\|_1 = 1$). Hence, by (1*),

$$inf p \cdot \Pi \geqq -M. \qquad (2).$$

Since $f(a)+u \gg f(a)$, the weak monotonicity property yields $f(a)+u \succ_a f(a)$; hence, by the definition of $\gamma(a)$,

$$f(a) + u - e(a) \text{ is in } \gamma(a), \text{ for all } a \text{ in } A.$$[11]

It follows that

$$inf p \cdot \gamma(a) \leqq p \cdot (f(a)+u-e(a)) \leqq p \cdot (f(a)-e(a))+1, \text{ for all } a \text{ in } A. \tag{5'.1}$$

But

$$\phi(a) = \{0\} \cup \gamma(a),$$

so

$$inf p \cdot \phi(a) \leqq inf p \cdot \gamma(a).$$

This, together with (5'.1), yields

$$inf p \cdot \phi(a) \leqq p \cdot (f(a) - e(a)) + 1, \text{ for all } a \text{ in } A.^{12} \tag{4'}$$

In what follows we shall have occasion to use the set of agents defined by

$$S = \{a \in A : p \cdot (f(a) - e(a)) < 0\}. \tag{7}$$

Since S is a subset of A, the inequality in (4') yields

$$\sum_{a \in S} [p \cdot (f(a) - e(a)) + 1] \geqq \sum_{a \in S} inf p \cdot \phi(a). \tag{8}$$

[11]This is the point where the distinction between divisible and indivisible commodity spaces is crucial in divisible commodity spaces it is the case that $f(a) + e(a) + \theta u$ belongs to $\phi(a)$ for any positive θ and not just for positive integer values of θ.

[12]In the divisible case, by letting θ tend to zero, Anderson obtains the counterpart of (4') without the 1 term on the right hand side which we shall refer to as (4).

Since $\sum_{a\in S} inf p \cdot \phi(a) \geqq \sum_{a\in A} inf p \cdot \phi(a) = inf p \cdot \Pi$, the inequalities in (8) and (2) yields

$$\sum_{a\in S}[p\cdot(f(a)-e(a))+1] \geqq inf p\cdot\Pi \geqq -M. \tag{7.2}$$

Therefore,

$$\sum_{a\in S}[p\cdot(f(a)-e(a))+1] \geqq -M, \tag{10.1}$$

and

$$\#S+\sum_{a\in S} p\cdot(f(a)-e(a)) \geqq -M, \tag{10.2}$$

i.e.,

$$-\sum_{a\in S} p\cdot(f(a)-e(a)) \leqq M+\#S. \tag{10.3}$$

But, by the definition (7) of S, the summation in the LHS of (10.3) is negative if S is non-empty and equals zero if S is empty. Hence (10.3) is equivalent to

$$\left|\sum_{a\in S} p\cdot(f(a)-e(a))\right| \leqq M+\#S, \tag{10.4}$$

and this inequality is valid whether S is empty or not.

Step 3. In this section we shall prove the inequality $(1.i')$ of Theorem 1′.

By feasibility of the core allocation f, we have

$$\sum_{a\in A} p\cdot(f(a)-e(a)) = p\cdot(\sum_{a\in A} f(a)-\sum_{a\in A} e(a)) = 0. \tag{13}$$

Using (13) and the definition (7) of S, we obtain

$$\sum_{a\in A} |p\cdot(f(a)-e(a))| = 2\sum_{a\in S} |p\cdot(f(a)-e(a))|. \tag{14}$$

Hence, from (10.4) we get

$$\sum_{a\in A} |p\cdot f(a)-e(a)| \leqq 2(M+\#S) \leqq 2(M+n) \tag{14.2}$$

where the second inequality follows from the fact that (since S is a subset of A), $\#S \leqq \#A = n$, and is valid whether S is empty or not.

Division by n then yields

$$\frac{1}{n}\sum_{a\in A} |p\cdot(f(a)-e(a))| \leqq 2(M/n)+2, \tag{15}$$

which is the inequality $(1, i')$ of Theorem 1′. (When S is empty, the left hand side of (15) vanishes.)

Step 4. In this section we prove the inequality (1.*i*′) of Theorem 1′.

Define

$$d(a) = \mathit{inf} p \cdot \gamma(a).$$

Then the inequality in (5′.1) becomes

$$d(a) \leqq p \cdot (f(a) - e(a)) + 1 \text{ for all } a \text{ in } A \tag{5'.2}$$

Also, since $\phi(a) = \{0\} \cup \gamma(a)$, we have (see (5*.1))

$$d(a) \geqq \mathit{inf} p \cdot \phi(a) \text{ for all } a \text{ in } A. \tag{6}$$

The proof of (1.*ii*′) involves the utilization of results obtained in step 2. In the notation we have adopted, the left hand side of the inequality in (1.*ii*′) becomes $\sum_{a \in A} |d(a)|$. Now define

$$A^* = \{a \in A : d(a) > 0\},$$

$$A^{**} = \{a \in A : d(a) < 0\}.$$

Then

$$\sum_{a \in A} |d(a)| = \sum_{a \in A^*} d(a) - \sum_{a \in A^{a**}} d(a). \tag{16}$$

Using (5′.2), we obtain for the first sum in the right hand side of (16) relations

$$\begin{aligned} \sum_{a \in A^*} d(a) &\leqq \sum_{a \in A^*} [p \cdot (f(a) - e(a)) + 1] \\ &= \#A^* + \sum_{a \in A^*} p \cdot (f(a) - e(a)). \end{aligned} \tag{17}$$

Define now two sets $A_1^* = A^* \cap (A \setminus S)$ and $A_2^* = A^* \cap S$, and the function $y(a) = p \cdot (f(a) - e(a))$. Then

$$\begin{aligned} \sum_{A^*} p \cdot (f(a) - e(a)) &= \sum_{A^*} \psi(a) = \sum_{A_1^*} \psi(a) + \sum_{A_2^*} \psi(a) \\ &= \sum_{A_1^*} \psi(a) - |\sum_{A_2^*} \psi(a)| \leqq \sum_{A_1^*} \psi(a) \\ &\leqq \sum_{A \setminus S} \psi(a) = -\sum_{S} \psi(a) \leqq M + \#S, \end{aligned}$$

where the last inequality follows from (10.4) and the preceding equality from (13).

So

$$S_{A^*} p \cdot (f(a) - e(a)) \leqq M + \#S \leqq M + n \tag{18}$$

and hence, using (17),

$$S_{a \in A^*} d(a) \leqq \#A + (M + n) = M + 2n. \tag{21}$$

As for the second sum in the right hand side of (16), we have

$$\sum_{a\in A^{**}} d(a) \geqq \sum_{A^{**}} infp \cdot \phi(a) \geqq \sum_{a\in A} infp \cdot \phi(a) = infp \cdot \Pi \geqq -M \qquad (22.1)$$

where the first inequality follows from (6) and the last one from (2). I.e.,

$$-\sum_{A^{**}} d(a) = \sum_{A^{**}} |d(a)| \leqq M. \qquad (22.2)$$

In turn, (16), (21), and (22.2) yield

$$\sum_{a\in A} |d(a)| \leqq (M+2n) + M = 2M + 2n. \qquad (23)$$

Dividing by n, we finally obtain

$$\frac{1}{n}\sum_{a\in A} |d(a)| \leqq 2(M/n) + 2, \qquad (24)$$

which is the inequality ($1.ii'$) in Theorem 1′. This completes the proof of Theorem 1′.

5. *A generalization.*[13] There is only one point in the preceding proof where the discrete case differs significantly from the continuum case: in the continuum case, commodity vectors of the form $f(a) + \theta u$ with any θ positive are in agent a's consumption set (assumed to be R^k_+), while in the discrete case this only holds when θ is a positive integer since the consumption set (as normalized in the preceding Theorem 1′) is assumed to be Z^k_+. To allow for alternative scales of measurement, and — more importantly — to cover in one theorem both the Anderson result and the above Theorem 1′, we proceed as follows. Let $C(a)$ denote the consumption set of agent a. Define $\beta = inf\{\theta : \theta$ nonnegative real such that, for all a in A and all x, x in $C(a)$ implies θx in $C(a)$ $\}$. In Anderson's case, $\beta = 0$, while in the (normalized) discrete case $\beta = 1$. A generalized result is obtained by inserting the β factor where integers enter in Theorem 1′ and its proof. In particular, formula (4′) becomes

$$infp \cdot \phi(a) \leqq p \cdot (f(a) - e(a)) + \beta. \qquad (4'')$$

Similarly, in the formulae (5′) and (7.2) the number β replaces 1. We thus obtain what may be labeled THEOREM 1″ in which the left-hand sides remain the same as in Theorem 1′, while the right-hand sides of ($1.i'$) and ($1.ii'$)

[13]This section is largely the result of a suggestion by Marcel K. Richter. He also pointed out that for Anderson's result it is sufficient to have a consumption set dense in the nonnegative reals, e.g., nonnegative rationals. This makes clear that the essential distinction is not that of a continuum vs. denumerable, but dense vs. discrete. (This can also be seen from Anderson's proof where $\theta = 1/m$, m a positive integer, with m tending to infinity.)

become $(2M/n) + 2\beta$. Anderson's Theorem 1 is the special case $\beta = 0$, while the above Theorem 1′ is the special case for $\beta = 1$. In what follows, further extensions are obtained: (i) for the case of unequal units for the various commodities and (ii) for mixed commodity spaces where some, but not necessarily all, goods are perfectly divisible (i.e., capable of assuming all real number values).

Section 5(i). Unequal units of different goods. All goods indivisible.

Here we shall replace z' with $z^\wedge = M \cdot (u_1, \ldots, u_k)$ with $u^\wedge = (u_1, \ldots, u_k)$ where u_j is the indivisible unit of good j. [Our treatment in sec. 4 normalized the units so that $u_j = 1$ for all $j = 1, \ldots, k$.] If we do not normalize, we must modify the argument following (2) in Step 2 by using $u^\wedge$ instead of u. As a result, the integer 1 in the right hand side of (5′.1) will be replaced by $u^{\wedge\wedge} = p \cdot u^\wedge$, i.e., the average of the u_j's. [Under normalization, $u^{\wedge\wedge} = 1$.] The same change must be made in formulae (8), (7.2) and (10.1). Then, starting with formula (10.2), the term $\#S$ must be multiplied by $u^{\wedge\wedge}$, and the right hand side of (14.2) becomes $2(M + nu^{\wedge\wedge})$, so that, in the right-hand side of (15), the integer 2 must also be multiplied by $u^{\wedge\wedge}$. The result, to be denoted by (15^^) is a generalized form of the first inequality in Theorem 1′. Analogous changes must be made in the second inequality of Theorem 2′, with $2u^{\wedge\wedge}$ replacing the integer 2.

Section 5(ii). The mixed case: some goods divisible.

In this section I consider the case where at least one good is divisible (i.e., its level can assume all nonnegative *real number* values) and there is an arbitrary number (zero, one, or more) of indivisible goods (whose levels can assume all nonnegative *integer* values). The divisible goods are $Y_1, Y_2, \ldots, Y_s$, $s \geq 1$, while the indivisible goods are denoted by X's. Thus the commodity space is $Z^h \times R^s$, where $h = 0, 1, \ldots$ is the number of indivisible goods. Note that this includes Anderson (1978) as a special case (when there are no X's). Each agent's consumption set is assumed to consist of all nonnegative points of the commodity space, i.e., it equals $Z^h_+ \times R^s_+$. We continue to assume that the set of agents, denoted by A, is finite, and that endowments are non-negative, i.e., that, for all a in A, $e(a) \in Z^h_+ \times R^s_+$.

For all a in A, the assumptions on the preference relation $\succ_a$ are: (a) local *non-satiation with respect to Y* (LNS-Y), and (b) free disposal (FD).

LNS-Y means that for every a in A, every $(x,y) \in Z^h \times R^s$, and every $\epsilon > 0$ there exists $y' \in R^s$ such that $\|y' - y\| < \epsilon$ and $y'y$.

Define *weak monotonicity with respect to Y (WM-Y)* by the following: for every a in A, every $(x,y) \in Z^h \times R^s$ and every $y' \in R^s$ the vector inequality $y' \gg y$ implies $(x;y') \succ_a (x;y)$. Define further *partial weak monotonicity with respect to a (non-empty) subset of the Y's (PWM-Y)* by the following: split the vector y into two subvectors, say $y = (y^*, y^{**})$, with y^* (of dimension $s^* > 0$) non-empty, and y^{**} of dimension s^{**}; then the economy is said to have the PWM-Y property if, for all in A, and all $x, y^*, y^{*\prime}, y^{**}$ in the appropriate spaces, it is the case that $y^{*\prime} \gg y^*$ implies $(x; y^{*\prime}, y^{**}) \succ_a (x; y^*, y^{**})$.

Observe that LNS-Y is implied by WM-Y, and even by PWM-Y. Hence, in particular, when there are no indivisible goods, our assumption WM-Y is identical with that made by Anderson, while for $s > 1$ (i.e., when the number of divisible goods is at least 2), PWM-Y is less restrictive than WM-Y. Hence the conclusion of this section inclides a generalization of Anderson's results as a special case.

Note also that when there is only one divisible good ($s = 1$), weak and strict monotonicity are identical.

Free disposal (FD) is defined as in section 3 above.

Reviewing our proof in section 4, we see that Step 1 and the inequality (2) remain valid in the present mixed case, and so does (5*.1). [Here it is iunderstood that $\gamma(a)$ is defined by $\gamma(a) = \{(x,y) - e(a) \in Z^h \times R^s : (x,y) \succ_a f(a)\}$.]

We must however modify the arguments in Step 2 following (2) and leading to (5′.1), and (4′).

By the local non-satiation property (LNS), there exists for every a in A, a sequence points (x, y^v) converging to $f(a)$, with $(x, y^v) \succ_a f(a)$ for all $v = 1, 2, \ldots,$ ad inf. Hence, for all a in A and all v, $(x, y^v) - e(a)$ belongs to $\gamma(a)$.

It follows that, with p fixed at its value obtained in Step 1, we have

$$inf p \cdot \gamma(a) \leqq p \cdot ((x, y^v) - e(a)) \leqq p \cdot (f(a) - e(a)). \qquad (5\text{^}.1)$$

This , together with (5*.1), yields

$$inf p \cdot \phi(a) \leqq p \cdot (f(a) - e(a)) \text{ for all } a \text{ in } A, \qquad (4\text{^})$$

which is the same as the inequality obtained by Anderson, p.1486, lines 11–12. An inspection of the remainder of the proof in Anderson shows that only (4^) and our inequality (2) are used.[14] Hence the inequalities (1) and (2) of his Theorems 1 and 2 respectively follow.

[14]Alternatively, this can be seen by looking at the formulae (8) and subsequent ones in sec.4 above, and removing the integer 1 or $\#S$, since we now have (4^) instead of (4′).

Thus Anderson's results are extended to the mixed case, with a possible weakening of assumptions in that his weak monotonicity is replaced by LNS. The possibility of such replacement suggests itself very naturally in the light of Koopmans' version (Proposition 4) of the First Theorem of Welfare Economics.

6. *The limit theorem.* A limit theorem for the discrete case is obtained exactly as in Anderson, by assuming a sequence of economies with $M(n)/n$ tending to zero, where $n = \#A$ in the n-th economy,[15] and $M(n)$ is the bound previously denoted by M in the n-th economy. However, because of the integer terms in the right-hand sides of the inequalities of Theorem 1′, it is no longer the case that the left-hand sides tend to zero. Instead, we have the following:
sc Theorem 2′

$$\lim sup \left[\frac{1}{n} \sum_{a \in A(n)} |p_n \cdot (f_n(a) - e_n(a))| \right] \leqq 2 \qquad (2.1')$$

and

$$\lim sup \left[\frac{1}{n} \sum_{a \in A(n)} |\, inf\{p_n \cdot (x - e_n(a)) : x \succ_a f_n(a)\}| \right] \leqq 2. \qquad (2, ii')$$

As in sec.5, the preceding result can be generalized so as to cover both the contimuum and discrete cases. We then have THEOREM 2″, with the left-hand sides the same as in Theorem 2′ while in the right-handed side 2 is replaced by 2β.

7. *An example* The following example shows that, in discrete spaces, the integer 1 cannot be, in general, omitted from the formula (5′.1).

There are two agents and two goods, both indivisible. Each agent has Leontief-type preferences (on the lattice of integer-valued points), represented by the utility function of the form

$$u_i(x_i, y_i) = min(x_i, y_i) \qquad i = 1, 2.$$

The respective endowments are

$$e(1) = (2, 0), \qquad e(2) = (0, 2).$$

Note that $u_1(e(1)) = u_2(e(2)) = 0$. Consider now the allocation $f = \langle f(1); f(2) \rangle = \langle (1,1); (1,1) \rangle$. We have $u_1(f(1)) = u_2(f(2)) = 1$. Hence f is individuality rational (IR). Furthermore, the only (feasible) allocation strictly

[15]A sequence of replicas satisfies this requirement.

preferred by agent 1 is $O_2 = \langle(2,2);(0,0)\rangle$. but O_2 i worse than f for agent 2, so O_2 is not Pareto Superior to f Neither is any other (feasible) allocation. Hence f is Pareto Optimal, as well as IR. Hence f is in the core of this economy. [For terminology, see footnote 16 below.]

To check the Anderson inequality (!!), we calculate:

$$f(1) - e(1) = (1,1) - (2,0) = (-1,1),$$
$$f(2) - e(2) = (1,1) - (0,2) = (1,-1),$$
$$\Gamma(1) = \{(x_1, y_1) : (x_1, y_1) \succ_1 f(1)\} = \{(2,2)\}$$
$$\gamma(1) = \{z_1 : z_1 = (x_1, y_1) - e(1), (x_1, y_1) \in \Gamma(1)\} = \{(2,2) - (2,0)\} = \{(0,2)\}$$
$$\Gamma(2) = \{(x_2, y_2) : (x_2, y_2) \succ_2 f(2)\} = \{(2,2)\}$$
$$\gamma(2) = \{z_2 : z_2 = (x_2, y_2) - e(1), (x_2, y_2) \in \Gamma(2)\} = \{(2,2) - (0,2)\} = \{(2,0)\}$$

Write $p = (p_x, p_y)$ where $p_x + p_y = 1$ and $p_x \geqq 0$, $p_y \geqq 0$. Then $inf p \cdot \gamma(1) = p \cdot (0,2) = 2p_y$, and $inf p \cdot \gamma(2) = p \cdot (2,0) = 2p_x$.

Now the Anderson inequality (!!) for agant a is

$$inf p \cdot \gamma(a) \leqq p \cdot [f(a) - e(a)]. \qquad (!!)$$

For agent 1, this becomes

$$2p_y \leqq p \cdot (-1,1) = -p_x + p_y = -(1 - p_y) + p_y = -1 + 2p_y,$$

i.e., $0 \leqq -1$, which is false *for any p*. So it is not case that there exists a price vector p for which (!!) holds when $a = 1$. In fact we have here

$$inf p \cdot \gamma(1) = p \cdot [f(1) - e(1)] + 1,$$

thus showing that, in general, the right hand side of this equation is the lowest upper bound obtainable for the left hand side, as obtained in the paper. The same result is obtained for agent 2, i.e., when $a = 2$.

8. *Individual vs. avarage.* The results reported in this note, as well as those in Anderson (1978) concern averages. But we do have an interesting (though elementary) inequality referring to individual agents, namely (5′.2) or in its generalized form

$$inf p \cdot \{x - e(a) : x \succ_a f(a)\} \leqq p \cdot (f(a) - e(a)) + \beta, \qquad (5''.2)$$

for any agent a in A, with $\beta = 0$ in the divisible case and $\beta = 1$ in the (normalized) indivisible case. In the classical (strongly convex differentiable) Edgeworth Box case this seems to correspond to the fact that when the marginal rate of substitution at a point of the contract curve is different from that at the Walrasian allocation , then the strict upper contour set of the core allocation "sticks out" below the equilibrium price line through the core allocation, and hence the value of the core allocation at the equilibrium price is higher than what one would have to spend to attain the same utility level. On the other hand, when the two marginal rates are equal, the expendure level is minimized.

Part Two. The Approximate Pareto Optimality of Competitive Allocations in Discrete Commodity Spaces.

9. Somewhat analogous phenomena are observed with regard to the Pareto optimality of Walrasian (i.e., competitive) allocations in discrete spaces (i.e., where all goods are indivisible). The proofs in Arrow and Hahn (1971) or Varian (1984), showing that in an exchange economy with selfish preferences a Walrasian allocation is *Weakly* Pareto Optimal[16] (Pareto *Efficient* in their terminology, here abbreviated as WPO), remain valid in discrete spaces. But as the following example[17] shows, a Walrasian allocation need not be Pareto Optimal.

An example. Consider a two-agent two-good (Edgeworth Box) exchange economy ($\#A = 2, k = 2$), with $e(1) = (1,0)$ and $e(2) = (0,1)$ Hence only the four corners of the 1 by 1 box are feasible. Preferences are defined by the values of utility fucntions on points with nonnegative integer coordinates. For agent 1, $u^1(x^1, y^1) = x^1 + (4/3)y^1$; for agent 2, $u^2 = x^2 + y^2$. (Here superscripts refer to agents. They are not exponents.) The core consists of the two points $e = \langle(1,0);(0,1)\rangle$ and $f = \langle(0,1);(1,0)\rangle$, but only f is Pareto Optimal. (Both are Weakly Pareto Optimal.) The pair (p^*, e) with $p^* = (1/3, 2/3)$ is a Walrasian equilibrium with a Walrasian allocation that is not Pareto Optimal. (There is also another Walrasian equilibrium in which f is the equilibrium allocation.)

It turns out, however, that — in a sense specified by Definition 1 below — in discrete spaces, Walrasian allocations are 'approximately Pareto Optimal.' In particular, as shown in Corollary 1 below, in a two-trader economy, ($n = 2$) with strictly monotonic preferences, given a Walrasian (hence, Weakly Pareto Optimal) allocation $z^* = (z^{*1}, z^{*2})$, there exist a Pareto Optimal allocation $z^{**} = (z^{**1}, z^{**2})$, agents i and j with j different from i, and a good r such that $z^{**i} \sim_i z^{*i}$ and $z^{**j} - u(r) \leqq_j z^{*j}$. [Here $u(r)$ in R^k is a commodity

[16]Although this terminology is widely used, let me state the distinction to avoid any misunderstandings. An allocation is *Weakly Pareto Optimal* (WPO) if it is impossible to make *everyone* better off. It is *Pareto Optimal* (PO) if it is impossible to make *even one* person better off without making some one else worse off. If an allocation is Pareto Optimal, it is necessarily Weakly Pareto Optimal. But, as seen in the aboved example, an allocation can be Weakly Pareto Optimal without being Pareto Optimal.

An allocation z' is *Pareto Superior* (PS) to z'' if z' is feasible and, in z', some persons are better off than under z'' and no one is worse off. An allocation z' is *Strongly Pareto Superior* (SPS) to z'' if z' it is feasible and everyone is better off under z' than under z''. So z' is Pareto Optimal if there is no Pareto Superior alternative; it is Weakly Pareto Optimal if there is no Strongly Pareto Superior alternative.

[17]This example is a slight modification of one constructed in the 1950's (but, I believe, never published) by Kenneth Arrow. (See also Quirk and Saposnik, 1968, p.134.)

space vector with 1 in the r-th place and zeros elsewhere. The symbol $\sim_i$ represents indifference for agent i.]

Thus when there are two traders with strictly monotonic preferences, a Walrasian allocation is, in satisfaction (or utility) terms for each trader, at most one unit-bundle away from a Pareto Optimal allocation. When an additional assumption (made on preferences assumption (A) below), it is possible to obtain analogous results for weakly monotone preferences (see Proposition 2 below).

When there are more than two traders ($n \geqq 3$), analogous results are obtained in Proposition 3 for strictly monotone preferences when other restrictions are imposed. Proposition 3(i) uses assumption (A) and requires that the number of goods k should be no less than $n-1$, where n is the number of agents. Proposition 3(ii) generalizes Corollary 1 without using (A) but imposing another restriction which happens to be automatically satisfied when $n=2$.

Definition 1. (i). Denote by $u(r)$ the k-vector with 1 in the r-th place and zeros elsewhere.

(ii). An allocation z is said to be *approximately Pareto Optimal* if either z is PO, or the following is true: there exists a PO allocation $z^\wedge$ that is Pareto Superior to z and, for every i in N, if $z^{\wedge i} \succ_i z^i$ then $z^i \geqq_i z^{\wedge i} - [u(r_1) + \cdots + u(r_{n-1})]$, with r_n in $\{1, \ldots, k\}$, not necessarily distinct. (Since $z^\wedge$ is PS to z, if it is not the case that $z^{\wedge j} \succ_j z^j$, then necessarily $z^{\wedge j} \sim_j z^j$.)

Remark. When $n=2$ and $k \geqq 1$, Definition 1 is equivalent to the following: either z^* is PO or there exists a PO allocation z^{**} such that z^{**} PS z^* and, for some i, j in $\{1,2\}$, $j \neq i$, and some r in $\{1, \ldots, k\}$, [1] $z^{*j} \sim_j z^{**j}$, and [2] $z^{*i} \geqq z^{**i} - u(r)$. (Since z^{**} PS z^* when z^* is not PO, it is then necessarily the case that $z^{**i} \succ_i z^*$.)

Proposition 1. Let the number of agent be $n=2$, and let the number of goods be $k \geqq 1$. Assume all commodities indivisible (normalized to integer-valued), consumption sets Z_+^K, endowments non-negative (integer-valued), (weak) preference relations transitive, reflexive, and total (hence complete), and also strictly monotone increasing (SM).

Let z^* be a WPO allocation. Then z^* is approximately Pareto Optimal.

Corollary 1. Under the assumptions of Proposition 1, any Walrasian allocation is approximately Pareto optimal.

Proof of the Corollary. A Walrasian allocation is WPO (see proofs in Arrow and Hahn or Varian). Hence, by Prop. 1, it is approximately Pareto Optimal.

Remark. It will be seen below that Corollary 1 may be viewed as a special case of Proposition 3 (ii) below.

Proof of Prop. 1. We may suppose that z^* is not PO. Then, because (by indivisibility) the feasible set is finite, there exists a z^{**} that is PS to z^*.

Then there exists an i such that z^{**i} is strictly preferred to z^{*i}. Then, since z^* is WPO, it must be that, for j unequal to i, $z^{*j} \sim_j z^{**j}$, so [1] holds. (The symbol $\sim_j$ represents indifference for agent j.) It will be shown below that the allocation

$z^{***} = \langle z^{***i}, z^{***j} \rangle = \langle z^{**i} - u(r), z^{**j} + u(r) \rangle$ is feasible. But suppose that [2] is false. Then, by strict monotonicity of preferences and because z^{**i} is strictly preferred to z^{*i}, it follows that z^{***} is Strongly Pareto Superior to z^*, hence — contrary to assumption — not WPO.

It remains to show that z^{***} is feasible. It is balanced by construction, so we must only show that it is individually feasible. Clearly z^{***j} is individually feasible since it is vectorially greater than individually feasible z^{**j}. As for z^{***i}, it is individually feasible if z^{**i} is non-negative and different from 0. For in this case, z^{**i} has at least one component, say the r-th, equal to 1 or more, so that $z^{**i} - u(r)$ is non-negative, hence feasible for agent i. But suppose that $z^{**i} = 0$. Since $z^{*i} \geq 0$, it follows that $z^{*i} \geq z^{**i}$, which is impossible by the assumption of strict monotonicity, since z^{**i} is strictly preferred by agent 1 to z^{*i}. Q.E.D.

The following Proposition 2, related to Corollary 1, asserts that a Walrasian allocation is approximately Pareto Optimal even when preferences are merely weakly monotone provided that the following assumption on preferences is added:

(A) for every consumer, the initial commodity bundle is weakly preferred to any bundle with a zero component; more formally, (A) holds if, for every agent i in $\{1,2\}$, if the commodity bundle z^i has a zero component, then $e(i) \geq_i z^i$.

(Admittedly, this is a rather high price to pay for weakening the strict monotonicity assumption, but the case is not vacuous: it is satisfied for Cobb-Douglas preferences.) Note that since Proposition 2 assumes (A), it is not comparable with Corollary 1.

Proposition 2. Retain the assumptions of Proposition 1, except that strict monotonicity is replaced by weak monotonicity, and assumption (A) is also adopted. Then any Walrasian allocation is approximately Pareto Optimal.

Proof. Since z^* is Walrasian, it is WPO. The first paragraph of the proof of Prop. 1 remains valid under weak monotonicity if $u(r)$ is replaced by $u = (1, \ldots, 1)$, a vector all of whose components are 1's, and z^{***} is replaced by $z\# = \langle z\#^i, z\#^j \rangle = \langle z^{**i} - u, z^{**j} + u \rangle$. It remains to show that $z\#$ is feasible. Again, only the individual feasibility of the bundle $z\#^i$ is a problem. Now clearly, $z\#^i$ is feasible for agent i if $z^{**i} \gg 0$. But if not, then z^{**i} has a zero component, hence ($^{\circ}$) $z^{**i} \leq_i e(i)$. But, as postulated in the first paragraph of the proof of Proposition 1, $z^{**i} \succ_i z^{*i}$. Also, since z^* is Walrasian, it is Individually Rational (IR), hence $z^{*i} \geq_i e(i)$, and so z^{**i} is strictly preferred by agent i to z^{*i}. This contradicts the inequality ($^{\circ}$). Q.E.D.

Proposition 3. (i) Let the set of agents $N = \{1, \ldots, n\}$ have at least two members ($n \geq 2$), and let the number of goods be $k \geq n-1$, with all commodities indivisible (normalized to integer-valued), consumption sets assumed to be Z_+^k, endowments non-negative (integer-valued), (weak) preference relations transitive, reflexive, and total (hence complete), are strictly monotone increasing (SM), and also satisfy assumption (A). Then every Walrasian allocation is approximately Pareto Optimal.

(ii) Furthermore, without assuming (A), if the number m of agents who strictly prefer the PO allocation z^{**} to the Walrasian allocation z^*, with z^{**} PS to z^*, is not less than $n/2$, then for every such agent i we have $z^{*i} \geq_i z^{**i} - u(r)$ for some good r.

Remark. Corollary 1 is a special case of Proposition 3 (ii) because in that case $m = n/2$.

Proof of Proposition 3. Let z^* be Walrasian and let the PO allocation z^{**} be PS to z^*. Without loss of generality, let [1] $z^{**1} \succ_1 z^{*1}$.

1.*Claim*: $z\#^1 = z^{**1} - [u(r_2) + \cdots + u(r_n)]$ is feasible for agent 1.

Proof. Suppose not. This means that, in z^{**1}, there are fewer than $n-1$ units of goods. But since $k \geq n-1$, it follows that z^{**1} has at least one zero component. Hence, by (A), $z^{**1} \leq_1 e(1)$. But since a Walrasian allocation is Individually Rational (IR), we have $e(1) \leq_1 z^{*1}$, hence $z^{**1} \leq_1 z^{*1}$. But this contradicts [1].

2.*Claim*: $z\#^1 \leq_1 z^{*1}$. *Proof.* Suppose not. Then $z\#^1 \succ_1 z^{*1}$. Consider the allocation

$$z\# = \langle z\#^1, z\#^2, \ldots, z\#^n \rangle$$

where, for $j > 1$,

$$[2] \quad z\#^j = z^{**j} + u(r_j).$$

Then, by strict monotonicity and [2], $z\#$ is Strongly Pareto Superior to z^*. But z^{**} is PS to z^*. Hence $z\#$ is Strongly Pareto Superior to Z^*, and therefore z^* is not WPO, contrary to our hypothesis. This completes the proof of (i).

Proof of (ii). Suppose now that $m \geq n-1$, and consider an allocation $z\#\#$ where m-n agents who strictly prefer z^{**} to z^* each give one unit of a good to one of the agents who are indifferent between z^{**} and z^*.

Suppose that giving up one unit of some good is infeasible for one of the m-n agents, say agent i. Then it must be that $z^{**i} = 0$. But, as in the proof of Proposition 1, since, by hypothesis, agent 1 strictly prefers z^{**1} to z^{*1}, and (by IR of a Walrasian allocation) z^{*1} is no worse for agent 1 to $e(1)$. But $e(1)$ is non-negative, so z^{**1} is strictly preferred to the 0 commodity vector, which yields a contradiction. Hence $z\#\#$ is feasible.

Now suppose that $z\#\#^1 \leq_1 z^{*1}$ is false. Then, as before, using strict monotonicity, we see that $z\#\#$ is Strongly Pareto Superior to z^*, so that z^* is not WPO, contrary to the hypothesis (because a Walrasian allocation is always WPO). This completes the proof.

References

[1] R.M. Anderson, *An Elementary Core Equivalence Theorem*, Econometrica **46** (1978), 1483–1487.

[2] R.M. Anderson, *Gap-minimizing Prices and Quadratic Core Convergence*, Journal of Mathematical Economics **16** (1987), 1–15.

[3] R.M. Anderson, *The Second Welfare Theorem with Nonconvex Preferences*, Econometrica **56** (1988), 361–382.

[4] K.J. Arrow and F.H. Hahn, *General Competitive Analysis*, Holden-Day, San Francisco, 1971.

[5] G. Debreu, *Theory of Value*, Wiley, New York, 1959.

[6] G. Debreu and H. Scarf, *A limit Theorem on the Core of an economy*, International Economic Review **5** (1963), 235–246.

[7] E. Dierker, *Gains and Losses at Core Allocations*, Journal of Mathematical Economics **2** (1975), 119–128.

[8] W. Hildenbrand, *Core of an Economy*, Ch.**18** in the Handbook of Mathematical Economics **II**, K.J. Arrow and M.D. Intriligator, North-Holland, Amsterdam, 1982.

[9] W. Hildenbrand and A.P. Kirman, *Equilibrium Analysis. Variations on Themes by Edgeworth and Walras*, North-Holland, Amsterdam, 1988.

[10] H. Keiding, *A Limit Theorem on the Cores of Large but Finite Economies*, preprint, April, 1974.

[11] T.C. Koopmans, *Three Essays on the State of Economic Science*, McGraw-Hill, New York, 1957.

[12] A.A. Lewis, *On the effective Content of Asymptotic Verifications of Edgeworth's Conjecture*, Mathematical Social Science **22** (1991), 275–324.

[13] A. Mas-Colell, *The Theory of General Economic Equilibrium*, Cambridge University Press, Cambridge, 1985.

[14] J. Quirk and R. Saposnik, *Introduction to General Equilibrium Theory and Welfare Economics*, McGraw-Hill, New York, 1968.

[15] H. Varian, *Microeconomic Analysis. Second Edition*, Norton, New York, 1984.

[16] K. Vind, *A Theorem on the Core of an Economy*, The Review of Economic Studies **5**, (1965), 165–177.

Cooperative Processing of Information

Tatsuro Ichiishi
Department of Economics, Ohio State University
Columbus, OH 43210-1172, U.S.A.

1. Introduction

The Nash equilibrium concept of a normal-form game (Nash [15]) has been extended in several directions in the past: One direction is to introduce asymmetric information explicitly into the normal-form game. The resulting model is Harsanyi's [9] *Bayesian game*. His *Bayesian equilibrium* concept extends the Nash equilibrium concept to this new framework.

Another direction is to introduce the possibility of coordination of strategies by several players. Recall that in regard to the noncooperative game theory on the one hand, a game in normal form (its Nash equilibrium, resp.) was generalized to Debreu's [5] abstract economy (its social equilibrium, resp.), the pure exchange economy (its competitive equilibrium, resp.) was generalized to Arrow and Debreu's [2] private-ownership economy (its competitive equilibrium, resp.), and the game-theoretical structure of the private-ownership economy (its competitive equilibrium, resp.) was summarized as Debreu's [5] abstract economy (its social equilibrium, resp.). Recall also that in regard to the classical cooperative game theory on the other hand, Bondareva [3] and Shapley's [18] work on the side-payment game (its core, resp.) was generalized by Scarf's [17] work on the non-side-payment game (its core, resp.), the pure exchange economy (its core allocation, resp.) was generalized to the coalition production economy (its core allocation, resp.), and the game-theoretical structure of the coalition production economy (its core allocation, resp.) was summarized as Scarf's [17] work on the non-side-payment game (its core, resp.). In order to analyze the simultaneous workings in the present-day economy of (1) the neoclassical price mechanism in a market where people are anonymous and (2) the cooperative resource allocation mechanism in a firm defined as an organization of human-resource holders, Ichiishi [10] proposed the model of society (social coalitional equilibrium, resp.) as a generalization of both (1) Debreu's [5] abstract economy (its social equilibrium, resp.) and (2) Scarf's [17] work on the non-side-payment game (its core, resp.). This synthesis was made possible by Fan's [7, 8] coincidence theorem, one of the highest achievements in the recent fixed-point theory. See Ichiishi [11, 12] for a summary of the development of the cooperative extension of noncooperative games and its underlying philosophy, motivated by the analytical theory of the firm.

The purpose of the present paper is to outline some results of Ichiishi and Idzik [13] and Ichiishi, Idzik and Zhao [14], in which a new model of the *Bayesian society* was formulated, both as a cooperative extension of the Bayesian game and as a Bayesian extension of the society, and in which a new solution concept of the *Bayesian strong equilibrium* was proposed for a Bayesian society, as an extention both of the Bayesian equilibrium and of the social coalitional equilibrium. There arise new issues intrinsic to this extension, so there are actually several versions of the Bayesian strong equilibrium concept. Three situations were discussed in the two papers, each giving rise to the associated specific equilibrium concept.

The first situation, called the *I-P case* (information pooling case), is one extreme case and postulates that coalition S can pool its members' private information, so a strategy bundle (contract) for S can be designed so as to take advantage of the pooled information. Bayesian strong equilibrium existence theorems were established for this case.

The second situation, called the *I-NP case* (information non-pooling case), is the other extreme case and postulates that nobody in coalition S can use the others' private information at the time of contract-execution. The members of S can design only strategy bundles such that player j's strategy is based solely on his own private information. In this case, j knows that the others do not have his private information at the time of contract-execution, so he may take a false action in his own interest by misrepresenting his private information; no strategy bundle leaving this possibility open can be agreed upon at the outset. Thus, the *Bayesian incentive compatible strong equilibrium* concept was introduced in this case, and its existence theorems were established.

The third situation is an intermediate case, in which player j can decide which part of his private information he will pass on to the other members of his coalition S. This is actually a challenging case to analyze, since j may not want to reveal his private information to the others, and even if he does, the others may not trust j's words, as j may manipulate his colleagues in his interest by telling them a lie. A specific structure was added to the model of Bayesian society in order to address endogenous determination of information structure in equilibrium. The associated (more involved) concept of *Bayesian incentive compatible strong equilibrium* was proposed. The pseudo-metric space (SPACE, d) of all logically conceivable Bayesian societies having this additional structure was constructed, and it was established that there exists an open and dense subset SPACE$'$ of SPACE so that every Bayesian society in SPACE$'$ possesses a Bayesian incentive compatible strong equilibrium.

To be precise, the above existence theorems and generic existence theorems were established for two classes of Bayesian societies: one the general class, and the other the specific class of Bayesian societies without externality defined by the condition that each player's utility function depends only on his own choice and his own type.[1] The emphasis was given on the latter specific class, because the assumptions for the existence and for the generic existence can then be succinctly stated in terms of the exogenously given data. This class, although too specific to include the normal-form game, still includes most of the centrally important economic models with organizations, such as the coalition production economy, a generalization of the pure exchange economy. The work outlined here should, therefore, be understood as a work to provide a game-theoretical setup of *economic* analysis – in particular, a game-theoretical setup of analytical theory of the firm.

For a detailed survey of related works, see Section 6 of literature discussions newly added to Ichiishi and Idzik [13].

2. Model

Let N be a finite set of players. The family of nonempty coalitions (nonempty subsets of N) is denoted by $\mathcal{N}$. Each player j has a choice set C^j, a type set T^j, and a von Neumann-Morgenstern utility function u^j. The type set T^j is assumed to be finite. For $S \in \mathcal{N}$, define $C^S := \prod_{j\in S} C^j$, $T^S := \prod_{j\in S} T^j$, and write $C := C^N$, $T := T^N$ for simplicity. Generic elements of sets C^S, T^S, C and T are denoted by c^S ($:= (c^j)_{j\in S}$), t^S, c and t, respectively. The function $u^j : C \times T \to \mathbf{R}$ associates j's utility level $u^j(c,t)$ to each choice bundle $c \in C$ and type-profile $t \in T$.

The set T^S gives rise to the partition of T: $\{\{t^S\} \times T^{N\setminus S} \mid t^S \in T^S\}$. Denote by $\mathcal{T}^S$ the algebra on T generated by this partition. At a certain stage of the game, player j has the private information, that is, he knows his true type in T^j, but does not know the others' true types. At this stage, his information structure is $\mathcal{T}^j$ ($:= \mathcal{T}^{\{j\}}$).

There is a strictly positive *ex ante* probability p on T. The probability on T held by player j of type $\bar{t}^j$ at the *interim* (or rather, *in mediis*) stage of the preceding paragraph is equal to the conditional probability of p given $\bar{t}^j$.

Harsanyi's [9] *Bayesian game* is defined as a list of specified data,

$$(\{C^j, T^j, u^j\}_{j\in N},\ p).$$

To formulate a cooperative extension of the Bayesian game, one needs to introduce a richer structure appropriate for analysis of cooperative behavior. First, one needs to distinguish between feasible and non-feasible coalitional choices. When the players' type-profile is given by t, each coalition S has a set of *feasible* joint choices, defined as subset $C_0^S(t)$ of C^S. Write for simplicity,

[1] For the I-P and I-NP cases, the treatment is somewhat more general.

$C_0^j(t) := C_0^{\{j\}}(t)$. Thus the feasible-choice correspondence $C_0^S : T \to C^S$ is defined for every $S \in \mathcal{N}$. Notice the possibilities that $C_0^S(t) \neq \prod_{j\in S} C_0^j(t)$, and that $C_0^S(t) \neq C_0^S(t')$ if $t \neq t'$. The first possibility is crucial for major economic applications, as the following example illustrates.

EXAMPLE 2.1 Given a type-profile t, the t-specific pure exchange economy is a list of specified data, $\mathcal{E}(t) := \{\mathbf{R}^l_+, u^j(\cdot,t), \omega^j(t)\}_{j\in N}$, where $\mathbf{R}^l_+$ is the consumption set of consumer j, $u^j(\cdot,t) : \mathbf{R}^l_+ \to \mathbf{R}$ is j's utility function, and $\omega^j(t)$ $(\in \mathbf{R}^l_+)$ is j's initial endowment vector. Then,

$$C^j := \mathbf{R}^l_+; \text{ and}$$

$$C_0^S(t) := \{x^S \in (\mathbf{R}^l_+)^S \mid \sum_{j\in S} x^j \leq \sum_{j\in S} \omega^j(t)\}.$$

Needless to say, $C_0^S(t) \neq \prod_{j\in S} C_0^j(t)$ in this case, and under the additional assumption that $u^j(\cdot,t)$ is a quasi-concave function only of c^j, this fact plays a crucial role in guaranteeing the existence of a core allocation of $\mathcal{E}(t)$ as a relative interior point of the Edgeworth box diagram,

$$\{x \in (\mathbf{R}^l_+)^N \mid \sum_{j\in N} x^j = \sum_{j\in N} \omega^j(t)\}.$$

A core allocation of $\mathcal{E}(t)$ is a strong equilibrium of $\mathcal{E}(t)$. □

The members of coalition S cannot distinguish the outsiders' type-profiles, so their choice set does not depend upon $t^{N\setminus S} \in T^{N\setminus S}$ $(C_0^S(t^S, t^{N\setminus S}) = C_0^S(t^S, t'^{N\setminus S})$ for any $t^{N\setminus S}, t'^{N\setminus S} \in T^{N\setminus S})$. Therefore, the feasible-choice correspondence $C_0^S : T \to C^S$ will be assumed to depend only upon $t^S \in T^S$, that is, $C_0^S(\cdot)$ *will be postulated to be* $\mathcal{T}^S$*-measurable.* For any $\mathcal{T}^S$-measurable function f on T, one may write $f(t^S)$ instead of $f(t)$.

A strategy of a player specifies his choice contingent upon a type-profile. Formally, a *strategy* of player j is a function $x^j : T \to C^j$. A choice $c^j \in C^j$ may be identified with the constant strategy $t \mapsto c^j$. Denote by X^j the set of all logically conceivable strategies of player j,

$$X^j := \{x^j : T \to C^j\}.$$

For $S \in \mathcal{N}$, define $X^S := \prod_{j\in S} X^j$, and write $X := X^N$ for simplicity. The model under construction explains which feasible strategy bundle $x^S := (x^j)_{j\in S}$ is agreed upon by the members of coalition S. Strategy bundle x^S may synonymously be called a *contract*.

The second additional structure clarifies which strategies are feasible. Feasibility is formulated by the feasible-strategy correspondence $F^S : X \to X^S$, which associates to each strategy bundle $\bar{x} \in X$ (which may not be feasible) the set $F^S(\bar{x})$ of all feasible strategy bundles available to coalition S. The correspondence is used as a description of coalition S's capacity when it tries to deviate from the grand coalition N. Suppose the grand coalition is entertaining possible contract $\bar{x} \in X$, which was proposed by its leaders. If coalition S is to deviate from N in this situation, the members of S have to know which strategy bundles x^S are feasible to them. The set $F^S(\bar{x})$ ($\subset X^S$) describes precisely these feasible strategies. *Any member of $F^S(\bar{x})$ will be postulated to be $\mathcal{T}^S$-measurable.* One could simply define the constant correspondence F^S by:

$$F^S(\bar{x}) := \{x^S \in X^S \mid x^S \text{ is a } \mathcal{T}^S\text{-measurable selection of } C_0^S.\}$$

However, the present approach is more general, and the correspondence F^S is arbitrarily given. Of course, there is then redundancy in definitions of C^S and F^S, but this demerit is dominated by the merits of wider economic applicability. Indeed, the present formal framework can be used to analyze the workings of an institution which allows the players to take advantage of the market mechanism when making cooperative choice of strategies; for detailed discussions of the following Example 2.2, in particular in relation to the balanced non-side-payment game, the ordinal convex non-side-payment game, the strong equilibria and the competitive equilibria, see Ichiishi [11, Example 2.A.14]. Let Δ^{l-1} denote the unit simplex in $\mathbf{R}^l$, $\{p \in \mathbf{R}^l_+ \mid \sum_{h=1}^l p_h = 1\}$. For two elements p, x in $\mathbf{R}^l$, let $p \cdot x$ denote their Euclidean inner product, $\sum_{h=1}^l p_h x_h$.

EXAMPLE 2.2 Let $\mathcal{E}(t) := \{\mathbf{R}^l_+, u^j(\cdot, t), \omega^j(t)\}_{j \in N}$ be a pure exchange economy given a type-profile t. Add one more hypothetical player, say the zeroth player or the *market participant*, whose choice set is the price simplex $C^0 := \Delta^{l-1}$. If players in N can use the market mechanism, the feasible joint choices $C_0^S(t)$ are enlarged. In this case, the feasible-strategy correspondence of coalition S ($\subset N$) is defined by: $x^S \in F^S(\bar{p}, \bar{x})$ iff x^S is $\mathcal{T}^S$-measurable, and

$$\forall\, t:\ \bar{p}(t) \cdot \sum_{j \in S} x^j(t) \le \bar{p}(t) \cdot \sum_{j \in S} \omega^j(t),\ x^j(t) \in \mathbf{R}^l_+. \qquad \square$$

DEFINITION 2.3 A *Bayesian society* is a list of specified data

$$\mathcal{S} := (\{C^j, T^j, u^j\}_{j \in N},\ \{C_0^S, F^S\}_{S \in \mathcal{N}},\ p)$$

of: (i) a finite set of players N; (ii) a choice set C^j, a finite set of types T^j, and a von Neumann-Morgenstern utility function $u^j : C \times T \to \mathbf{R}$ for each player j; (ii) a $\mathcal{T}^S$-measurable feasible-choice correspondence $C_0^S : T \to C^S$, and a feasible-strategy correspondence $F^S : X \to X^S$ for each coalition S, such that each element of $F^S(\bar{x})$ is a $\mathcal{T}^S$-measurable function; and (iii) a strictly positive *ex ante* objective probability p on T.

In an attempt to design and agree upon some contract, the players play a cooperative game, entertaining a host of possible joint strategies. Eventually a feasible and group incentive compatible strategy bundle is voluntarily chosen and voluntarily sustained. This particular strategy bundle is agreed upon, that is, the contract is signed. Strictly speaking, as soon as a contract (strategy bundle) is signed, the game is over. The rest of the time sequence is spent for contract-execution (choices at subsequent information sets – to use the terminology of the extensive game). The agreed upon contract is binding in the subsequent stages; this is the case, regardless whether the agreement is made noncooperatively (as in the Stackelberg-type mechanism-design *à la* principal-agent theory) or cooperatively (as in the coalitional mechanism-design of the present work).

The final issue of this section is timing of contract execution. Two extreme cases are now defined (for the intermediate case, see Section 4). Suppose coalition S is to form, and the members are to agree upon contract $x^S(\cdot)$. The I-P case (the information pooling case) is defined as the situation in which the members of a realized coalition S can eventually *pool* information, that is, if the members' true types are given by $t^S = \{t^j\}_{j \in S}$, each member j will know the sharper conditional probability $p(\cdot \mid t^S)$ on T given t^S at the time of contract execution. The I-NP case (the information non-pooling case) is defined as the situation in which no coalition S can pool its members' information, so that each member j only knows the conditional probability given t^j, even after coalition-formation. In the I-P case, any contract can be feasible (as long as it is a member of $F^S(\bar{x})$ – in view of the measurability postulate in Definition 2.3 (ii)). In the I-NP case, only those contracts x^S in $F^S(\bar{x})$ for which each x^j depends solely upon t^j are feasible. Therefore, define the feasible-strategy correspondence for the I-NP case, $F^S_{i\text{-}np} : X \to X^S$ by:

$$F^S_{i\text{-}np}(\bar{x}) := \{x^S \in F^S(\bar{x}) \mid \forall\, j \in S : \ x^j \text{ is } \mathcal{T}^j\text{-measurable.}\}.$$

DEFINITION 2.4 The *I-P case* is the situation in which for every $\bar{x} \in X$, members of coalition S can choose any strategy bundle from $F^S(\bar{x})$. The *I-NP case* is the situation in which for every $\bar{x} \in X$, members of coalition S can choose strategy bundles only from $F^S_{i\text{-}np}(\bar{x})$.

The I-NP case creates the incentive for players to misrepresent their true types after a strategy is chosen, specifically during the time a particular coalitional choice is made according to the contract. If members of coalition S foresee at the outset that a particular contract x^S may later induce such misrepresentation, they do not agree on the contract x^S. Therefore, the feasible-strategy set is further restricted to those strategies that are Bayesian incentive compatible in the sense of d'Aspremont and Gérard-Varet [4]. Suppose that players S agreed on $x^S \in F^S_{i\text{-}np}(\bar{x})$ when the players N were discussing $\bar{x}$. Let j be any member of S, whose true type is $\bar{t}^j$. If he reports to S that his type is $\tilde{t}^j$, then he makes the choice $x^j(\tilde{t}^j)$ according to the contract, so his conditional expected utility given his true type is

$$Eu^j(x^j(\tilde{t}^j), x^{S\setminus\{j\}}, \bar{x}^{N\setminus S} \mid \bar{t}^j),$$

where $x^j(\tilde{t}^j)$ is the constant strategy, $t \mapsto x^j(\tilde{t}^j)$. The *Bayesian incentive compatibility* says that he has the incentive to report his true type $\bar{t}^j$ and enjoy the conditional expected utility $Eu^j(x^S, \bar{x}^{N\setminus S} \mid \bar{t}^j)$.

Let $\mathcal{S}$ be a Bayesian society in the I-NP case. The set $F^S_{i\text{-}np,ic}(\bar{x})$ is defined as the set of those $x^S \in F^S_{i\text{-}np}(\bar{x})$ for which

$$\forall\, j \in S:\ \forall\, \bar{t}^j, \tilde{t}^j \in T^j:\ Eu^j(x^S, \bar{x}^{N\setminus S} \mid \bar{t}^j) \geq Eu^j(x^j(\tilde{t}^j), x^{S\setminus\{j\}}, \bar{x}^{N\setminus S} \mid \bar{t}^j).$$

3. Equilibria for the I-P and I-NP Cases

DEFINITION 3.1 Let $\mathcal{S}$ be a Bayesian society in the I-P case. A strategy bundle $x^* \in X$ is called a *Bayesian strong equilibrium,* if
(i) $x^* \in F^N(x^*)$; and
(ii) it is not true that there exist $S \in \mathcal{N}$ and $x^S \in F^S(x^*)$ such that for each $j \in S$,

$$\forall\, t^S \in T^S:\ Eu^j(x^S, x^{*N\setminus S} \mid t^S) \geq Eu^j(x^* \mid t^S),$$

with strict inequality for at least one t^S.

DEFINITION 3.2 Let $\mathcal{S}$ be a Bayesian society in the I-NP case. A strategy bundle $x^* \in X$ is called a *Bayesian incentive compatible strong equilibrium,* if
(i) $x^* \in F^N_{i\text{-}np,ic}(x^*)$; and
(ii) it is not true that there exist $S \in \mathcal{N}$ and $x^S \in F^S_{i\text{-}np,ic}(x^*)$ such that for each $j \in S$,

$$\forall\, t^j \in T^j:\ Eu^j(x^S, x^{*N\setminus S} \mid t^j) \geq Eu^j(x^* \mid t^j),$$

with strict inequality for at least one t^j.

Condition (i) in Definitions 3.1 and 3.2 says that the contract x^* is feasible in the grand coalition N. One may introduce a coalition structure instead of the grand coalition, but such generalization is straightforward (see Ichiishi [10]). Condition (ii) is the coalitional stability in that no coalition can improve upon x^*.

Ichiishi and Idzik [13] provides Bayesian strong equilibrium existence theorems for the I-P and I-NP cases, and Bayesian incentive compatible strong equilibrium existence theorems for the I-NP case: First, existence thoerems are established for the general class of Bayesian societies with externality in which each utility function u^j depends fully on $(c,t) \in C \times T$. The assumptions in these theorems are stated, however, partly in terms of derivative concepts such as the parameterized non-side-payment games. Second, as applications of the theorems for the general class, existence theorems are established for the specific class of Bayesian societies without externality in which each utility function u^j depends only on $(c^j,t) \in C^j \times T$. Assumptions of the theorems for this specific class are stated only in terms of the exogenously given data $\mathcal{S}$; these theorems are presented below (Theorems 3.3 and 3.4). Notice that in spite of the terminology "no-externality", the feasible-strategy correspondences F^S depend fully on $x \in X$, and to this extent externality is still considered.

Of course, given the no-externlity condition, the normal-form games or the Bayesian games are no longer included in the analysis, but vertually all the major economic models (extended to cover asymmetric information) are included: the pure exhange economy, the coalition production economy, and the economic models of Ichiishi [11, 12] with organizations. In the literature the set of strong equilibria of the pure exchange economy, and the set of strong equilibria of the coalition production economy have been called the *core*. Moreover, due to the full dependence of F^S on $x \in X$ (the preceding paragraph), the core of a production economy with public goods can also be analyzed.

A subfamily $\mathcal{B}$ of $\mathcal{N}$ is called *balanced* if there exists $\{\lambda_S\}_{S\in\mathcal{B}} \subset \mathbf{R}_+$ such that $\sum_{S\in\mathcal{B}:S\ni j} \lambda_S = 1$ for every $j \in N$.

THEOREM 3.3 (Ichiishi and Idzik [13]) *Let $\mathcal{S}$ be a Bayesian society in the I-P case. Assume that each von Neumann-Morgenstern utility function u^j depends only on $C^j \times T$. Then there exists a Bayesian strong equilibrium, if:*
(i) *for any j, C^j is a nonempty, compact, convex subset of a Hausdorff locally convex topological vector space over $\mathbf{R}$;*
(ii) *for any j and any t, $u^j(\cdot,t)$ is continuous and concave on $C_0^j(t)$;*
(iii) *for any S and any t, $C_0^S(t)$ is nonempty, closed and convex;*
(iv) *for any S, correspondence F^S is both upper and lower semicontinuous in X, and has nonempty, closed and convex values; and*
(v) *for any $\bar{x} \in X$ and any balanced family $\mathcal{B}$ with the associated balancing coefficients $\{\lambda_S\}_{S\in\mathcal{B}}$, it follows that $\sum_{S\in\mathcal{B}} \lambda_S \tilde{F}^S(\bar{x}) \subset F^N(\bar{x})$, where $\tilde{F}^S(\bar{x}) := \{x \mid x^S \in F^S(\bar{x}),\ x^{N\setminus S} = \mathbf{0}\}$.*

THEOREM 3.4 (Ichiishi and Idzik [13]) *Let $\mathcal{S}$ be a Bayesian society in the I-NP case which satisfies all the assumptions of Theorem 3.3. Assume furthermore that each C^j is metrizable;*
(ii′) *for any j and any t, $u^j(\cdot,t)$ is linear affine in C^j; and*
(vi) *for any S, either F^S is a constant correspondence, or there exists $\hat{x}^S \in X^S$*

such that for all $\bar{x} \in X$, $\hat{x}^S \in F^S(\bar{x})$, and for all $j \in S$ and all $\bar{t}^j, \tilde{t}^j \in T^j$ for which $\bar{t}^j \neq \tilde{t}^j$,

$$Eu^j(\hat{x}^j \mid \bar{t}^j) > Eu^j(\hat{x}^j(\tilde{t}^j) \mid \bar{t}^j).$$

Then, there exists a Bayesian incentive compatible strong equilibrium of $\mathcal{S}$.

The linearity condition (ii′) on $u^j(\cdot, t)$ requires some comments: If choices here are interpreted as *pure choices*, then this assumption imposes the strong condition of risk-neutrality on the players' preference relations. If, on the other hand, choices are interpreted as mixed choices, then the utility here should be interpreted as the expected utility. Of course, the expected utility is linear in probabilities, so the assumption is automatically satisfied under the second interpretation of the choices (that is, the affine linearity does not have to be stated as an assumption). Indeed, in much of the existing literature on Bayesian analysis of games with asymmetric information, mixed choices are explicitly assumed, and the linearity of the utility functions is fully utilized.

4. Additional Structure for Endogenous Determination of Information Structure

It would be virtually impossible to address endogenous determination of the information structure in equilibrium within the general framework of Bayesian society, so one needs to introduce an additional specific structure to the model. It should be acknowledged that the microeconomic works on the rational expectations equilibrium by Radner [16] and by Dubey, Geanakoplos and Shubik [6] were fundamentally helpful in formulating the required specific structure here.

The additional structure for a Bayesian society $\mathcal{S}$ can now be introduced (Postulates 4.1 and 4.2). First, a scenario to motivate this structure is in order: Suppose a firm (coalition of human-resource holders) is going to form. The signed contract specifies a [type-profile]-contingent choice to each member. Suppose that the production activity takes two months, and that each member makes a choice each month (that is, each member has two information sets, one information set for each month). Each month, choices of all members, one choice for each member, are made simultaneously. It is only after the second month that the firm completes its production activity, sells its products, and distributes its value-added net of dividend-payment to its members according to the contract.

POSTULATE 4.1 For each player $j \in N$, his choice set is of the form, $C^j = C_1^j \times C_2^j$.

Set $c^j = (c_1^j, c_2^j)$. Player j makes choice c_1^j (c_2^j, resp.) at his information set of the first month (the second month, resp.).

For two information structures $\mathcal{B}$ and $\mathcal{C}$ (algebras on T), denote by $\mathcal{B} \vee \mathcal{C}$ the algebra generated by $\mathcal{B} \cup \mathcal{C}$ (the smallest algebra on T that contains both $\mathcal{B}$ and $\mathcal{C}$). For any set Z and any function $f : T \to Z$, denote by $\mathcal{A}(f)$ the algebra generated by f (the smallest algebra on T that contains the sets $\{f^{-1}(z) \mid z \in Z\}$).

A strategy of player j in coalition S can also be denoted by $(x_1^j(\cdot), x_2^j(\cdot))$. Given a strategy bundle x^S, information is processed within coalition S in the following way: In the first month, each player is *assumed* to have only his own private information. So, the component $x_1^j(\cdot)$ has to be $\mathcal{T}^j$-measurable. *If it is common knowledge in S that player j has the incentive to make a choice (say, c_1^j) in the first month according to his true type*, then by the beginning of the second month the occurrence of event

$$\{t^j \in T^j \mid x_1^j(t^j) = c_1^j\}$$

has become common knowledge in S. Let $\bar{t}$ be the true type profile, and suppose choice bundle $c_1^S \in C_1^S$ is made in the first month. Then each player j has the information that event

$$\{t \in T \mid t^j = \bar{t}^j,\ x_1^S(t) = c_1^S\}$$

has occurred with probability 1. In a nutshell, the members of coalition S process and update information *by taking action* rather than *by talking*.

When designing the other component of the strategy bundle $x_2^S = (x_2^i)_{i \in S}$ at the time of contract agreement, the members can anticipate the information structure available to i at the beginning of the second month, $\mathcal{T}^i \vee (\vee_{j \in S} \mathcal{A}(x_1^j))$, and make each x_2^i measurable with respect to it. To simplify notation, set

$$\hat{\mathcal{T}}^i(x_1^S) := \mathcal{T}^i \vee (\vee_{j \in S} \mathcal{A}(x_1^j)).$$

Thus, one can make the following postulate of information-pooling rule:

POSTULATE 4.2 Given any strategy bundle $\bar{x} \in X$, coalition S designs only those $x^S \in F^S(\bar{x})$ such that for all $j \in S$ it follows that
(i) x_1^j is measurable with respect to $\mathcal{T}^j$, and
(ii) x_2^j is measurable with respect to $\hat{\mathcal{T}}^j(x_1^S)$.

Denote by $\hat{F}^S(\bar{x})$ the set of those feasible strategies x^S that satisfy Postulate 4.2 (information-pooling rule):

$$\hat{F}^S(\bar{x}) := \left\{ x^S \in F^S(\bar{x}) \;\middle|\; \begin{array}{l} \forall\, j \in S: \\ x_1^j \text{ is measurable w. r. t. } \mathcal{T}^j, \text{ and} \\ x_2^j \text{ is measurable w. r. t. } \hat{\mathcal{T}}^j(x_1^S) \end{array} \right\}.$$

Recall that in order for the present information-processing mechanism to work, the members of coalition S need to have the common knowledge that each player has the incentive to make a choice in the first month according to his true type. After all, the contract will not be enforced, if some member has the incentive to make a choice with false pretension about his true type either during the first month or during the second month. If the members of S foresee at the time of contract design that a particular contract x^S may later induce such false pretension, they do not agree on the contract x^S. Instead of the strategy set $\hat{F}^S(\bar{x})$, therefore, they will consider only the restricted subset $\hat{F}^S_{ic}(\bar{x})$ of those strategy bundles that are Bayesian incentive compatible. The definition of Bayesian incentive compatibility in the context of Postulates 4.1 and 4.2 is much involved; see the Appendix for the detail.

5. Equilibria for the Intermediate Case

DEFINITION 5.1 Let $\mathcal{S}$ be a Bayesian society which satisfies Postulates 4.1 and 4.2. A strategy bundle $x^* \in X$ is called a ***Bayesian incentive compatible strong equilibrium***, if
(i) $x^* \in \hat{F}^N_{ic}(x^*)$; and
(ii) it is not true that there exist $S \in \mathcal{N}$ and $x^S \in \hat{F}^S_{ic}(x^*)$ such that for all $j \in S$,

$$\forall\, t \in T:\ Eu^j(x^S, x^{*N\setminus S} \mid \hat{T}^j(x^S_1))(t) \geq Eu^j(x^* \mid \hat{T}^j(x^*_1))(t),$$

with strict inequality for at least one $t \in T$.

The difficulty in establishing an existence theorem for Bayesian incentive compatible strong equilibrium (Definition 5.1) lies in the fact that even if F^S is well-behaved, $\hat{F}^S_{ic}$ is not convex-valued or upper semicontinuous, so the standard existence techniques do not apply. Ichiishi, Idzik and Zhao [14] provides generic existence theorems for a Bayesian incentive compatible strong equilibrium. Again, a generic existence thoerem is established first for the general class of Bayesian societies with externality in which each utility function u^j depends fully on $(c, t) \in C \times T$, and then as its application a generic existence theorem is established for the specific class of Bayesian societies without externality[2] in which each utility function u^j depends only on $(c^j, t^j) \in C^j \times T^j$. Assumptions of the theorem for this specific class are stated only in terms of the exogenously given data $\mathcal{S}$'s; this theorem is presented below (Theorems 5.9). Notice again that in spite of the terminology "no-externality", the feasible-strategy correspondences F^S depend fully on $x \in X$, and to this extent externality is still considered.

A Bayesian society studied here is a specified list of data,

$$\mathcal{S} := (\{C^j, T^j, u^j\}_{j \in N}, \{C^S_0, F^S\}_{S \in \mathcal{N}},\ p)$$

[2]The concept of no-externality is more stringent here than in Ichiishi and Idzik [13].

(Definition 2.3), with a rich structure (Postulates 4.1 and 4.2) and the further assumption of no-externality. One part of the data $(\{C^j, T^j, u^j\}_{j\in N}, \{C_0^S\}_{S\in\mathcal{N}}, p)$ will be fixed throughout. By changing the other part of the data from $\{F^S\}_{S\in\mathcal{N}}$ to $\{F'^S\}_{S\in\mathcal{N}}$, one obtains *another* Bayesian society

$$\mathcal{S}' := (\{C^j, T^j, u^j\}_{j\in N},\ \{C_0^S, F'^S\}_{S\in\mathcal{N}},\ p).$$

By varying $\{F^S\}_{S\in\mathcal{N}}$, one obtains the *space* of Bayesian societies, $\mathrm{SPACE}_{\mathrm{ne}}$. The space will be endowed with a natural pseudo-metric d. The pseudo-metric space $(\mathrm{SPACE}_{\mathrm{ne}}, d)$ of Bayesian societies will thus be constructed.

A property $\mathcal{P}$ is called a *generic property* of a Bayesian society *in* $\mathrm{SPACE}_{\mathrm{ne}}$, if there exists an open and dense subset $\mathrm{SPACE}'_{\mathrm{ne}}$ of $(\mathrm{SPACE}_{\mathrm{ne}}, d)$ such that every $\mathcal{S} \in \mathrm{SPACE}'_{\mathrm{ne}}$ satisfies $\mathcal{P}$. The following theorem (Theorem 5.9) clarifies conditions on $\mathrm{SPACE}_{\mathrm{ne}}$ under which the following is a generic property of a Bayesian society: There exist multitude of Bayesian incentive compatible strong equilibria, and there exists a Bayesian incentive compatible strong equilibrium x^* such that x_1^{*j} is 1-1 on T^j (so the equilibrium strategy fully reveals private information by the end of the first month).[3]

Two assumptions (Assumptions 5.2 and 5.3) on the fixed data $(\{C^j, T^j, u^j\}_{j\in N}, \{C_0^S\}_{S\in\mathcal{N}}, p)$, hence on the fixed strategy-spaces $X^j := \{x^j : T \to C^j\}$, are now stated:

ASSUMPTION 5.2 For every $j \in N$,
(i) C^j is a nonempty, compact, convex and metrizable subset of a real Hausdorff locally convex topological vector space;
(ii) for each $S \in \mathcal{N}$ and each $(c^{N\setminus S}, t) \in C^{N\setminus S} \times T$, $u^j(\cdot\ , c^{N\setminus S}, t)$ is linear affine and continuous on C_0^S.

Condition (i) of Assumption 5.2 is standard in economic theory. Condition (ii) is interpreted in two different ways: First, if each of spaces C_1^j and C_2^j consists only of *pure* (or, *deterministic*) *choices*, then it means the risk-aversion. Second, in the no-externality case, if each of spaces C_1^j and C_2^j consists only of *mixed choices* (or *probabilities on pure-choices*) for j, if $u^j(\cdot\ , t^j)$ is interpreted as the expected utility as a function of j's mixed-choice pairs, and if the underlying von Neumann-Morgenstern utility function is additively separable with respect to the pure choice of the first month and the pure choice of the second month, then condition (ii) is automatically satisfied.

ASSUMPTION 5.3 Assumption for the no-externality case: For each $j \in N$, there exist $c_m^j \in C^j$ and a finite subset C_f^j of C^j such that

[3]Recall that the type space T^j is defined so that the private information structure $\mathcal{T}^j$ can distinguish any two distinct members of T^j. Function x_1^{*j} is $\mathcal{T}^j$-measurable, iff it may be considered a function only of $t^j \in T^j$. A $\mathcal{T}^j$-measurable function is called 1-1 on T^j, if it is 1-1 as a function defined on T^j.

(i) for all $c^j \in C^j_f$ and all $t^j \in T^j$, $u^j(c^j, t^j) > u^j(c^j_m, t^j)$;
(ii) for all $c^j, c'^j \in C^j_f$ for which $c^j \neq c'^j$, it follows that $c^j_1 \neq c'^j_1$;
(iii) for all $t^j, t'^j \in T^j$ for which $t^j \neq t'^j$, there exist $c^j, c'^j \in C^j_f$ such that $u^j(c^j, t^j) > u^j(c'^j, t^j)$, and $u^j(c'^j, t'^j) > u^j(c^j, t'^j)$.

Assumption 5.3 and Assumption 5.7 (ii) (to be presented below) are made so that they guarantee existence of a *strictly* Bayesian incentive compatible strategy bundle (like $\hat{x}^S$ in condition (vi) of Theorem 3.4); see Abreu and Matsushima [1].

In the context of the present structure (Postulates 4.1 and 4.2), the essential role of the feasible-strategy correspondences $\{F^S\}_{S \in \mathcal{N}}$ in Bayesian society $\mathcal{S}$ is described by the correspondences $\{G^S\}_{S \in \mathcal{N}}$ defined by

$$G^S(\bar{x}) := \left\{ x^S \in F^S(\bar{x}) \;\middle|\; \begin{array}{l} \forall\, j \in S: \\ x^j_1 \text{ is } \mathcal{T}^j\text{-measurable, and} \\ x^j_2 \text{ is } \mathcal{T}^S\text{-measurable} \end{array} \right\}.$$

The following Assumption 5.4 is standard.

ASSUMPTION 5.4 For each $S \in \mathcal{N}$, correspondence $G^S : X \to C^S$ is upper and lower semicontinuous in X, and for each $\bar{x} \in X$, $G^S(\bar{x})$ is nonempty, closed and convex.

For the grand coalition, a stronger assumption on G^N is imposed. A subset A of space X is called *relatively strictly convex*, if the strict convex combinations of any two distinct members of A are in the relative interior of A. For each strategy bundle $x \in X$, denote by $Eu^j(x)$ and $Eu(x)$ player j's unconditional *ex ante* expected-utility of $x \in X$ and the unconditional *ex ante* expected-utility allocation $(Eu^j(x))_{j \in N}$, respectively.

ASSUMPTION 5.5 The correspondence $G^N(\cdot)$ is a constant correspondence on X, so one may write $G^N := G^N(\bar{x})$. The set G^N is relatively strictly convex. There exist $x, x' \in G^N$ such that $Eu(x) \ll Eu(x')$.

ASSUMPTION 5.6 Assumption for the no-externality case: Choose any $\bar{x} \in X$ and any balanced subfamily $\mathcal{B}$ of $\mathcal{N} \setminus \{N\}$ with the associated balancing coefficients $\{\lambda_S\}_{S \in \mathcal{B}}$. For each $S \in \mathcal{B}$ choose any $(x^{(S)j})_{j \in S} \in G^S(\bar{x})$, and define $\mathcal{T}^j$-measurable strategy $x^j : T^j \to C^j$ by

$$x^j(\bar{t}^j) := \sum_{S \in \mathcal{B}: S \ni j} \lambda_S(x^{(S)j}_1(\bar{t}^j), \sum_{t^{S \setminus \{j\}}} p(t^{S \setminus \{j\}} \mid \bar{t}^j) x^{(S)j}_2(\bar{t}^j, t^{S \setminus \{j\}})).$$

Then, $x \in G^N$.

Assumption 5.6 is a version of the balancedness condition on the sets $\{G^S(\bar{x})\}_{S\in\mathcal{N}}$, and makes explicit the extent to which the grand coalition has a large feasible-strategy set G^N. It means (1) that for each j the combination of the strategies $\{x^{(S)j}\}_{S\in\mathcal{B}:S\ni j}$ with the convex coefficients $\{\lambda_S\}_{S\in\mathcal{B}:S\ni j}$ is feasible in the grand coalition; and (2) that each member j is insured in the grand coalition to be able to choose $x^j(\bar{t}^j)$ at any state $t \in T^{N\setminus\{j\}} \times \{\bar{t}^j\}$.

The convex hull of a subset A of a vector space is denoted by co A.

ASSUMPTION 5.7 Assumption for the no-externality case: For each $S \in \mathcal{N}$ one of the following two conditions holds true:
(i) $G^S(\cdot)$ is a constant correspondence; or
(ii) For each $j \in S$, there exists a finite subset C^j_f of C^j such that for every $\bar{x} \in X$, $\prod_{j\in S}\{x^j : T^j \to \text{co } C^j_f\} \subset F^S(\bar{x})$, and such that for all $t^j, t'^j \in T^j$ for which $t^j \neq t'^j$, there exist $c^j, c'^j \in C^j_f$ so that $u^j(c^j, t^j) > u^j(c'^j, t^j)$, and $u^j(c'^j, t'^j) > u^j(c^j, t'^j)$.

Since each C^j is a metric space (Assumption 5.2 (i)) and T is finite, X^j is also a metric space. Denote by ρ_S the Hausdorff distance on the closed subsets of X^S.

DEFINITION 5.8 Definition for the no-externality case: Let $(\{C^j, T^j, u^j\}_{j\in N}, \{C^S_0\}_{S\in\mathcal{N}}, p)$ be a list of given data, and assume that each u^j depends only upon $(c^j, t^j) \in C^j \times T^j$, and that Assumptions 5.2 and 5.3 are satisfied. The *space of Bayesian societies without externalities* SPACE_{ne} is the set of all Bayesian societies obtained by adding to this list all $\{F^S\}_{S\in\mathcal{N}}$ which satisfy Assumptions 5.4-5.7. The pseudo-metric d on SPACE_{ne} is defined by: $d(\mathcal{S}, \mathcal{S}') := \max_{S\in\mathcal{N}} \max_{\bar{x}\in X} \rho_S(G^S(\bar{x}), G'^S(\bar{x}))$.

Notice that d may not be a metric, since two distinct sets, $F^S(\bar{x})$ and $F'^S(\bar{x})$, may give rise to the identical sets, $G^S(\bar{x}) = G'^S(\bar{x})$.

THEOREM 5.9 (Ichiishi, Idzik and Zhao [14]) *Let* $(\text{SPACE}_{\text{ne}}, d)$ *be the pseudo-metric space of Bayesian societies without externalities that satisfy Assumptions 5.2-5.7. The following is a generic property of a Bayesian society in* $(\text{SPACE}_{\text{ne}}, d)$*: There exist multitude of Bayesian incentive compatible strong equilibria, and there exists a Bayesian incentive compatible strong equilibrium* x^* *such that* x^{*j}_1 *is 1-1 on* T^j.

Appendix

This Appendix presents the definition of Bayesian incentive compatibility for the intermediate case of Sections 4-5. Given a strategy bundle $x \in X$ and an algebra $\mathcal{B}$ on T, player j's *conditional expected utility of x given $\mathcal{B}$* is the function $Eu^j(x \mid \mathcal{B}) : T \to \mathbf{R}$ defined as

$$Eu^j(x \mid \mathcal{B})(t) := \frac{1}{p(B(t))} \sum_{s \in B(t)} u^j(x(s), s)p(s),$$

where $B(t)$ is the minimal element of $\mathcal{B}$ that contains t. A *pretension function* of player j is a function $\sigma : T^j \to T^j$, which says that when his true type is t^j, he acts (makes a choice) as though his type were $\sigma(t^j)$. Given any algebra $\mathcal{B}^j$ on T^j, denote by $\text{endo}(T^j, \mathcal{B}^j)$ the set[4] of all functions that map each t^j into the minimal set of $\mathcal{B}^j$ that contains t^j. If every member of coalition S has information structure $\mathcal{B}^j$ about player j's type, then j can only choose a pretension function $\sigma \in \text{endo}(T^j, \mathcal{B}^j)$.

In order to define the Bayesian incentive compatibility, one needs to clarify first which choice of player j is *legal* in the sense that the other members of the coalition S cannot catch j's false pretension about his true type. Suppose the members of S are deliberating on whether or not to sign a contract $x^S \in \hat{F}^S(\bar{x})$. At the beginning of the first month, player j's information structure is given as $\mathcal{T}^j$, and no other member has any part of this information (that is, if $i \in S \setminus \{j\}$, then $\mathcal{T}^i \cap \mathcal{T}^j = \{\emptyset, T\}$). So, player j is not caught in the first month no matter which choice he makes from $\{x_1^j(t^j) \mid t^j \in T^j\}$; that is, he can make choice according to any pretension function $\sigma \in \text{endo}(T^j, \{\emptyset, T^j\})$, so that when player j's true type is t^j, he makes the choice $x_1^j(\sigma(t^j))$. By acting according to the function $x_1^j \circ \sigma$, player j having his true type t^j passes on to all the other members of S the information that event $A := (x_1^j)^{-1}(x_1^j \circ \sigma(t^j))$ has occurred. This information may be false, that is, t^j may not be a member of A, but the other members take it as j's testimony about himself and expect that j will act according to this information in the second month, that is, j will have to make a choice from $x_2^j(A)$ in the second month. Therefore, j's pretension function in the second month has to be of the form $\tau \circ \sigma$ for some $\tau \in \text{endo}(T^j, \mathcal{A}(x_1^j))$.

When j chooses such a pair of pretension functions, $\sigma \in \text{endo}(T^j, \{\emptyset, T^j\})$ and $\tau \in \text{endo}(T^j, \mathcal{A}(x_1^j))$, the other members $i \in S \setminus \{j\}$, acting honestly, would make choices $(x_1^i(t^i), x_2^i(\sigma(t^j), t^{S\setminus\{j\}}))$, because x_2^i is $\hat{\mathcal{T}}^i(x_1^S)$-measurable and $(\sigma(t^j), t^{S\setminus\{j\}}) \in ((x_1^j)^{-1}(x_1^j \circ \sigma(t^j)), (x_1^{S\setminus\{j\}})^{-1}(x_1^{S\setminus\{j\}}(t^{S\setminus\{j\}})))$.

The present concept of *Bayesian incentive compatibility* says that player j cannot benefit from any pair of pretension functions that are not caught. Thus, the set of Bayesian incentive compatible feasible strategy set $\hat{F}_{\text{ic}}^S(\bar{x})$ is defined by: $x^S \in \hat{F}_{\text{ic}}^S(\bar{x})$, iff $x^S \in \hat{F}^S(\bar{x})$, and for all $j \in S$, all $\sigma \in \text{endo}(T^j, \{\emptyset, T^j\})$, all $\tau \in \text{endo}(T^j, \mathcal{A}(x_1^j))$ and all $t \in T$,

[4] Let $\mathcal{P}$ be the partition of T^j that consists of the minimal nonempty members of $\mathcal{B}^j$. Then, $\text{endo}(T^j, \mathcal{B}^j) := \{\sigma : T^j \to T^j \mid \forall\, P \in \mathcal{P} : \sigma(P) \subset P\}$.

$$
\begin{aligned}
&Eu^j(x^S,\ \bar{x}^{N\setminus S} \mid \hat{T}^j(x_1^S))(t) \\
&\quad \geq\ Eu^j((x_1^j \circ \sigma,\ x_2^j \circ (\tau \circ \sigma,\ \mathrm{id})), \\
&\qquad\qquad (x_1^{S\setminus\{j\}},\ x_2^{S\setminus\{j\}} \circ (\sigma,\ \mathrm{id})),\ \bar{x}^{N\setminus S} \mid \hat{T}^j(x_1^S))(t),
\end{aligned}
$$

where id is the identity map on $T^{S\setminus\{j\}}$.

Many of the above inequalities of the Bayesian incentive compatibility condition are redundant; for example, the inequality for $(t^S, t^{N\setminus S})$ and the inequality for $(t^S, t'^{N\setminus S})$ are identical. For notational simplicity, however, the inequalities will be given for *all* $t \in T$.

References

[1] D. Abreu and H. Matsushima, *Virtual implementation in iteratively undominated strategies: Complete information*, Econometrica, **60** (1992), 993–1008.

[2] K. J. Arrow and G. Debreu, *Existence of an equilibrium for a competitive economy*, Econometrica, **22** (1954), 265–290.

[3] O. N. Bondareva, *Teoriia iadra v igre n lits*, Vestnik Leningrad University Math., **13** (1962), 141–142 (in Russian).

[4] C. d'Aspremont and L.-A. Gérard-Varet, *Incentives and incomplete information*, Journal of Public Economics, **11** (1979), 25–45.

[5] G. Debreu, *A social equilibrium existence theorem*, Proceedings of the National Academy of Sciences of the U.S.A., **38** (1952), 886–893.

[6] P. Dubey, J. Geanakoplos and M. Shubik, *The revelation of information in strategic market games*, Journal of Mathematical Economics, **16** (1987), 105–137.

[7] K. Fan, *Extensions of two fixed point theorems of F. E. Browder*, Mathematische Zeitschrift, **112** (1969), 234–240.

[8] K. Fan, *A minimax inequality and applications*, in: O. Shisha (ed.), *Inequalities III*, pp. 103–113, Academic Press, New York (1972).

[9] J. C. Harsanyi, *Games with incomplete information played by 'Bayesian' players*, Management Science: Theory, **14** (1967/1968), 159–182 (Part I), 320–334 (Part II), 486–502 (Part III).

[10] T. Ichiishi, *A social coalitional equilibrium existence lemma*, Econometrica, **49** (1981), 369–377.

[11] T. Ichiishi, *The Cooperative Nature of the Firm*, Cambridge University Press, Cambridge, U.K. (1993).

[12] T. Ichiishi, *The cooperative nature of the firm: Narrative*, Managerial and Decision Economics, **14** (1993), 387–407. In: Special Issue edited by Koji Okuguchi on Labor-Managed Firms under Imperfect Competition (and Related Problems).

[13] T. Ichiishi and A. Idzik, *Bayesian cooperative choice of strategies*, mimeo., Department of Economics, The Ohio State University (July 1991).

[14] T. Ichiishi, A. Idzik and J. Zhao, *Cooperative processing of information via choice at an information set*, mimeo., Department of Economics, The Ohio State University (May 1993). Forthcoming in International Journal of Game Theory.

[15] J. F. Nash, Jr., *Equilibrium points in n-person games*, Proceedings of the National Academy of Sciences of the U.S.A., **36** (1950), 48–49.

[16] R. Radner, *Rational expectations equilibrium generic existence and the information revealed by prices*, Econometrica, **47** (1979), 655–678.

[17] H. Scarf, *The core of an n-person game*, Econometrica, **35** (1967), 50–69.

[18] L. S. Shapley, *On balanced sets and cores*, Naval Research Logistics Quarterly, **14** (1967), 453–460.

A Survey of Stochastic Differential Equations

Kiyosi Itô

RIMS, Kyoto University, Kyoto, 606, JAPAN

Introduction

Let us consider a system, dynamical,biological or economical, that is determined by a finite number of parameters:

$$\underline{x} = (x^1, x^2, \ldots, x^r) \in \boldsymbol{R}^r.$$

The time evolution of the system is governed by an ordinary differential equation:

$$dx_t^i = a^i(\underline{x}_t)dt, \qquad i = 1, 2, \ldots, r. \tag{1}$$

If the system is perturbed by infinitesimal random causes at every time point, then it is reasonable to replace (1) by the following stochastic differential equation:

$$dX_t^i = a^i(\underline{X}_t)dt + \sum_{j=1}^{n} b_j^i(\underline{X}_t)dB_t^j, \tag{2}$$

where $(B_t^j,\ t \geq 0)$, $j = 1, 2, \ldots, n$, are independent Wiener processes and $(X_t^i,\ t \geq 0)$, $i = 1, 2, \ldots, r$ are sample continuous stochastic processes adapted to the filtration generated by the Wiener processes.

In this note we will give a brief survey of basic facts on stochastic differential equations without bothering with technical details. Throughout in the discussion below we denote the base probability space by $(\Omega, \mathcal{F}, P)$.

1. Wiener processes

A stochastic process $B_t = B_t(\omega)$, $\omega \in (\Omega, \mathcal{F}, P)$, is called **Wiener process**, if (i) it has independent increments, (ii) it is normalized: $E(B_t) = 0$, $E(B_t^2) = t$, and (iii) it is sample continuous, *i.e.* $B_t(\omega)$ is continuous in t for every ω. This is a rigorous mathematical model of the socalled Brownian motion of a particle suspended in water discovered by R. Brown in 1827 and discussed mathematically by A. Einstein in 1905. Also in L. Bachlier's theory of random fluctuation of stock price one can find a prototype of Wiener processes.

2. Filtration

An increasing family of σ–subalgebras of $\mathcal{F}$, $\boldsymbol{B} = \{\mathcal{B}_t,\ t \geq 0\}$ is called a **filtration.** Intuitively $\mathcal{B}_t$ means the information obtained up to time t that determines the value of a random variable $X = X(\omega)$ if and only if X is $\mathcal{B}_t$-measurable. A random time $T = T(\omega) \in [0, \infty)$ is called a **$\boldsymbol{B}$-stopping time** if the event $\{T \leq t\}$, *i.e.* the set $\{\omega : T(\omega) \leq t\}$ belongs to $\mathcal{B}_t$ for every $t \geq 0$.

A real-valued stochastic process $X_t = X_t(\omega)$ is called **$\boldsymbol{B}$-adapted,** if $\{X_t \leq a\} \in \mathcal{B}_t$ for every $t \geq 0$ and every $a \in \boldsymbol{R}$. The family of all sample continuous $\boldsymbol{B}$-adapted processes is denoted by $\mathcal{C}$. Let

$$\mathcal{A} := \left\{A \in \mathcal{C},\ {}^{\exists}T_n : \int_0^{T_n} |dA_t| < \infty\right\},$$

and

$$\mathcal{M} := \{M \in \mathcal{C},\ {}^{\exists}T_n : E(M_{t\wedge T_n}|\mathcal{B}_s) = M_{s\wedge T_n}\ (s < t)\},$$

where $T_n(\omega)$ is an increasing sequence of $\boldsymbol{B}$-stopping times tending to ∞ for every ω. Let

$$\mathcal{Q} := \{Q \in \mathcal{C},\ Q = M + A,\ M \in \mathcal{M},\ A \in \mathcal{A}\}.$$

Every process $A \in \mathcal{A}$ is of bounded variation on every bounded interval, every $M \in \mathcal{M}$ is called a **local $\boldsymbol{B}$-martingale** and every $Q \in \mathcal{Q}$ is called a **local $\boldsymbol{B}$-semimartingale.**

3. Stochastic differentials[P.1]

For $X \in \mathcal{C}$ we define a random interval function:

$$(dX)(s,t] := X(t) - X(s) \qquad (s < t)$$

and let

$$d\mathcal{C} := \{dX : X \in \mathcal{C}\}.$$

Similarly we define $d\mathcal{Q}$, $d\mathcal{M}$ and $d\mathcal{A}$. We will introduce some operations on $d\mathcal{Q}$.

Definition 3.1 The $\mathcal{C}$-**multiplication** in $d\mathcal{Q}$ is defined as follows:

$$(Y \cdot dX)(s,t] := \underset{|\Delta|\to 0}{\text{l.i.p.}} \sum_{i=1}^{n} Y(\tau_{i-1})dX(\tau_{i-1}, \tau_i] \qquad (dX \in d\mathcal{Q},\ Y \in \mathcal{C}),$$

where $\Delta = \{s = \tau_0 < \tau_1 < \cdots < \tau_n = t\}$, $|\Delta| = \max_i(\tau_i - \tau_{i-1})$ and l.i.p. abbreviates limit in probability. $(Y \cdot dX)(s,t]$ is often denoted by $\int_s^t Y(\tau)dX(\tau)$ and called the **stochastic integral** of Y relative to dX over $(s,t]$.

Theorem 3.2
(a) $d\mathcal{Q}$ is a module over the ring $\mathcal{C}$.
(b) $\mathcal{C} \cdot d\mathcal{M} \subset d\mathcal{M}$ (*i.e.* $Y \cdot dX \in d\mathcal{M}, {}^\forall Y \in \mathcal{C}, {}^\forall dX \in d\mathcal{M}$).
(c) $\mathcal{C} \cdot d\mathcal{A} \subset d\mathcal{A}$.
(d) $d\mathcal{Q} = d\mathcal{A} \oplus d\mathcal{M}$ (as modules over $\mathcal{C}$).

Definition 3.3 The product in $d\mathcal{Q}$ is defined as follows:

$$(dX \cdot dY)(s,t] := \underset{|\Delta|\to 0}{\text{l.i.p.}} \sum_{i=1}^{n} dX(\tau_{i-1}, \tau_i] dY(\tau_{i-1}, \tau_i], \quad dX, dY \in d\mathcal{Q};$$

see **Definition 3.1** for Δ and $|\Delta|$.

Theorem 3.4
(a) $d\mathcal{Q}$ is an algebra over $\mathcal{C}$.
(b) $d\mathcal{Q} \cdot d\mathcal{Q} \subset d\mathcal{A}$.
(c) $d\mathcal{Q} \cdot d\mathcal{A} = 0$.

Definition 3.5 For $Y \in \mathcal{Q}$ and $dX \in d\mathcal{Q}$ we define the **symmetric $\mathcal{Q}$-multiplication** $Y \circ dX \in d\mathcal{Q}$ by

$$Y \circ dX = Y \cdot dX + \frac{1}{2} dY \cdot dX.$$

$(Y \circ dX)(s,t]$ is often denoted by $\int_s^t Y(\tau) \circ dX(\tau)$ or $(S)\int_s^t Y(\tau) \circ dX(\tau)$ and called the **Stratonovich- Fisk** integral.

Theorem 3.6 $(Y \circ dX) \cdot dZ = (Y \cdot dX) \cdot dZ$ (obvious by **Theorem 3.4** (b),(c)).

4. Chain rules

Suppose that $f \in C(\boldsymbol{R}^n)$ and $X^i \in \mathcal{Q}$, $i = 1, 2, \ldots, n$. Then it is trivial that $f(\underset{\sim}{X}) = \{f(X_t^1, X_t^2, \ldots, X_t^n), t \geq 0\} \in \mathcal{C}$.

Theorem 4.1 (**Stochastic chain rule** due to Itô-Kunita-Watanabe) If $f \in C^2(\boldsymbol{R}^n)$, then $f(\underset{\sim}{X}) \in \mathcal{Q}$ and

$$d(f(\underset{\sim}{X})) = \sum_i \partial_i f(\underset{\sim}{X}) dX^i + \frac{1}{2} \sum_{i,j} \partial_i \partial_j f(\underset{\sim}{X}) dX^i dX^j.$$

Note that the second term of the right hand side belongs to $d\mathcal{A}$.

Theorem 4.2 (Symmetric stochastic chain rule due to Stratonovich-Fisk) If $f \in C^3(\boldsymbol{R})$, then

$$d(f(\underline{X})) = \sum_i \partial_i f(\underline{X}) \circ dX^i.$$

This follows immediately from **Theorem 4.1** by observing that $\partial_i f \in C^2(\boldsymbol{R})$ and

$$\begin{aligned}
\sum_i \partial_i f(\underline{X}) \circ dX^i &= \sum_i \partial_i f(\underline{X}) \cdot dX^i + \frac{1}{2}\sum_i d(\partial_i f(\underline{X})) \cdot dX^i \\
&= \sum_i \partial_i f(\underline{X}) \cdot dX^i + \frac{1}{2}\sum_i (\sum_j \partial_j \partial_i f(\underline{X}) dX^j + dA^i) \cdot dX^i \\
&\qquad\qquad (dA^i \in d\mathcal{A}) \\
&= d(f(\underline{X})) \qquad \text{(by Theorems 4.1 and 3.4(c))}.
\end{aligned}$$

This chain rule takes the same form as in the chain rule in the ordinary analysis. Hence it is very useful in some cases; see the examples in **Section 4**.

5. Stochastic differential equations

Let $B^i = \{B^i_t\}$, $i = 1, 2, \ldots, n$ be independent Wiener processes. Then

$$\boldsymbol{B} := \{\mathcal{B}_t \equiv \sigma[B^i_s,\ s \le t,\ i = 1, 2, \ldots, n],\ 0 \le t < \infty\}$$

is a filtration to which B^i is adapted for every i. Using the notation introduced above, we obtain

$$dB^i \in d\mathcal{M},\ dB^i dB^j = \delta_{ij} dt \text{ and } dt \in d\mathcal{A}.$$

Suppose that $X^i \in Q$, $i = 1, 2, \ldots, r$ and denote $(X^i,\ i = 1, 2, \ldots, r)$ by $\underline{X}$. Consider the following stochastic differential equation:

$$\begin{aligned}
dX^i_t &= a^i(\underline{X}_t)dt + \sum_{j=1}^n b^i_j(\underline{X}_t) dB^j_t,\ i = 1, 2, \ldots, r, \\
\underline{X}_0 &= \boldsymbol{x} \in \boldsymbol{R}^r.
\end{aligned}$$

Theorem 5.1 If $a^i : \boldsymbol{R}^r \to \boldsymbol{R}$ and $b^{ij} : \boldsymbol{R}^r \to \boldsymbol{R}$ are Lipschitz continuous, then the equation has a unique solution.

Proof Apply the Picard approximation method.

Theorem 5.2 X_t is a Markov process whose transition semigroup $\{T_t\}$ is generated by the following elliptic operator

$$\mathcal{G} = \sum_{i=1}^{r} a^i \partial_i + \frac{1}{2} \sum_{i,j=1}^{r} B^{ij} \partial_i \partial_j,$$

where $B^{ij} = \sum_{k=1}^{n} b_k^i b_k^j$.

Proof Using the stochastic chain rule, we obtain

$$d(\varphi(\underline{X}_t)) = \sum_{i=1}^{r} \partial_i \varphi(\underline{X}_t) dX_t^i + \frac{1}{2} \sum_{i,j=1}^{r} \partial_i \partial_j \varphi(\underline{X}_t) dX_t^i \cdot dX_t^j.$$

But

$$dX_t^i = a^i(\underline{X}_t)dt + \sum_{k=1}^{n} b_k^i(\underline{X}_t) dB_t^k.$$

Since $dt \cdot dB_t^k = 0$ and $dB_t^k \cdot dB_t^l = \delta_{kl} dt$, we obtain

$$dX_t^i \cdot dX_t^j = \sum_{k,l=1}^{n} b_k^i(\underline{X}_t) b_l^j(\underline{X}_t) \delta_{kl} dt = B^{ij}(\underline{X}_t) dt,$$

which implies

$$d(\varphi(\underline{X}_t)) = \mathcal{G}\varphi(\underline{X}_t)dt + \sum_{i=1}^{r} \sum_{k=1}^{n} b_k^i(\underline{X}_t) \partial_i \varphi(\underline{X}_t) dB_t^k.$$

Noting that $dB \in d\mathcal{M}$, we obtain

$$E(\varphi(\underline{X}_t)) - E(\varphi(\underline{x})) = \int_0^t E(\mathcal{G}\varphi(\underline{X}_s))ds,$$

from which our theorem follows.

6. Application of the symmetric stochastic chain rule

6.1

Let us solve the equation

$$dX_t = X_t \cdot dB_t.$$

$$\begin{aligned} dX_t &= X_t \circ dB_t - \frac{1}{2} dX_t \cdot dB_t \\ &= X_t \circ dB_t - \frac{1}{2}(X_t \cdot dB_t) \cdot dB_t \\ &= X_t \circ dB_t - \frac{1}{2} X_t dt \\ &= X_t \circ dB_t - \frac{1}{2} X_t \circ dt \qquad (\text{by } dX_t \cdot dt = 0) \\ &= X_t \circ d(B_t - \frac{t}{2}). \end{aligned}$$

Hence

$$X_t = X_0 \exp(B_t - \frac{t}{2}).$$

6.2 Stochastic parallel displacement[P.2]

Let M^r be an r-dimensional Riemannian manifold. A deterministic curve γ is expressed by a local coordinate system as follows:

$$x_t^i, \ t \geq 0, \ i = 1, 2, \ldots, r.$$

If we use another coordinate system $y^i = f^i(x^1, x^2, \ldots, x^r)$, $i = 1, 2, \ldots, r$, then the same curve is expressed by

$$y_t^i = f^i(x_t^1, x_t^2, \ldots, x_t^r) = f^i(\underline{x}_t), \ t \geq 0, \ i = 1, 2, \ldots, r.$$

Hence

$$dy_t^i = \sum_j \partial_j f^i(\underline{x}_t) dx_t^j.$$

Now we consider a stochastic curve Γ in M^r. Let the x-coordinates of Γ be $\underline{X}_t = (X_t^1, X_t^2, \ldots, X_t^r)$. If $X^i = (X_t^i, t \geq 0) \in \mathcal{Q}$ for every i in the sense explained in 2, then we say that $\Gamma \in \mathcal{Q}$. This concept is well-defined independently of the choice of the local coordinates. In fact, if we take the y-coordinates, then Γ is represented by

$$Y_t^i = f^i(X_t^1, X_t^2, \ldots, X_t^r) = f^i(\underline{X}_t)$$

and

$$dY_t^i = \sum_j \partial_j f^i(\underline{X}_t) \circ dX_t^i.$$

The parallel displacement of a vector along a deterministic curve (x_t^i) is given by a differential equation

$$dv_{i,t} = \sum_{j,k} \Gamma_{ji}^k(\underline{x}_t) v_{k,t} dx_t^j.$$

The stochastic counterpart of this equation is

$$dV_{i,t} = \sum_{j,k} \Gamma^k_{ji}(\underline{X}_t) V_{k,t} \circ dX^j_t.$$

Since the symmetric stochastic chain rule takes the same form as the ordinary one, stochastic parallel displacement is well-defined independently of the choice of local coordinates and the inner product of the vectors remains constant by stochastic parallel displacement.

7. Stochastic control[B.4][B.7]

Consider a **stochastic differential equation with a control process** $U = U_t$:

$$dX^i_t = a^i(\underline{X}_t, U_t)dt + \sum_{j=1}^{n} b^i_j(\underline{X}_t, U_t)dB^j, \quad i = 1, 2, \ldots, r,$$

where U is a stochastic process taking values in a compact subset Γ of $\boldsymbol{R}^m$ and adapted to $\mathcal{B}_t = \sigma[B_s, s \le t]$, $t \ge 0$. Let $f(\underline{x}, \underline{u})ds$ be the gain in ds when $\underline{X}_s = \underline{x}$ and $\underline{U}_s = \underline{u}$, $\varphi(\underline{x})$ the last gain in the period $[0, t]$ when $\underline{X}_t = \underline{x}$, and c the interest rate.

Then the **maximum average gain** in $[0, t]$ under the initial condition $\underline{X}_0 = \underline{x}$ is given as follows:

$$v(t, \underline{x}) := \sup_U E\left[\int_0^t e^{-cs} f(\underline{X}_s, \underline{U}_s)ds + e^{-ct}\varphi(\underline{X}_t) | \underline{X}_0 = \underline{x}\right].$$

and $v(t, \underline{x})$ satisfies the well-known **Bellman equation** :

$$\frac{\partial v}{\partial t}(t, \underline{x}) = \mathcal{G}v(t, \underline{x}), \quad v(0, \underline{x}) = \varphi(\underline{x}),$$

where $\mathcal{G}$ is the **Bellman operator** :

$$\mathcal{G}\varphi(\underline{x}) = \sup_{\underline{u} \in \Gamma} \left[\sum_{i,j} B^{ij}(\underline{x}, \underline{u})\partial_i\partial_j\varphi(\underline{x}) + a^i(\underline{x}, \underline{u})\partial_i\varphi(\underline{x}) - c\varphi(\underline{x}) + f(\underline{x}, \underline{u})\right],$$

where $B^{i,j} = \sum_{k=1}^n b^i_k b^j_k$.

In view of the fact that $v(t, \underline{x})$ is determined by t and φ we obtain a one-parameter family of non-linear operators :

$$T_t\varphi(\underline{x}) = v(t, \underline{x}), \ t \ge 0, \qquad T_0\varphi(\underline{x}) = \varphi(\underline{x}).$$

Since $T_t T_s = T_{t+s}$, the family $\{T_t, \ t \ge 0\}$ turns out to be a non-linear semigroup generated by the Bellman operator above. This semigroup interpretation of the Bellman theory is due to M.Nisio [B.13].

8. The Stroock-Varadhan martingale problem[B.19]

The problem to determine a diffusion process (sample continuous Markov process) whose transition semigroup is generated by an elliptic operator :

$$\mathcal{G} = \sum_{i=1}^{r} a^i \partial_i + \frac{1}{2} \sum_{i,j=1}^{r} B^{ij} \partial_i \partial_j$$

goes back to A.Kolmogorov[P.3]. Hereafter we call such diffusion process a **$\mathcal{G}$-process.**

This problem was first treated by appealing to the theory of partial differential equations by W.Feller[P.4], R.Fortet [P.5], K.Yosida[B.22] and others. The method of stochastic differential equations that was explained in Section 5 is probabilistic. Another probabilistic method is to solve the Stroock-Varadhan martingale problem, which will be explained below.

Let $\{\underline{X}_t\}$ be a $\mathcal{G}$-process. For every smooth $\varphi : \boldsymbol{R}^r \to \boldsymbol{R}$ with compact support we obtain

$$T_t\varphi(\underline{x}) = E[\varphi(\underline{X}_t)|\underline{X}_0 = \underline{x}], \tag{1}$$

$$T_t\varphi(\underline{x}) - \varphi(\underline{x}) = \int_0^t T_u\mathcal{G}\varphi(\underline{x})du, \tag{2}$$

and

$$E[\varphi(\underline{X}_{t+s})|\mathcal{B}_s] = T_t\varphi(X_s) \tag{3}$$

by the Markov property of $\{\underline{X}_t\}$, where $\mathcal{B}_s = \sigma[X_u,\ u \le t]$.

Let

$$M_t^\varphi := \varphi(\underline{X}_t) - \int_0^t \mathcal{G}\varphi(\underline{x}_u)du. \tag{4}$$

Then the process is a $\boldsymbol{B}$-martingale ($\boldsymbol{B} = \{\mathcal{B}_s,\ s \ge 0\}$) by (1),(2) and (3). Conversely, if M_t^φ is a $\boldsymbol{B}$-martingale for every φ, then $\{\underline{X}_t\}$ is a $\mathcal{G}$-diffusion. In view of this fact finding a $\mathcal{G}$-diffusion is equivalent to determining the process $\{\underline{X}_t\}$ for which $\{M_t^\varphi\}$ is a martingale for every φ. The latter problem is called the Stroock-Varadhan problem.

The Stroock-Varadhan method is sometimes more direct and powerful than the stochastic differential equation method, because the latter involves the Wiener processes, while the former does not. It should be noted that if $\{\underline{X}_t\}$ satisfies the stochastic differential equation (Section 5), then the stochastic chain rule ensures that

$$d\varphi(\underline{X}_t) = \mathcal{G}\varphi(\underline{X}_t)dt + \sum_{i=1}^{r}\sum_{k=1}^{n} b_k^i \partial_i\varphi(\underline{X}_t)dB^k,$$

and so

$$dM^{\varphi} \in d\mathcal{M},$$

i.e. $\{M_t^{\varphi},\ t \geq 0\}$ is a local martingale.

9. Malliavin calculus[B.12][B.18][B.20][B.21]

The stochastic differential equation method is more convenient to prove the existence and the uniqueness of the $\mathcal{G}$-process mentioned in the last section; for example, this method is applicable to the case where $\mathcal{G}$ is weakly elliptic. However, in order to prove the existence of the density of the transition probability (under certain stronger conditions) that satisfies the Kolmogorov forward equation we had to use the results on partial differential equations.

It was P.Malliavin that succeeded in establishing a purely probabilistic theory to deal with this problem. Remember that the sample space of the n-dimensional Wiener process is the Banach space of all $\boldsymbol{R}^n$-valued continuous functions endowed with the so-called Wiener measure, denoted by (E, μ). Representing the Wiener process $\{B_t^i,\ i = 1, 2, \ldots, n\}$ in Section 5 on (E, μ) (the coordinate representation), we see that the solution X_t of the stochastic differential equation is Borel measurable on the space E. Extending the Sobolev differentiation on the finite dimensional spaces to the infinite dimensional space (E, μ), Malliavin defined **stochastic calculus of variation**[P.6] on (E, μ) and proved that X_t is infinitely differentiable on (E, μ) if the coefficients a^i and b_k^i are smooth. This splendid idea of Malliavin produced many interesting and useful results that turned out to be a beautiful theory, now called **Malliavin calculus**.

References

The list below is not complete; see [B.19],[B.21] for more information.

Books on stochastic differential equation and related topics

[B.1] J-M. Bismut, *Mécanique aléatoire, Springer Lecture*, Notes in Math. **866**, Springer, 1981.

[B.2] K. I. Chung and R. J. Williams, *Introduction to stochastic integration*, Birkhäuser, 1983.

[B.3] K. D. Elworthy, *Stochastic differential equations on manifolds*, Cambridge Univ. Press, 1982.

[B.4] W. H. Fleming and R. W. Rishel, *Deterministic and stochastic optimal control*, Springer, 1975.

[B.5] A. Friedman, *Stochastic differential equations and applications*, Acad. Press, Vol.**1**, 1975, and Vol.**2**, 1976.

[B.6] I. Karatzas and S. E. Shreve, *Brownian motion and stochastic calculus*, Springer, 1988.

[B.7] N. V. Krylov, *Controlled diffusion processes*, Springer, 1980.

[B.8] H. Kunita, *Stochastic flows and stochastic differential equations*, Cambridge Univ. Press, 1990.

[B.9] H. Kunita, *Stochastic flows and applications*, Tata Inst. of Fund. Research, Bombay, 1986.

[B.10] R. S. Liptser and A. N. Shiryayev, *Statistics of random processes* **I**, general theory, 1977 and **II**, applications, 1978, Springer.

[B.11] H. P. Mckean, *Stochastic integrals*, Acad. Press, 1969.

[B.12] P. Malliavin, *Stochastic calculus*, Springer, to appear.

[B.13] M. Nisio, *Stochastic control theory*, Indian Stat. Institute Lecture Note Series **9**, 1980.

[B.14] L. C. G. Rogers and D. Williams, *Diffusions, Markov processes, and martingales, Vol. 2 Itô calculus*, J.Waley & Sons, 1987.

[B.15] Z. Schuss, *Theory and applications of stochastic differential equations*, J.Wiley & Sons, 1980.

[B.16] L. Schwartz, *Semimartingales and their stochastic calculus on manifolds*, Presses de l'univ. Montréal, 1984.

[B.17] D. W. Stroock, *Lectures on stochastic analysis : diffusion theory*, Cambridge Univ. Press, 1987.

[B.18] D. W. Stroock, *Topics in stochastic differential equations*, Tata Inst. of Fund. Research, Bombay, 1982.

[B.19] D. W. Stroock and S. R. S. Varadhan, *Multidimensional diffusion processes*, Springer, 1979.

[B.20] S. Watanabe, *Stochastic differential equations and Malliavin calculus*, Tata Inst. of Fund. Research, Bombay, 1984.

[B.21] N. Ikeda and S. Watanabe, *Stochastic differential equations and diffusion processes*, North-Holland, Kodansha, 1981(1st ed.) and 1989(2nd ed.).

[B.22] K. Yosida, *Functional Analysis*, Springer, 1964.

Papers cited in this note

[P.1] K. Itô, *Stochastic differentials*, Appl. Math. and Opt. **1**(1974), 374-381.

[P.2] K. Itô, *Stochastic parallel displacement*, Springer Lecture Notes in Mathematics **451**(1974), 1-7.

[P.3] A. Kolmogorov, *Über die analytischen Methoden im der Wakrscheinlichkeitsrechnung*, Math. Ann. **104** (1931), 415-458.

[P.4] W. Feller, *Zur Theorie der stochastischen Prozesse (Existenz und Eindentingkaits Sätze)*, Math. Ann. **113** (1936), 113-160.

[P.5] R. Fortet, *Les fonctions aléatoires du types de Markoff associées à certaines equations linéares aux derivées partielles du type paraboliques*, J. Math. Pures Appl. **22**(1943), 177-243.

[P.6] P. Malliavin, *Stochastic calculus of variation and hypo-elliptic operators*, Proc. Intern. Symp. SDE Kyoto, 195-263, Kinokuniya, Tokyo, 1978.

Stability and Oscillations in a Dynamical Kaldorian Model

Kunio Kawamata
Department of Economics
Keio University
Mita, Minato-ku Tokyo, 108, JAPAN

1. Introduction

In a classic paper N.Kaldor (1940) presented a model of a closed economy and demonstrated the existence of a periodic orbit which may be interpreted as a trade cycle in an autonomous system.[1] In establishing this result he assumed not only that investment and saving decisions are made without any time lag, but also that investment is adjusted to saving instantaneously so far as the adjustment is possible by a small change in output. Thus, in his model, non-negligible adjustment period is taken into account only in installing capital and making it ready for production.

In this paper we modify Kaldorian model to allow for a delay in the adjustment of investment to saving and suppose that this delay is not, in general, negligible compared with the gestation period of capital. In other words, we suppose that the adjustment speed k of output is not necessarily infinitely larger than that of capital in productive use which we may suppose to be equal to 1.

Besides the problems on the existence and uniqueness of equilibrium, we discuss broadly four separate but related topics. In this way we exemplify how various mathematical techniques have been applied to the trade cycle theories,using the Kaldorian model as a basic framework.

In Section 4 we investigate conditions under which the uniquely determined equilibrium is stable in the large, while in Section 5 we give conditions guaranteeing the existence of a non-degenerate periodic orbit. Needless to say, these results depend on the magnitude of k as well as on the forms of investment and saving functions. The existence of a periodic orbit was demonstrated by Chang and Smith (1971) in a similar model, and Torre

[1] A related classical models of trade cycle include Goodwin (1951) and Kalecki (1968).

(1977) discusses the occurrence of the Poincare–Hopf bifurcation in a related model.

In Section 7, we consider the situation where there is an exogenous periodic disturbance in the final demand and show the existence of a periodic solution of the adjustment process under appropriate conditions. The method of proof is new in the economic analysis literature and could be applied to show the existence of periodic solution in many other situations.

2. Basic Assumptions

Let us consider a closed economy without governmental and foreign activities. We indicate the state of the economy at time t by the following symbols:

$$\begin{aligned} Y(t) &: \text{ gross national product} \\ C(t) &: \text{ consumption} \\ I(t) &: \text{ gross investment} \\ K(t) &: \text{ capital stock} \end{aligned}$$

These are all real numbers for a fixed t. In the following we shall describe how these variables are related and make suitable assumptions to determine their behavior.

We first assume that the consumption function can be expressed in the following simple form[2]

$$C = \tilde{C}(Y) \tag{2.1}$$

and satisfies the standard conditions:

$$\begin{aligned} &\tilde{C}(0) \geq 0 \text{ and} \\ &0 < \tilde{C}'(Y) < 1 \text{ for any } Y > 0. \end{aligned} \tag{2.2}$$

The saving function is defined by

$$\tilde{S}(Y) = Y - \tilde{C}(Y) \tag{2.3}$$

which, in view of (2.2), satisfies

$$\begin{aligned} &\tilde{S}(0) \leq 0 \text{ and} \\ &0 < \tilde{S}'(Y) < 1 \text{ for any } Y > 0. \end{aligned} \tag{2.4}$$

For the most of paper we consider a smooth invest function of the form

$$I = \tilde{I}(Y, K) \tag{2.5}$$

which satisfies

[2]In general K may enter as an independent variable of the saving function, so long as the influence is small compared with its influence on the investment function.

$$\tilde{I}(0, \underline{K}) \geq 0 \quad \text{for some } \underline{K} > 0$$
$$\tilde{I}(Y, 0) \geq 0 \quad \text{for any } Y > 0$$

and

$$\begin{aligned} &\tilde{I}_Y(Y, K) > 0 \\ &\tilde{I}_K(Y, K) < 0 \quad \text{for any } (Y, K) \in R^2_+ \end{aligned} \tag{2.6}$$

where R^2_+ denotes the non-negative orthant of the two dimensional Euclidean space R^2. In the sequel we shall assume (2.2) (hence (2.4)) and (2.6) without mentioning it.

A typical form of I is depicted in Fig.1 for K_1, K_2 and K_3 where $K_3 > K_2 > K_1$. This type of the investment function is referred to as the Kaldorian investment function.

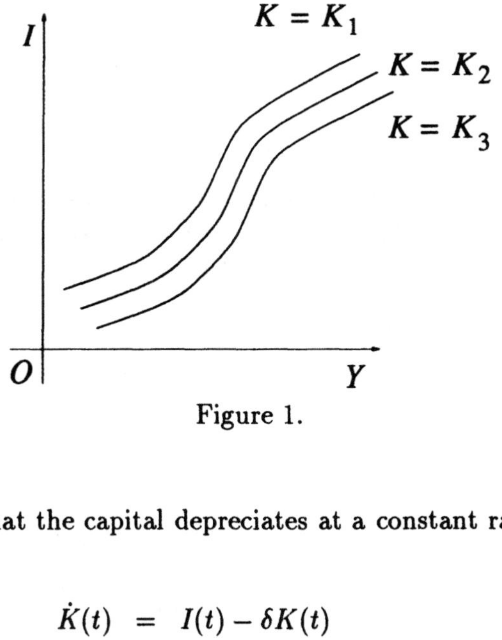

Figure 1.

We assume that the capital depreciates at a constant rate $\delta > 0$, and so $\dot{K}(t)$ satisfies

$$\begin{aligned} \dot{K}(t) &= I(t) - \delta K(t) \\ &= \tilde{I}(Y(t), K(t)) - \delta K(t) \end{aligned} \tag{2.7}$$

where $K(0)$ and $Y(0)$ are given initial conditions.

To complete the model we finally assume that the output level is adjusted according to the rule

$$\dot{Y}(t) = k(\tilde{I}(Y(t), K(t)) - \tilde{S}(Y(t))) \tag{2.8}$$

where k is a positive number. As will be explained below, the adjustment process of Kaldor (1940) can be described as the limiting case where k is infinitely large.

In what follows we shall often delete tilde sign representing consumption, saving and investment functions, and delete t to simplify notation. Hence the differential equations (2.8) and (2.7) describing our dynamic process can be rewritten as

$$\begin{aligned} \dot{Y} &= k(I(Y,K) - S(Y)) \\ \dot{K} &= I(Y,K) - \delta K \end{aligned} \tag{P}$$

where $Y(0)$ and $K(0)$ are exogenously given as initial conditions.[3]

3. Basic Properties of the Model

By definition, an equilibrium (Y^*, K^*) of the dynamic process (P) is the point $(Y, K) \in R^2_+$ which satisfies

$$I(Y,K) = S(Y) \tag{3.1}$$

and

$$I(Y,K) = \delta K. \tag{3.2}$$

Since $I_K < 0$, for each $Y > 0$ there exists at most one K which satisfies (3.1), and similarly for (3.2). We note that this does not rule out the possibility that for some K there are more than one Y satisfying (3.1) (see, Fig.2 (ii) below).

From (3.1) it follows that

$$\left.\frac{dK}{dY}\right|_{\dot{Y}=0} = \frac{S_Y - I_Y}{I_K} \tag{3.3}$$

and so

$$\left.\frac{dK}{dY}\right|_{\dot{Y}=0} \gtreqless \text{ depending on } S_Y \lesseqgtr I_Y. \tag{3.4}$$

Typical forms of the curve (3.1) or $\dot{Y} = 0$ together with the signs of $\dot{Y}$ can be depicted as in Fig. 2. We consider two cases. In case (A) it is assumed that $S_Y > I_Y$ holds for any $Y > 0$, and in case (B), $S_Y < I_Y$ holds for intermediate levels of Y, while $S_Y > I_Y$ holds for low and high levels of Y. The latter case corresponds to the Kaldorian investment function as depicted in Fig.1.[4] In either case we shall assume that for each K , $S_Y > I_Y$ holds for small and large Y's.

Similarly, from (3.2) we obtain

[3]Slightly different, but related dynamical models occur in other economic situations. For example, see, Nikaido(1980),(1985) .

[4]We are assuming that the slope of saving function lies between the upper and lower bounds of the slope of investment function.

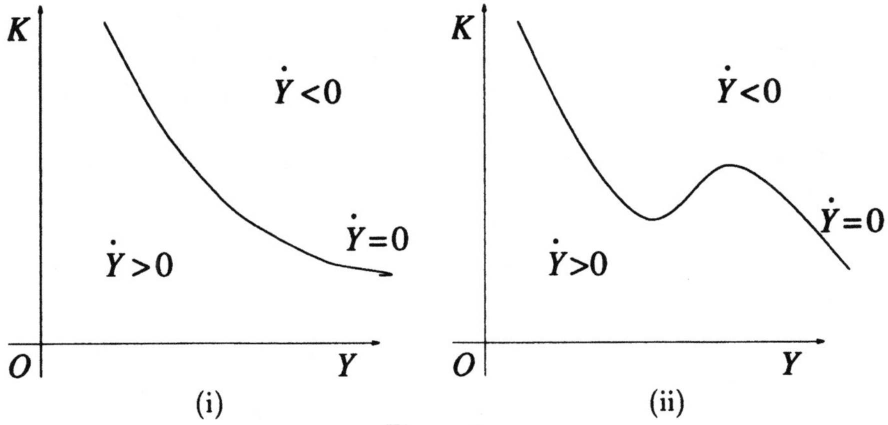

Figure 2.

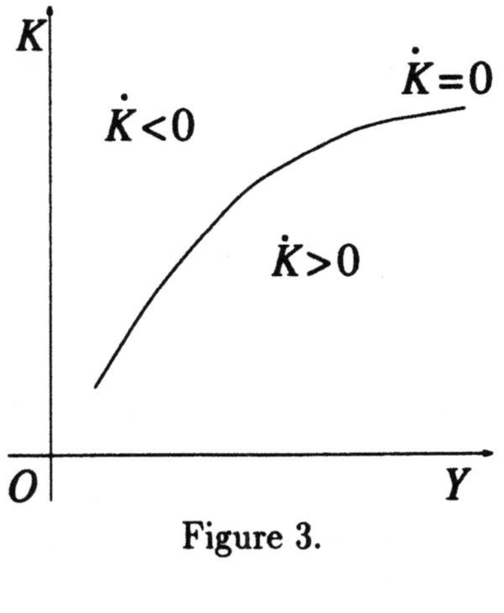

Figure 3.

$$\left.\frac{dK}{dY}\right|_{\dot{K}=0} = \frac{I_Y}{\delta - I_K} \tag{3.5}$$

hence, $dK/dY > 0$ along the curve (3.2) (i.e. the curve $\dot{K} = 0$). A typical form of the curve together with the signs of $\dot{K}$ are indicated in Fig.3.

It is clear that the dynamic process (P) has an equilibrium point if and only if two curves $\dot{Y} = 0$ and $\dot{K} = 0$ have a point in common. For this it is sufficient that

(a) There exists Y close to zero such that the curve $\dot{Y} = 0$ lies above the curve $\dot{K} = 0$ and

(b) For any sufficiently large Y the absolute value of the slope of curve $\dot{Y} = 0$ (or that of $\dot{K} = 0$) exceeds a positive constant.

We first prove the existence of equilibrium.

PROPOSITION 1. The dynamic system (P) has an equilibrium (Y^*, K^*) if (a)′ there exist K_1, $K_2 > 0$ such that $I(0, K_1) = S(0)$ and $I(0, K_2) = \delta K_2$ and (b)′ there exist $\alpha_1 > 0$ and $\beta_1 > 0$ such that $S_Y - I_Y > \alpha_1$ for any K and sufficiently large Y and $-I_K < \beta_1$ for any $(Y, K) \in R^2_+$. The conclusion also holds if condition (b)′ is replaced by (b)″ there exist α_2, $\beta_2 > 0$ such that $I_Y > \alpha_2$ and $-I_K < \beta_2$ for any $(Y, K) \in R^2_+$.

Proof) Let K_1 and K_2 be such that $I(0, K_1) = S(0)$ and $I(0, K_2) = \delta K_2$. Then condition (a)′ implies that

$$\begin{aligned} I(0, K_2) &= \delta K_2 \\ &> 0 \\ &\geq S(0) \\ &= I(0, K_1). \end{aligned} \tag{3.6}$$

It then follows $K_1 > K_2$ and so condition (a) is satisfied by continuity.

On the other hand, if (b)′ holds, then, in view of (3.3), we know that the absolute value of the slope of $\dot{Y} = 0$ is larger than α_1/β_1. Therefore condition (b) is satisfied.

Similarly, if condition (b)″ is satisfied, then, by (3.5), the slope of $\dot{K} = 0$ is larger than $\alpha_2/(\delta + \beta_2)$. Therefore, Proposition 1 holds under conditions (a)′ and (b)″. Q.e.d.

Henceforward, we shall suppose that an equilibrium exists often without mentioning it.

As to the number of equilibria, there are various possibilities; we have shown several representative cases in Fig.4. The arrows in figures indicate the directions of the change of (Y, K).

For each $(Y, K) \in R^2_+$ let $J(Y, K)$, or simply J , be defined by

$$J = \begin{pmatrix} k(I_Y - S_Y) & kI_K \\ I_Y & I_K - \delta \end{pmatrix} \tag{3.7}$$

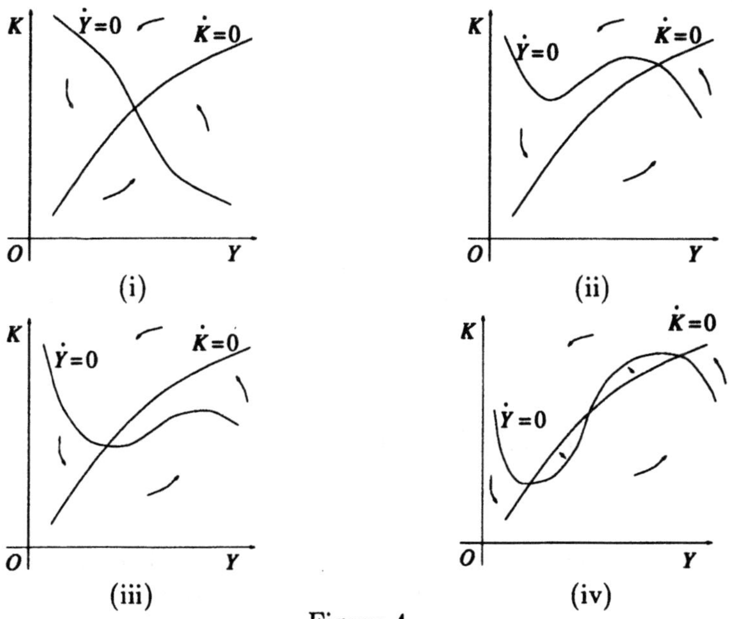

Figure 4.

which is the Jacobian matrix of the system (P) evaluated at (Y, K). We also define

$$\operatorname{tr} J = k(I_Y - S_Y) + I_K - \delta \tag{3.8}$$

and

$$\det J = k[-S_Y I_K + \delta(S_Y - I_Y)] \tag{3.9}$$

REMARK 1. Kaldor (1940) assumes in Appendix that the slope of saving function is larger than that of the line RR which is defined by (3.2). Since $\left.\frac{dI}{dY}\right|_{\dot{K}=0} = \delta \left.\frac{dK}{dY}\right|_{\dot{K}=0}$, this condition reduces to

$$\begin{aligned} S_Y &> \left.\frac{dI}{dY}\right|_{\dot{K}=0} \\ &= \frac{\delta I_Y}{\delta - I_K} \end{aligned} \tag{3.10}$$

using (3.5). This is easily seen to be equivalent to $J > 0$.

We will prove under the above Kaldorian assumption there can not be more than one equilibrium of the system (P).

PROPOSITION 2. Under the Kaldorian condition (i.e.. if $\det J > 0$ for each (Y, K)), the dynamical process (P) has at most one equilibrium.

Proof) The condition $\det J > 0$ can be restated as

$$\delta(S_Y - I_Y) > S_Y I_K,$$

and, since $I_K < 0$, as

$$\frac{I_Y}{\delta - I_K} > \frac{S_Y - I_Y}{I_K}. \tag{3.11}$$

In view of (3.3) and (3.5), the last condition implies that the slope of the curve $\dot{K} = 0$ is algebraically greater than that of $\dot{Y} = 0$. Hence these two curves do not interest more than once.

4. Behavior of the Dynamical Process When k is Infinitely Large

Assume that k is much larger than 1. Then arrows in Fig.4 are almost horizontal except on the curve $\dot{Y} = 0$. In this case the stability of the dynamical process (P) can be analyzed quite accurately using only diagrams. Similar analysis can be applied when k is close to zero, in which case the arrows are almost vertical except on the curve $\dot{K} = 0$.

(A) Consider the case where $S_Y > I_Y$ holds for any Y and K. Then the curve $\dot{Y} = 0$ is downward sloping and we have a situation as in Fig.5. Hence given any initial point (Y_0, K_0) the system converges to the equilibrium point E. Following Zeeman (1973) the curve $\dot{Y} = 0$ will be referred to as the slow manifold, emphasizing the fact that the movement of the arrows along the curve is slow.

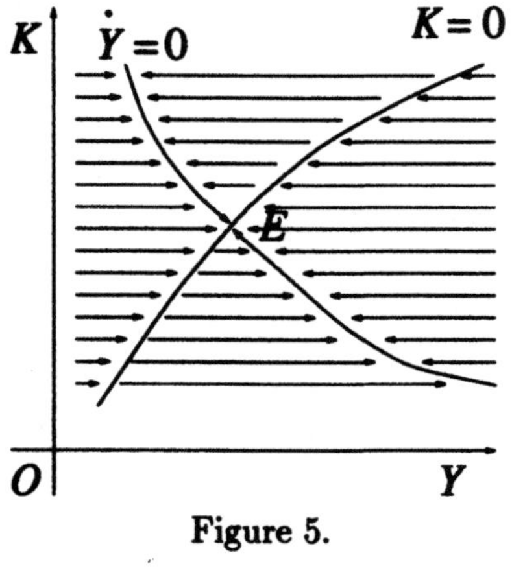

Figure 5.

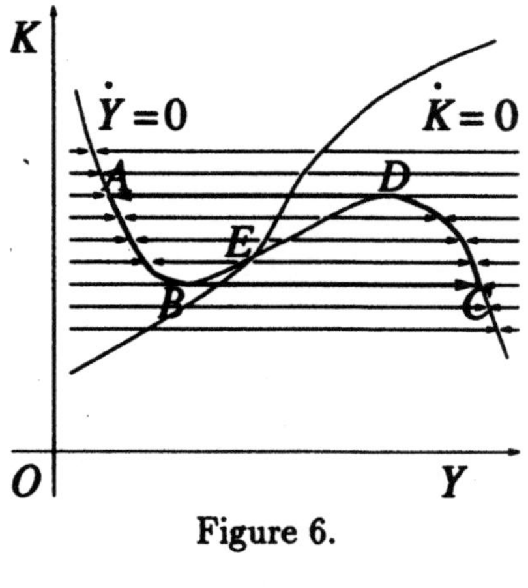

Figure 6.

(B) In the case of Kaldorian investment function, the curve $\dot{Y} = 0$ is as depicted in Fig.2 (ii). In this case we may have multiple equilibria as in Fig.4(iv).

Multiple equilibria is ruled out under the Kaldorian assumption that $\det J > 0$ (cf. Proposition 2 and Remark 1). If we assume further that the equilibrium occurs at an intermediate level of Y (namely if $I_Y > S_Y$ at E) as in Fig.6, then the equilibrium point E is unstable and the Kaldorian periodic orbit $ABCD$ appears starting from any non-equilibrium position. It may easily be checked that a similar cycle appears for the case of Fig.4 (iv) if all the equilibrium points lie in the intermediate region, namely if all equilibria occur at the points where $\dot{Y} = 0$ is increasing (that is where $S_Y < I_Y$).

Notice that the Kaldorian cycle may not appear if two curves intersect in different positions. In particular, in the case of Fig.4 (ii) the equilibrium is stable so long as k is sufficiently large.

5. Sufficient Conditions for Global Stability

Let us turn to the general case where k is no longer assumed to be infinitely large. In this section we shall derive sufficient conditions under which the dynamic process (P) has a unique stable equilibrium point. Conditions guaranteeing existence and uniqueness were given in Propositions 1 and 2.

Before stating the stability theorem, we give a brief discussion on a basic concept.

Given $f : R^n \to R^n$, assume that the solution of the system of differential equations

$$\dot{y} = f(y) \tag{5.1}$$

is uniquely determined by initial conditions. We say that a equilibrium y^* is *globally asymptotically stable* if $y(t)$ is a solution for (5.1), then $y(t)$ exists for all t and $y(t) \to y^*$ as $t \to \infty$. (See, e.g., Hartman (1982) or Wiggins (1990)).

PROPOSITION 3. Assume that for all $(Y, K) \in R^2$, $S_Y > I_Y$. Then the dynamical system (P) has a unique equilibrium (Y^*, K^*) which is globally asymptotically stable.

Proof) By Olech's Theorem, Olech (1963) for $n = 2$ (5.1) is asymptotically stable if

(i) $\det J(y) > 0$

(ii) $\operatorname{tr} J(y) < 0$

and either

(iii) $f_{11}(y) \cdot f_{22}(y) \neq 0$ or $f_{12}(y) \cdot f_{21}(y) \neq 0$

for each y, where f_{ij} $(i,j = 1,2)$ denotes the i,j-th element of Jacobian matrix of f.

Condition (iii) is satisfied by assumption and for each (Y, K) the Jacobian of the system (P) is expressed as in (3.7). Since under the assumptions of Theorem 3

$$\operatorname{tr} J = k(I_Y - S_Y) + I_K - \delta < 0 \tag{5.2}$$

$$\det J = k[-S_Y I_K + \delta(S_Y - I_Y)] > 0 \tag{5.3}$$

and

$$kI_Y I_Y \neq 0. \tag{5.4}$$

Proposition 3 is established.

6. Existence of Trade Cycles

In this section we shall investigate sufficient conditions for the existence of a non-degenerate periodic orbit of the solution path of the dynamical process (P). A similar study was made by Chang and Smith (1971). Also in a related economic model Torre (1977) discusses the occurrence of the Poincaré-Andronov-Hopf bifurcation. Our analysis will be mainly concerned with the Kaldorian case characterized by the following properties:

(a) Investment function is of Kaldorian type and there is a unique equilibrium point in the range of Y where $I_Y > S_Y$ (see, Fig.4 case (iii))

(b) The curve $\dot{K} = 0$ has a steeper slope than the curve $\dot{Y} = 0$ everywhere.

We remarked earlier that this condition (b) is equivalent to the condition $\det J > 0$ and also to the Kaldorian assumption that the saving function is everywhere steeper than RR curve defined in the Appendix of Kaldor (1940).

The next condition is to guarantee that the curve $\dot{Y} = 0$ takes on all non-negative values of K, and that the curve $\dot{K} = 0$ is bounded above and goes through the origin (see, Fig.7 below).

(c) For any $K > 0$ there is $Y > 0$ such that $I(Y,K) = S(Y)$. Moreover, $I(Y,0) \geq 0$ for all $Y \geq 0$, the equality holding only at $Y = 0$. Also there is $\overline{K}$ so large that for any $K \geq \overline{K}$, $\delta K > I(Y,K)$ holds for all Y. This condition can be weakened as in Remark 2 below.

We are now ready to state the main result on the existence of periodic orbit in the autonomous system (P). We notice that condition (d) in the following Theorem is satisfied when k is large as in the original Kaldor model.

Proposition 4. If conditions (a), (b) and (c) are satisfied and if (d) $I_Y - S_Y > (\delta - I_K)/k$ in some neighborhood of the equilibrium, then the dynamical system (P) has a non-trivial periodic solution.

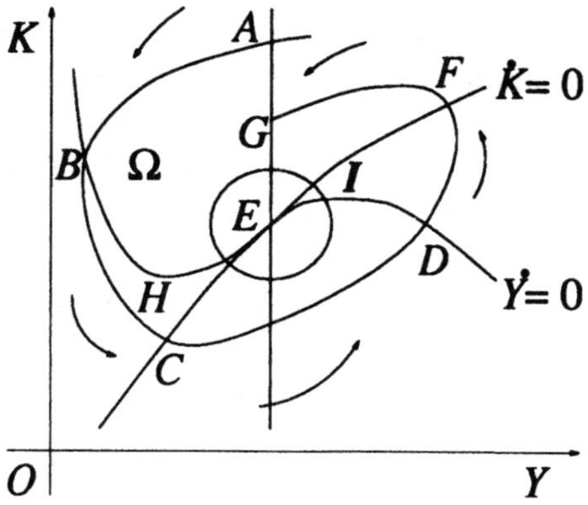

Figure 7.

Proof) We first consider the linear approximation of the system (P). For each (Y, K) the Jacobian of (P) can be expressed as in (3.7). We will denote by $\overline{J}$ the matrix J evaluated at the equilibrium point.

It is well known[5] that if all the characteristic roots of $\overline{J}$ have positive real parts then the equilibrium point is either an unstable focus or an unstable node, hence[6] the path that starts from any non-equilibrium point stays away from some neighborhood of the equilibrium for all $t > 0$.

We claim that a sufficient condition that $\overline{J}$ has this property is that

$$\operatorname{tr} \overline{J} = -(k(S_Y - I_Y) + \delta - I_K) > 0. \tag{6.1}$$

To show this we first note that the sum of the two characteristic roots of $\overline{J}$ is given by $\operatorname{tr} \overline{J}$, and the product, by $\det \overline{J}$, which is positive by assumption (b). Hence if the two roots are real both of them are positive; and if two of them are complex, the real part of the two conjugate roots is given by the half of $\operatorname{tr} \overline{J}$ and so is positive.

Let us next consider the global behavior of the dynamical process (P). In Fig.7, E is the equilibrium point (Y^*, K^*).

By assumptions (a) and (b) of the theorem , the curves $\dot{Y} = 0$ and $\dot{K} = 0$ are shaped and intersect as in the case of Fig.4(iii). Moreover, by assumption (c) of the proposition, the curve $\dot{Y} = 0$ takes on all non-negative values of K and the curve $\dot{K} = 0$ is bounded above and goes through the origin.

Now draw a line perpendicular to the horizontal axis that goes through the equilibrium point E. Since the orbits in Fig.7 are defined everywhere and

[5]See, e.g., Hale (1969) pp.101-104.

[6]If $\overline{J}$ has no purely imaginary characteristic roots (including zero), then there is a homeomorphism locally taking orbits of the system near equilibrium to the orbits of its linear approximation. See, Theorem 7.1 of Hartman(1973) for more rigorous statement.

do not intersect, we may find a point A so high on the perpendicular line that the orbit starting from A encloses stationary points H and I of $\dot{Y} = 0$ before it meet the same perpendicular line in the same direction at G. Since the curve $\dot{K} = 0$ is bounded above we may suppose that A lies above G.

Let us consider the region bounded by $ABCDFGA$ and delete an open neighborhood of the equilibrium point E where the path through A does not enter. Call the remaining compact region Ω.

The dynamic process (P) starting from A remains in Ω and there is no equilibrium point there. Hence Poincare=Bendixon Theorem[7] implies the existence of a periodic orbit of the system (P).

REMARK 2. Condition (c) of Theorem 4 is a rather strong requirement. The following condition (c)′ is sufficient for the proposition.

(c)′ let $\overline{Y} > 0$ be the solution of $I(Y,0) = S(0)$ and $\overline{K} > 0$ be defined by the solution of $I(\overline{Y}, K) = \delta K$. Then for any $0 \leq K \leq \overline{K}$ there exists $Y > 0$ such that $I(Y,K) = S(Y)$. Moreover, $I(Y,0) \geq 0$ for all $Y \geq 0$, the equality holding only at $Y = 0$.

The proof of theorem in this version can be conducted in the following way. Since $I(\overline{Y},0) = S(\overline{Y})$ and I is increasing in Y and decreasing in K it follows that $\dot{Y} < 0$ for all $Y > \overline{Y}$. We note that this implies that the curve $\dot{Y} = 0$ intersects with the horizontal axis at $Y = \overline{Y}$ and does not extend beyond the intersection point. Since the curve $\dot{K} = 0$ is increasing in Y and $\dot{K} < 0$ above it, any orbit of the solution of the system that goes through the region $\dot{Y} < 0$ (does not go beyond $\overline{Y}$ and so) does not go through the region $K \geq \overline{K}$. Hence G must lie in the region $K < \overline{K}$ and so we may find a point A which lies in the same region and above G. Since the curve $\dot{K} = 0$ is defined for any $K \leq \overline{K}$ the rest of the proof can be conducted as before.

7. Exogenous Disturbance and Trade Cycle

That an economic system behaves harmoniously with the exogenous fluctuation was implicitly assumed in the economic literature of trade cycles. Theories of Schumpeter (1939) and Frish (1933) among others, stand on the basis of this supposition. In this section we assume that there is an exogenous disturbance in the final demand of period ω and demonstrate that under appropriate conditions the adjustment system has the ω-periodic solution.

[7]Poincare-Bendixon Theorem states that a non-empty compact limit set of a C^1 planar dynamical system which contains no equilibrium point is a closed orbit. Uzawa (1963) is an early article which applied this theorem in a different economic context.

REMARK 3. The period of solution of a system is not in general the same as the period of disturbance term. For example the system

$$\begin{aligned}\frac{dx}{dt} &= y + (x^2 + y^2 - 1)\sin ax \\ \frac{dy}{dt} &= -x\end{aligned}$$

has the periodic solution $x = -\cos t, y = \sin t$ with period 2π, although the period of the right sides is $2\pi/a$.

To simplify the exposition we assume that there is an exogenous disturbance $\beta(t)$ of saving at time t for $t \geq 0$.It is assumed that $\beta(t)$ has period ω (i.e., $\beta(t) = \beta(t+\omega)$ for any t) and continuously differentiable. The modified dynamical process (P)$'$ may now be written as

$$\begin{cases} \dot{Y} = k(I(Y,K) - S(Y) - \beta(t)) \\ \dot{K} = I(Y,K) - \delta K \end{cases} \tag{P$'$}$$

where $Y(0)$ and $K(0)$ are given.

Now let α be a positive real number, and rewrite (P)$'$ in an equivalent form as

$$\begin{cases} \dot{Y} + \alpha Y = k(I(Y,K) - S(Y) - \beta(t)) + \alpha Y \\ \dot{K} + \alpha Y = I(Y,K) - (\alpha - \delta)K. \end{cases} \tag{P$''$}$$

For each α and t define a map $g_{\alpha,t} : R^2_+ \to R^2$ by

$$g_{\alpha,t}(Y,K) = \frac{1}{\alpha}\begin{pmatrix} k(I(Y,K) - S(Y) - \beta(t)) + \alpha Y \\ I(Y,K) + (\alpha - \delta)K \end{pmatrix}. \tag{7.1}$$

The Jacobian matrix $J_{\alpha,t}$ of $g_{\alpha,t}$ does not contain t explicitly, and so we write it simply as J_α. Thus J_α is given by

$$J_\alpha = \frac{1}{\alpha}\begin{pmatrix} k(I_Y - S_Y) + \alpha & kI_K \\ I_Y & I_K + (\alpha - \delta) \end{pmatrix}. \tag{7.2}$$

For a moment we fix α and write $g_{\alpha,t}$ simply as g_t. In the sequel we give a sufficient condition under which g_t is a contraction map uniformly in t.The condition means that there exists a constant $\kappa < 1$ satisfying

$$|g_t(z) - g_t(u)| < \kappa|z - u| \quad \text{for any } z, u \in R^2_+ \text{ and } t. \tag{7.3}$$

The following result is established in the Appendix in more general situation.

LEMMA 1. In order that g_t be a contraction map[8] for some norm, it is sufficient that for each $(Y,K) \in R^2_+$, the absolute values of the characteristic roots of its Jacobian matrix are all less than 1.

[8]Strictly speaking, we only prove that g_t is a contraction map in the wide sense as defined in the Appendix of this section. This is sufficient for the remaining discussion.

For each fixed (Y, K) let $f\alpha(\lambda)$ be the characteristic polynomial of αJ_α. We will look for conditions under which all roots of $f\alpha(\lambda) = 0$ have absolute values less than α for a suitable choice of α.

By definition (see, (3.8) and (3.9)), we have

$$\begin{aligned} f_\alpha(\lambda) &= \begin{vmatrix} \lambda - k(I_Y - S_Y) - \alpha & -kI_k \\ -I_Y & \lambda + (\delta - I_k) - \alpha \end{vmatrix} \\ &= (\lambda - \alpha)^2 - (\lambda - \alpha)(\operatorname{tr} J) + \det J. \end{aligned} \tag{7.4}$$

For each $(Y, K) \in R^2_+$ the discriminant Δ of $f_\alpha(\lambda) = 0$ may be written as

$$\Delta = (\operatorname{tr} J)^2 - 4 \det J.$$

If $\Delta < 0$, $f_\alpha(\lambda) = 0$ has two complex roots and their product, which is the square of the absolute value of each characteristic root, is given by

$$\lambda_1 \lambda_2 = \alpha^2 + \alpha \operatorname{tr} J + \det J \tag{7.5}$$

If $\Delta > 0$, $f_\alpha(\lambda) = 0$ has two real roots. We will proceed to study their magnitude.

By putting $\lambda = \alpha$ in (7.4) we obtain

$$f_\alpha(\alpha) = \det J. \tag{7.6}$$

Similarly, by putting $\lambda = -\alpha$ in (7.4) we have

$$f_\alpha(-\alpha) = 4\alpha^2 + 2\alpha \operatorname{tr} J + det J. \tag{7.7}$$

Hence, in particular, we have,

$$f_1(-1) = 4 + 2 \operatorname{tr} J + \det J \tag{7.8a}$$

and

$$f_k(-k) = 4k^2 + 2k \operatorname{tr} J + \det J. \tag{7.8b}$$

From (7.6) and (7.8) we can easily infer that

$$\text{if } 2 > -\operatorname{tr} J \text{ and } \det J > 0 \text{ then } f_1(1) > 0 \text{ and } f_1(-1) > 0 \tag{7.9a}$$

and for $k > 0$

$$\text{if } 2k > -\operatorname{tr} J \text{ and } \det J > 0 \text{ then } f_k(k) > 0 \text{ and } f_k(-k) > 0 \tag{7.9b}$$

Next, define

$$\lambda_\alpha = \alpha + \frac{1}{2} \operatorname{tr} J. \tag{7.10}$$

It is clear from this that

$$\text{if } 4\alpha > -\operatorname{tr} J > 0 \text{ then } \alpha > \lambda_\alpha > -\alpha. \tag{7.11}$$

From (7.4) and (7.10) we obtain

$$\begin{aligned} f_\alpha(\lambda_\alpha) &= -\frac{1}{4}((\operatorname{tr} J)^2 - 4\det J) \\ &= -\frac{1}{4}[(k^2(S_Y - I_Y)^2 \\ &\quad +2k(S_Y - I_Y)(I_K - \delta) + (\delta - I_K)^2 + 4kI_Y I_K] \end{aligned} \tag{7.12}$$

Now consider the following two sets of conditions

$$\begin{aligned} &\det J > 0,\ 2 > -\operatorname{tr} J \\ \text{and}\quad &\operatorname{tr} J + \det J < 0 \end{aligned} \tag{C1}$$

and

$$\begin{aligned} &\det J > 0,\ 4 + \operatorname{tr} J > 0 \\ \text{and}\quad &k(S_Y - I_Y)^2 > \max(-2(I_K - \delta), I_Y I_K). \end{aligned} \tag{C2}$$

where $\det J$ and $\operatorname{tr} J$ are defined by (3.8) and (3.9).

We note that the last two conditions in (C1) are likely to be satisfied when k is small and the last two conditions in (C2) are likely to be satisfied when k is large.

It is now easy to prove

LEMMA 2. If condition (C1) is satisfied then $f_1(\lambda) = 0$ has two roots whose absolute values are less than 1, and if condition (C2) is satisfied then $f_k(\lambda) = 0$ has two real roots between $-k$ and k.

Proof) If condition (C1) is satisfied, then $f_1(1) > 0$ and $f_1(-1) > 0$ by (7.9a), and also if $\Delta \geq 0$ where Δ is the discriminant defined above, then $f_1(\lambda_1) < 0$ by (7.12) and λ_1 lies between -1 and 1 by (7.11). Hence $f_1(\lambda) = 0$ has real roots whose absolute values are less than 1. On the other hand if $\Delta < 0$ then (7.5) and the last condition of (C1) implies the desired conclusion.

When condition (C2) is satisfied, the same proof applies for $\alpha = 1$ using (7.6), (7.9b), (7.11) and (7.12). In this case two roots are real.

We now state

PROPOSITION 5. If either condition (a) or (b) is satisfied for any $(Y, K) \in R_+^2$, then the dynamical system (P)′ has a non-generate periodic solution with the same period as the exogenous disturbance $\beta(t)$.

The proof of Proposition 5 follows from Lemmas 1, 2 and the following result which guarantees the existence of a periodic solution. We only have to define $g_{\alpha,t}$ by (7.1) for $\alpha = 1$ and k and note that $\beta(t)$ is an ω-periodic function of t. Lemma 3 is then applicable to (P)″.

LEMMA 3. Let X be a convex subset of R^n and assume that for a given $\alpha > 0$, $g_t : X \to X$ satisfies (7.3) for some $\kappa \leq 1$ and $\forall z, u \in R^n$. Also assume that as a function of t $g_t(x)$ is continuous and ω-periodic. Then given $x(0)$

$$\frac{dx(t)}{dt} + \alpha x(t) = \alpha g_t(x(t))$$

has a solution satisfying $x(0) = x(w)$.

For a proof of this lemma, see, for example, Brezis (1973) pp.10–15. The basic idea of the proof is as follows. To each point x in X associate the point $f(x)$ where x arrives at ω period later along the solution path of the system, and apply a suitable fixed point theorem.

Appendix to Section 7

Let X be a convex subset of R^n with a norm $|\ |$ and $g : X \to R^n$ be continuously differentiable. We say that g is a contraction mapping (in the wide sense) if there exists a positive number $\kappa \leq 1$ such that

$$|g(x) - g(y)| < \kappa|x - y| \quad \forall x, y \in X \tag{A.1}$$

In this Appendix we prove the following result we used in Section 7.

THEOREM A. If for each $x \in X$ the characteristic roots of the Jacobian matrix $Dg(x)$ are all less than 1 in absolute values then g is a contraction mapping in the wide sense for some norm $|\ |$.

Proof) For each $z \in X$ let the characteristic roots of the linear map $T = Dg(z)$ have absolute values less than 1. Then it is known that there is $\kappa < 1$ (which may depend on z) and a norm $|\ |$ such that

$$|Tx| \leq \kappa|x| \quad \forall x \in X \tag{A.2}$$

(See, for example Hirsh and Smale (1974) pp.279–280)

Now if $x, y \in X$ then for each $0 \leq t \leq 1$, $z = x + t(y - x) \in X$ by the convexity of X. Hence if we define a mapping $\phi : [0,1] \to R^n$ by $\phi(t) = g(x + t(y - x))$, then

$$\begin{aligned} g(y) - g(x) &= \phi(1) - \phi(0) \\ &= \int_0^1 \phi'(t)dt \\ &= \int_0^1 \phi Dg(x + t(y - x))(y - x))dt \end{aligned} \tag{A.3}$$

Hence there is $\overline{\kappa} \leq 1$ such that

$$\begin{aligned} |g(y) - g(x)| &\leq \int_0^1 \overline{\kappa}|y - x|dt \\ &= \overline{\kappa}|y - x|. \end{aligned}$$

References

[1] H. Brezis, *Operateurs Maximaux Monotones*, North-Holland, 1973.

[2] W. W. Chang and D.J. Smyth. *The Existence and Persistence of Cycles in a Non-linear Model: Kaldor's 1940 Model Re-examined*, The Review of Economic Studies, January (1971), 37–44.

[3] R. Frish, *Propagation Problem and Impulse Problem in Dynamic Economics*, Economic Essays in Honor of G. Cassel, 1933.

[4] R. M. Goodwin, *The Nonlinear Accelerator and the Persistence of Business Cycles*, Econometrica, **19** (1951), 1–17.

[5] J. K. Hale, *Ordinary Differential Equations*, Wiley-Interscience, 1969.

[6] P. Hartman, *Ordinary Differential Equations*, The Johns Hopkins University, 1973.

[7] M. W. Hirsch and S. Smale, *Differential Equations, Dynamical Systems and Linear Algebra*, Academic Press, 1974.

[8] N. Kaldor, *A Model of the Trade Cycle*, Economic Journal, March, 1940.

[9] M. Kalecki, *Trend and Business Cycles Reconsidered*, Economic Journal, **78** (1968), 263–76.

[10] H. Nikaido, *Harrodian Pathology of Neoclassical Growth: The Irrelevance of Smooth Factor Substitution*, Zeitschrift für Nationalökonomie (Journal of Economics),**40**, No.1-2 (1980),111–134.

[11] H. Nikaido, *Dynamics of Growth and Capital Mobility in Marx's Scheme of Reproduction*, Zeitschrift für Nationalökonomie (Journal of Economics), **45**, No.3 (1985), 197–218.

[12] C. Olech, Z. Opial and T, Wazenski, *On the global stability of autonomous systems in the plane*, Contributions to Differential Equations **1** (1963), 389–400.

[13] J. Schumpeter, *Business Cycles : A Theoretical, Historical and Statistical Analysis of the Capitalist Process*, McGraw-Hill, 1939.

[14] V. Torre, *Existence of Limit Cycles and Control in Complete Keynesian System by Theory of Bifurcations*, Econometrica, September (1977),1457–66

[15] H. Uzawa, *On a Two Sector Model of Economic Growth: II*, The Review of Economic Studies, Vol. 30(2), 1963.

[16] S. Wiggins, *Introduction to Applied Nonlinear Dynamical Systems and Chaos*, Springer-Verlag, 1990.

[17] E. C. Zeeman, *Differential Equations for the Heartbeat and Nerve Impulse*, in C.H.Waddington, Toward a Theoretical Biolozy, Vol.4 (E.U.P) also in Peixoto, Dynamical Systems, 1973.

[18] E. C. Zeeman, *On the Unstable Behavior of Stock Exchanges*, Journal of Mathematical Economics, Vol.1, 1974.

On Large Games with Finite Actions: A Synthetic Treatment*

M. Ali Khan

Department of Economics,
The Johns Hopkins University, Baltimore, MD 21218, U.S.A.

Yeneng Sun

Department of Mathematics,
National University of Singapore, SINGAPORE 0511.

Abstract

In this expository paper, we untangle the relationship between anonymous and non-anonymous versions of the theory of large games. Our treatment shows that the two formulations collapse to one essentially equivalent theory in the case of finite action spaces, and exhibit rich differences only when this finiteness is dispensed with.

1. Introduction

There are two formulations of the theory of large games, both dating to the seventies. The first, due to Schmeidler (1973), is based on a measure space of player's names, and hence is a formulation of large non-anonymous games; while the second, due to Mas-Colell (1978; but also see the more mature 1984 formulation), is based on a distribution of player's characteristics, and hence is a formulation of large anonymous games. The distinction between these two formulations can also be made on the ground of how they conceive of

*A Preliminary version of this paper was presented as an invited talk at the K.E.S/T.I.Tech Conference on *Nonlinear and Convex Analysis in Economic Theory* held in Tokyo in July 1993. The current version was presented at the Microeconomics Workshops at Indiana University, both at Bloomington and at Indianapolis, and at Johns Hopkins. Both authors thank Professors Bob Anderson, Subir Chakrabarti, Joe Harrington, Leo Hurwicz, Sung Kim, Kali Rath, Neil Rothman, Bob Sandy, Akira Yamazaki, Shinji Yamashige and Makoto Yano for their questions and encouragement, but retain sole responsibility for errors. The work was completed during the visit of Yeneng Sun to the Department of Economics at Johns Hopkins in October-November 1993; both authors thank Lou Maccini for making this visit possible.

a player's dependence on *societal responses;* in the first, her payoff depends on the average (integral) responses, while in the second, it depends on the distribution (probability measure) of responses. All of this underscores the fact that the two formulations are technically and substantively different. The first is primarily measure-theoretic in its conception, the space of players' names insures this at the very outset, while the second is primarily topological inspite of the fact that it is enunciated in terms of probability measures; see Khan (1986, 1989) for a fuller articulation of these observations. What has not been equally well-appreciated is the essential equivalence of these two formulations when the space of players' names and the space of players' characteristics are both atomless, and the action sets are restricted to a finite number of objects, the case that served as the focus of both Schmeidler and Mas-Colell in the theorems that are our primary concern here. The two formulations of *societal responses* are then essentially identical – essential that is with respect to a homeomorphism. This paper provides an elaboration of this equivalence, and also supplements it by some observations on the case when the action sets are countably infinite.

We focus then on the existence of pure-strategy Cournot-Nash equilibria in the case of Schmeidler, and on the existence of symmetric Cournot-Nash equilibrium distributions in the case of Mas-Colell. If we now see Schmeidler's theorem in the setting of an abstract measure space rather than on the Lebesgue space $[0, 1]$, as he originally did, we can show that his existence theorem implies that of Mas-Colell simply by looking on the space of players' characteristics as the space of players' names, and consequently connecting these two spaces by the identity map. Armed with this elementary observation, the rest of the argument is few routine steps of verification. However, if one insists on setting Schmeidler's theorem on $[0, 1]$, one has to use Skorokhod's lemma whereby any measure on a complete separable metric space can be viewed as the measure induced by a random variable defined on an atomless probability space and proceed with essentially identical computations; see the hint in Mas-Colell (1984) and its elucidation in Rath (1991).

The other way round has proved more elusive simply on account of the fact that an atomless game of the Schmeidler type with all identical players induces an atomic game of the Mas-Colell type concentrated at one point in the space of characteristics. In the parlance of probability theory, this is simply the observation that the measure induced by a random variable on an atomless probability space need not be atomless. To put this in a less technical way, a non-anonymous game in which all players are negligible may be one in which a non-negligible proportion of the players may have the same characteristic. This observation seems to preclude the use of Mas-Colell's theorem (his Theorem 2) which takes as its basic hypothesis an atomless measure on the space of players' characteristics. Rath (1991) overcomes this difficulty through the general existence theorem of Mas-Colell, his Theorem

1, and the "disjointification" technique of Khan-Sun (1987). Rath's lemma has obvious independent interest, but his approach still does not enable a proof of Schmeidler's theorem from our primary vantage point, Mas-Colell's Theorem 2.

We show that the theorem of Mas-Colell implies that of Schmeidler simply by rescaling the payoffs in such a way that identical agents with different names get, through this rescaling, apparently different payoffs. The mechanics behind this rescaling use the scalar version of Lyapunov's theorem, and draw on an elementary lemma from probability theory, whose use in economic theory goes back at least to Gabszewicz-Mertens (1971). Since nothing essential of a game-theoretic nature changes as a result of this rescaling, we bypass the difficulty mentioned above, and squarely faced in the work of Rath. Once we have the rescaling, the rest of the argument is again few routine steps of verification. Of course, if it is the "Lebesgue interval" version of Schmeidler's result that one is after, even this elementary lemma is not necessary.

Given the ease with which one can deduce Mas-Colell's theorem from that of Schmeidler's, one is tempted to ask whether a somewhat stronger result can be obtained through this route. We present such a strengthening of Mas-Colell's theorem, one that applies to a situation where the space of characteristics is conceived of in a bipartite way, one part essential and the other non-essential. To put the matter another way, our treatment allows us to consider a situation where a dummy index *and* payoffs jointly constitute characteristics. Of course, we stretch the interpretation of *anonymous* here, but it is of some interest that Schmeidler's theorem can be used to prove this strengthened version, and also that this strengthened version allows a proof of his theorem without even appealing to the minimal machinery we used earlier. The argument is simply an exploitation of the observation that the induced measure on the characteristic space is now always atomless.

This testimony to the strength of Schmeidler's theorem notwithstanding, there is of course an element of circularity in all of this – we use a result to prove another which is then used to prove the first result we began with. Even though an elaboration of this circularity is really the point of this paper – we are after all showing the essential equivalence of two different formulations of the theory – for ease of reference we recall some standard results on integration of correspondences taking values in finite-dimensional Euclidean space, and sketch how they can be used to provide independent existence proofs of the results we relate.

We conclude the paper by showing how this theory fares when the space of actions is countably infinite. Here we expose the recent work of the authors, see Khan-Sun (1993), which generalizes the folklore that the extreme point of the set of all probability measures with given marginals is concentrated on the graph of a measurable function; see Douglas (1964), Lindenstrauss (1965) and the survey by Yor (1978).

2. Some Straightforward Observations

$\mathbb{R}^k$ is the k-dimensional Euclidean space, with $\mathbb{R}^k_+$ its non-negative orthant and $\Delta_k = \{x \in \mathbb{R}^k_+ : \sum_{i=1}^k x_i = 1\}$ its $(k-1)$-dimensional simplex. Until the last but one section of this paper, we shall reserve the symbol A for a finite set of k objects $\{a_1, \cdots, a_k\}$, each of which is different. The set of probability measures on the measurable space $(A, \mathcal{A})$, $\mathcal{A}$ the power set of A is denoted by $\mathcal{M}(\{a_1, \cdots, a_k\})$ or by $\mathcal{M}(A)$.

We now recall for the reader the following well known remarks.

Remark 1 *Δ_k endowed with its relativized Euclidean topology and $\mathcal{M}(\{a_1, \cdots, a_k\}$ endowed with the weak*topology are homeomorphic.*

Proof of Remark 1: Let $\phi : \Delta_k \longrightarrow \mathcal{M}(\{a_1, \cdots, a_k\})$ be such that for any $x \in \Delta_k$, and any $V \subseteq \{a_1, \cdots, a_k\}$ $\phi(x)(V) = \sum_{a_j \in V} x_j$. Certainly $\phi(x) \in \mathcal{M}(\{a_1, \cdots, a_k\})$ for any $x \in \Delta_k$. It is easy to see that ϕ is bijective, and that Δ_k is compact. Thus all that remains to be shown is that ϕ is continuous. Towards this end, pick any sequence $\{x^n\}_{n \in \mathbf{N}}$ from Δ_k such that $x_n \longrightarrow x$. Now pick any continuous function f on $\{a_1, \cdots, a_k\}$, which is to say a k-tuple of real numbers $(f_1, \cdots f_k)$. We have to show that

$$\sum_{j=1}^{k} f_j x_j^n \longrightarrow \sum_{j=1}^{k} f_j x_j,$$

where x_j^n is the j-th coordinate of x^n. But this assertion is obvious given our hypothesis. ∎

We leave it to the reader to show

Remark 2 *Let $\psi : A \longrightarrow \mathbb{R}^k$ be a mapping such that for any $i \in (1, \cdots, k)$, $\psi(a_i) = e_i$ where e_i is the ith unit vector in $\mathbb{R}^k$. Then ψ is an embedding of A endowed with its discrete topology which is to say that there is a homeomorphism between A and $\psi(A)$ endowed with its relativized Euclidean topology.*

Remark 3 *If two compact topological spaces are homeomorphic, then the spaces of real-valued continuous functions on them, endowed with the sup-norm topology, are homeomorphic.*

Let $\mathcal{U}_A^{\Delta}$ and $\mathcal{U}_A^{M}$ be the spaces of real-valued continuous functions on $A \times \Delta_k$, and on $A \times \mathcal{M}(A)$, each endowed with the sup-norm topology. By Remarks 1 and 3, these two spaces are homeomorphic, and we shall move between them without warning. Because of this, we shall also obscure the notational distinction and denote either of them by $\mathcal{U}_A$. We shall also identify A with the set $\{e_1, \cdots, e_k\}$.

3. Two Benchmarks

Theorem 1 [**Schmeidler**] *Let* $(T, \mathcal{T}, \lambda)$ *be an atomless probability space and* $\mathcal{G}$ *a measurable map from* T *to* $\mathcal{U}_A$ *with the Borel* σ*-algebra* $\mathcal{B}(\mathcal{U}_A)$. *Then there exists a measurable function* $f : T \longrightarrow \mathbb{R}^k$; $f(t) \in \{e_1, \cdots, e_k\}$ *such that for* λ*-almost all* $t \in T$,

$$u_t(f(t), \int_{t \in T} f(t) d\lambda) \geq u_t(e_i, \int_{t \in T} f(t) d\lambda), \ i = 1, \cdots, k,$$

where $u_t = \mathcal{G}(t) \in \mathcal{U}_A$.

We remind the reader that Schmeidler states his theorem in the setting of the Lebesgue measure space $[0, 1]$. As we emphasized in the introduction this is of no essential consequence for the points that we want to make here.

Theorem 2 [**Mas – Colell**] *Let* μ *be an atomless probability measure on* $(\mathcal{U}_A, \mathcal{B}(\mathcal{U}_A))$. *Then there exists* $\tau \in \mathcal{M}(\mathcal{U}_A \times A)$ *with marginal measures* τ_A *and* $\tau_{\mathcal{U}_A}$ *such that*
(i) $\tau_{\mathcal{U}_A}$ *is* μ,
(ii) $\tau(B_\tau) = \tau(\{(u, a) \in (\mathcal{U}_A \times A) : u(a, \tau_A) \geq u(x, \tau_A)$ *for all* $x \in A\}) = 1$,
(iii) there exists a measurable function $h : \mathcal{U}_A \longrightarrow A$ *with* $\tau(Graph_h) = \tau(\{(u, h(u)) \in (\mathcal{U}_A \times A) : u \in \mathcal{U}_A\}) = 1$.

In the sequel, we shall refer to Theorems 1 and 2 as Theorems S and M respectively.

4. Some Useful Notation

Let $(X, \mathcal{X})$ and $(Y, \mathcal{Y})$ be measurable spaces. For any $f : X \longrightarrow Y$ a $(\mathcal{X}, \mathcal{Y})$-measurable map, let

$$f_* : \mathcal{M}(X) \longrightarrow \mathcal{M}(Y) \text{ such that } f_*(\mu)(B) = \mu(f^{-1}(B)), \ \ \forall B \in \mathcal{Y}, \ \forall \mu \in \mathcal{M}(X),$$

where $\mathcal{M}(X)$ and $\mathcal{M}(Y)$ are the space of all probability measures on $(X, \mathcal{X})$ and $(Y, \mathcal{Y})$ respectively.

Let $(Z, \mathcal{Z})$be another measurable space and $g : Y \longrightarrow Z$ a $(\mathcal{Y}, \mathcal{Z})$-measurable map. Then

$$(g \circ f)_* = g_* \circ f_* \text{ where } (g \circ f)_* : \mathcal{M}(X) \longrightarrow \mathcal{M}(Z).$$

This follows simply from the facts that for any $\lambda \in \mathcal{M}(X)$, and any $B \in \mathcal{Z}$,

$$(g \circ f)_*(\lambda)(B) = \lambda((g \circ f)^{-1}(B)) = \lambda(f^{-1}(g^{-1}(B))) = f_*(\lambda)(g^{-1}(B)) = (g_* \circ f_*)(\lambda)(B).$$

5. Theorem M as a Consequence of Theorem S

We shall prove the theorem as a consequence of Theorem S. Towards this end, consider the space $(\mathcal{U}_A, \mathcal{B}(\mathcal{U}_A), \mu)$ also as the space of players' names. It is certainly an atomless probability space. Let i be the identity map on $\mathcal{U}_A$ – it stands for the mapping $\mathcal{G}$ in Theorem S and is certainly measurable. Then Theorem S applies and we can assert the existence of a measurable function $f : \mathcal{U}_A \longrightarrow \mathbb{R}^k$; $f(u) \in \{e_1, \cdots, e_k\}$, such that for μ-almost $u \in \mathcal{U}_A$,

$$u(f(u), \int_{u \in \mathcal{U}_A} f(u) d\mu) \geq u(e_i, \int_{u \in \mathcal{U}_A} f(u) d\mu), \; i = 1, \cdots, k.$$

Certainly $\int_{u \in \mathcal{U}_A} f(u) d\mu \in \Delta_k$ and therefore by the definition of ϕ in the proof of Remark 1, $\phi(\int_{u \in \mathcal{U}_A} f(u) d\mu) \in \mathcal{M}(A)$ and such that $\phi(\int_{u \in \mathcal{U}_A} f(u) d\mu)) = f_*(\mu)$.

Since the map $(i, f) : \mathcal{U}_A \longrightarrow (\mathcal{U}_A \times A)$ is measurable, we can consider $\tau \equiv (i, f)_*(\mu)$. μ is a probability measure, and therefore τ is a probability measure. We shall show that τ satisfies all of the conditions required by Theorem M. First,

$$\begin{aligned} \tau(\mathrm{Graph}_f) &= \tau(\{(u,a) \in \mathcal{U}_A \times A : a = f(u), u \in \mathcal{U}_A\}) = \tau(\{(u, f(u)) : u \in \mathcal{U}_A\}) \\ &= (i,f)_*(\mu)((i,f)(\mathcal{U}_A)) = \mu((i,f)^{-1}((i,f)(\mathcal{U}_A))) = \mu(\mathcal{U}_A) = 1. \end{aligned}$$

Next, we show that

$$\tau_{\mathcal{U}_A} = (\mathrm{proj}_1)_*(\tau) = (\mathrm{proj}_1)_*((i,f)_*(\mu)) = (\mathrm{proj}_1 \circ (i,f))_*(\mu) = i_*(\mu) = \mu.$$

Third, we observe that

$$\tau_A = (\mathrm{proj}_2)_*(\tau) = (\mathrm{proj}_2)_*((i,f)_*(\mu)) = (\mathrm{proj}_2 \circ (i,f))_*(\mu) = f_*(\mu).$$

Finally, we observe that there exists $\mathcal{U}_A^F \in \mathcal{B}(\mathcal{U}_A), \mu(\mathcal{U}_A^F) = 1$, such that

$$\begin{aligned} B_\tau = B_{\tau_A} = B_{f_*(\mu)} &= \{(u,a) \in (\mathcal{U}_A \times A) : u(a, f_*(\mu)) \geq u(x, f_*(\mu)) \text{ for all } x \in A\} \\ &\supseteq \{(u, f(u)) : u \in \mathcal{U}_A^F\} = (i,f)(\mathcal{U}_A^F), \end{aligned}$$

which implies that $(i,f)^{-1}(B_\tau) \supseteq (i,f)^{-1}((i,f)\mathcal{U}_A^F)) \supseteq \mathcal{U}_A^F$, and hence we obtain

$$\tau(B_\tau) = (i,f)_*(B_\tau) = \mu((i,f)^{-1}(B_\tau)) \geq \mu(\mathcal{U}_A^F) = 1.$$

Since τ is a probability measure, we are done. ∎

As Rath's (1991) work makes clear, if Theorem S is set on the Lebesgue measure space $[0,1]$ instead of $(T, \mathcal{T}, \lambda)$, we need to work with the random variable furnished by Skorokhod's lemma rather than with the identity map.

6. Theorem S as a Consequence of Theorem M

We shall need a preliminary result concerning the so-called *universality* property of atomless probability spaces; see Gabszewicz-Mertens (1971; Lemma 1) for a statement and a proof based on Caratheodory's theorem on the extension of measures.

Lemma 1 *Let $(T, \mathcal{T}, \lambda)$ be an atomless probability space. Then there exists a measurable function $g : T \longrightarrow [0,1]$ which induces the Lebesgue measure.*

We now turn to the proof of Theorem S. Since $\mathcal{G}$ is measurable, we can consider the probability measure $\mu \equiv \mathcal{G}_*(\lambda)$. It is clear that inspite of λ being atomless, μ is not necessarily atomless. However, since μ is a probability measure, the number of its atoms is at most countable. Denote them by $\{u_i\}_{i\in I}$, where I is a countable index set. Let

$$T^i \equiv \mathcal{G}^{-1}(\{u_i\}) \quad \text{and} \quad T^0 \equiv T/\cup_{i\in I} T^i.$$

For all $i \in I$, let $\mu_i = \mu(\{u_i\})$. It is clear that $\lambda(T^i) = \mu_i$. Now consider the measure space $([1, 1+\mu_i], \nu)$, where ν denotes the Lebesgue measure on the corresponding interval. Lemma 1 yields the existence of a measure-preserving mapping

$$f_i : (T^i, \mathcal{T} \cap T^i, (\lambda \mid_{T^i})) \longrightarrow ([1, 1+\mu_i], \nu).$$

Now define a mapping $\mathcal{F}$ from T to $\mathcal{U}_A$ such that

$$\mathcal{F}(t) = \begin{cases} \mathcal{G}(t) \text{ for all } t \in T^0 \\ f_i(t)\mathcal{G}(t) \text{ for all } t \in T^i,\ i \in I. \end{cases}$$

Certainly, $\mathcal{F}$ is a measurable mapping with respect to the relevant σ-algebras. Also, by construction $\mu^{\mathcal{F}} \equiv \mathcal{F}_*(\lambda)$ is an atomless measure.

The hypotheses of Theorem M are fulfilled, and we can now apply it to assert the existence of a probability measure $\tau \in \mathcal{M}(\mathcal{U}_A \times A)$ and a measurable function $h : \mathcal{U}_A \longrightarrow A$ such that $\tau_A = h_*(\mu^{\mathcal{F}})$, and that $\mu^{\mathcal{F}}$-almost everywhere in $\mathcal{U}_A$,

$$u(h(u), \tau_A) \geq u(x, \tau_A) \text{ for all } x \in A.$$

But this means that for λ-almost all $t \in T$,

$$\mathcal{F}(t)(h(\mathcal{F}(t)), \tau_A) \geq \mathcal{F}(t)(x, \tau_A) \text{ for all } x \in A.$$

For all $t \in T$, let $g(t) = h(\mathcal{F}(t))$. Then

$$\tau_A = h_*(\mu^{\mathcal{F}}) = h_*\mathcal{F}_*(\lambda) = (h\mathcal{F})_*(\lambda) = g_*(\lambda).$$

Now for each $i \in I$ and for λ-almost all $t \in T^i$

$$f_i(t)\mathcal{G}(t)(g(t), g_*(\lambda)) \geq f_i(t)\mathcal{G}(t)(x, \tau_A) \text{ for all } x \in A.$$

Since for all $i \in I$, and for all $t \in T^i$, $f_i(t) > 0$, we obtain for λ-almost everywhere in T,

$$\mathcal{G}(t)(g(t), g_*(\lambda)) \geq \mathcal{G}(t)(x, g_*(\lambda)) \text{ for all } x \in A.$$

On appealing to Remark 1, we observe that $g_*(\lambda) = \int_{t \in T} g(t) d\lambda$, and complete the proof. ∎

It is worth pointing out that if Theorem S is set on the Lebesgue measure space $[0,1]$ instead of $(T, \mathcal{T}, \lambda)$, it can be derived from Theorem M even without the recourse to Lemma 1.

7. Anonymous Games with Essential and Dummy Characteristics

We present a straightforward generalization of Theorem M in which the characteristics space is a product of the space $\mathcal{U}_A$, as in Theorem M, and of an abstract probability space.

Theorem 3 *Let μ be an atomless probability measure on $(\mathcal{U}_A \times T, \mathcal{B}(\mathcal{U}_A) \otimes \mathcal{T})$. Then there exists $\tau \in \mathcal{M}((\mathcal{U}_A \times T) \times A)$such that*
(i) $\tau_{(\mathcal{U}_A \times T)}$ is μ,
(ii) $\tau(B_\tau) = \tau(\{((u,t),a) \in ((\mathcal{U}_A \times T) \times A) : u(a, \tau_A) \geq u(x, \tau_A)$ for all $x \in A\}) = 1$,
(iii) there exists a measurable function $h : \mathcal{U}_A \times T \longrightarrow A$ with $\tau(Graph_h) = \tau(\{((u,t), h(u,t)) \in ((\mathcal{U}_A \times T) \times A) : (u,t) \in (\mathcal{U}_A \times T)\}) = 1$, and with $\tau_A = h_*(\mu)$.

7.1 Theorem 3 as a Consequence of Theorem S

Let proj_1 be the first projection map on $\mathcal{U}_A \times T$ – it stands for the mapping $\mathcal{G}$ in Theorem S and is certainly measurable. Since $(\mathcal{U}_A \times T, \mathcal{B}(\mathcal{U}_A) \otimes \mathcal{T}, \mu)$ is an atomless probability, Theorem S applies to furnish a measurable function $f : \mathcal{U}_A \times T \longrightarrow \mathbb{R}^k$; $f(u,t) \in \{e_1, \cdots, e_k\}$, such that for μ-almost all $(u,t) \in \mathcal{U}_A \times T$,

$$u(f(u,t), \int_{(u,t)\in(\mathcal{U}_A \times T)} f(u,t) d\mu) \geq u(e_i, \int_{(u,t)\in(\mathcal{U}_A \times T)} f(u,t) d\mu), \; i = 1, \cdots, k.$$

Certainly $\int_{(u,t)\in(\mathcal{U}_A \times T)} f(u,t) d\mu \in \Delta_k$ and by the definition of ϕ in the proof of Remark 1, $\phi(\int_{(u,t)\in(\mathcal{U}_A \times T)} f(u,t) d\mu)) \in \mathcal{M}(A)$ and such that $\phi(\int_{(u,t)\in(\mathcal{U}_A \times T)} f(u,t) d\mu$ $f_*(\mu)$.

Since the map $(i, f) : (\mathcal{U}_A \times T) \longrightarrow ((\mathcal{U}_A \times T) \times A)$ is measurable, we can consider $\tau \equiv (i, f)_\star(\mu)$, a probability measure. We now leave it to the reader to show, by substituting $(\mathcal{U}_A \times T)$ and $f(u,t)$ for $\mathcal{U}_A$ and $f(u)$ respectively in the Proof of Theorem M, that τ satisfies all of the conditions required by Theorem 3. ∎

7.2 Theorem S as a Consequence of Theorem 3

Consider the mapping $(\mathcal{G}, i) : T \longrightarrow (\mathcal{U}_A \times T)$ with $(\mathcal{U}_A \times T)$ equipped with the product σ-algebra $\mathcal{B}(\mathcal{U}_A) \otimes \mathcal{T}$. It is easy to show that $(\mathcal{G}, i)$ is measurable, and that the measure $\mu \equiv (\mathcal{G}, i)_\star(\lambda)$ is an atomless probability measure. Then Theorem 3 furnishes us with a measureable function $h : \mathcal{U}_A \times T \longrightarrow A$ and a μ-null set $B \in \mathcal{B}(\mathcal{U}_A) \otimes \mathcal{T}$ such that for all $(u,t) \in (\mathcal{U}_A \times T/B)$,

$$u(h(u,t), h_\star(\mu)) \geq u(x, h_\star(\mu)) \text{ for all } x \in A.$$

Let $f = h \circ (\mathcal{G}, i)$. We can now assert that for λ-almost all $t \in T$,

$$\mathcal{G}(t)(f(t), \int_{t\in T} f(t)d\lambda) \geq \mathcal{G}(t)(e_i, \int_{t\in T} f(t)d\lambda),\ i = 1, \cdots, k.$$

Suppose not; then there exists $K \in \mathcal{T}, \lambda(K) > 0$ such that

$$\mathcal{G}(t)(x, \int_{t\in T} f(t)d\lambda) > \mathcal{G}(t)(f(t), \int_{t\in T} f(t)d\lambda) \text{ for some } x \in A, \text{ and for all } t \in K.$$

By Remark 1, this implies

$$\mathcal{G}(t)(x, f_\star(\lambda)) > \mathcal{G}(t)(f(t), f_\star(\lambda)) \text{ for some } x \in A, \text{ and for all } t \in K.$$

But $f = h \circ (\mathcal{G}, i)$ implies $f_\star(\lambda) = (h \circ (\mathcal{G}, i))_\star(\lambda) = h_\star \circ (\mathcal{G}, i)_\star(\lambda) = h_\star(\mu)$, and hence we can rewrite the above expression as

$$\mathcal{G}(t)(x, h_\star(\mu)) > \mathcal{G}(t)(h(\mathcal{G}(t), t), h_\star(\mu)) \text{ for some } x \in A, \text{ and for all } t \in K.$$

On substituting u for $\mathcal{G}(t)$, and on noting that $B \supseteq (\mathcal{G}, i)(K)$ and $(\mathcal{G}, i)^{-1}(B) \supseteq K$, we have

$$\mu(B) = \lambda((\mathcal{G}, i)^{-1}(B)) \geq \lambda(K) > 0,$$

which contradicts the fact that B is a μ-null set. ∎

8. A Note on Existence Proofs

In 1965, Aumann developed a theory of integration of correspondences which take their values in a finite dimensional Euclidean space. Thus, for a correspondence $F : T \longrightarrow \mathbb{R}^k_+$ on an atomless probability space $(T, \mathcal{T}, \lambda)$, Aumann sharpened the Lyapunov-Richter theorem on the convexity of the integral to the assertion that

$$\int_{t\in T} co(F(t))d\lambda = \int_{t\in T} F(t)d\lambda,$$

where $co(A)$ is the convex hull of the set A. Furthermore, for a correspondence $F : T \times X \longrightarrow \mathbb{R}^k$ on a probability space $(T, \mathcal{T}, \lambda)$, and on a metric space X, he also showed that the upper semicontinuity of $F(t, \cdot)$ on X for λ-almost all t in T implies the upper semicontinuity of the integral on X; see Aumann (1965, 1976) for details.

As is well known, Schmeidler proved his theorem on the existence pure strategy Cournot-Nash equilibria in atomless non-anonymous games by "purifying" mixed strategy equilibria obtained in a more general set-up; see his Theorem 1. For this purification, Schmeidler used the theory of integration as developed by Aumann, and in particular the two theorems quoted above. In 1991, Rath furnished a direct proof of Schmeidler's result based on the integral of the correspondence

$$F : T \times \Delta_k \longrightarrow \Delta_k \text{ where } F(t,x) = \text{Argmax}_{a\in A} u_t(a,x),$$

and A is identified, as in Section 2 onwards, with $\{e_1, \cdots, e_k\}$. Rath looks for, and finds through the use of Kakutani's theorem, a fixed point of the correspondence

$$\phi : \Delta_k \longrightarrow \Delta_k \text{ where } \phi(x) = \int_{t\in T} F(t,x)d\lambda.$$

In addition to Aumann's measurable selection theorem, and the Lyapunov-Richter convexity theorem, Rath also needed the result on the upper semicontinuity of the integral. Because of Rath's work, we now have both direct and indirect proofs of Theorem S.

In 1984, Mas-Colell had already based his direct proof of the existence of symmetric Cournot-Nash equilibrium distributions on the integral of the correspondence

$$G : \mathcal{U}_A \times \Delta_k \longrightarrow \Delta_k \text{ where } G(u,x) = \text{Argmax}_{a\in A} u(a,x),$$

and A is again identified with the set $\{e_1, \cdots, e_k\}$. Mas-Colell also finds through the use of Kakutani's theorem, a fixed point of the correspondence

$$\phi : \Delta_k \longrightarrow \Delta_k \text{ where } \phi(x) = \int_{u\in \mathcal{U}_A} G(u,x)d\mu,$$

and in this, he too has to rely on the full power of the integration theory developed by Aumann.

In 1987, the authors showed that every Cournot-Nash equilibrium distribution of an atomless finite action game can be symmetrized without appealing to any results on integration. In this way, they could derive Mas-Colell's result by "symmetrizing" a Cournot-Nash equilibrium distribution obtained in a more general set-up, as for example in Mas-Colell's Theorem 1. Thus, we now have both direct and indirect proofs of Theorem M.

In the context of this expository paper, all that remains is a proof of Theorem 3. We leave it to the reader to provide a proof based, for example, on a suitable modifications of the mappings considered in Mas-Colell (1984).

9. A Countably Infinite Action Space

We now turn to the case where A is a countably infinite compact metric space. We shall limit ourselves to a few cursory remarks that lay the background for the principal result in Khan-Sun (1993).

To begin with, it is clear that a naive approach in which the set $\{e_1, \cdots, e_k, \cdots\}$ is substituted for the set $\{e_1, \cdots, e_k\}$ in the statement of Theorem S will not work. For one thing, one has to specify the ambient space in which the set $\{e_1, \cdots, e_k, \cdots\}$ is situated. Since such a space is necessarily infinite dimensional, one is directly led to integrals over an infinite-dimensional space. It is a well known fact that the Lyapunov-Richter theorem on the convexity of the integral does not obtain for a general infinite dimensional space, and one is then led to approximate Cournot-Nash equilibria, as in Khan (1986).

On the other hand, the statement of Theorem M with $\{e_1, \cdots, e_k, \cdots\}$ substituted for $\{e_1, \cdots, e_k\}$ has meaning, at least as a conjecture. Here again a naive approach to existence will not work. If, for example, in which the set $\{e_1, \cdots, e_k, \cdots\}$ is embedded in the Hilbert space ℓ^2, we lose compactness in the norm topology of the action set, and even weak closedness if we choose to work in the weak topology. Thus the space of probability measures on A is no longer compact, and hence precludes the use of the standard fixed point theorems.

In recent work, the authors have reported the existence of symmetric Cournot-Nash equilibrium distributions for atomless games with any countable compact metric action space. (Note that such a space could be much more complicated than the set $\{e_1, \cdots, e_k, \cdots\} \cup \{0\}$ endowed with the weak topology on ℓ^2). They proceed through "symmetrization," but since the set of all subsets of a countable action space is no longer countable, they cannot appeal to their earlier elementary (but nevertheless measure-theoretic) argument. Instead, they observe that a Cournot-Nash equilibrium distribution is symmetric if and only if it is an extreme point of a particular set of measures with given marginals. This observation then brings the Krein-Milman theorem directly to the fore, and allows a satisfactory extension of the results; see Khan-Sun (1993) for details.

10. Concluding Remark

We conclude by pointing out that even in the case of a finite number of actions, the two formulations of large games are not necessarily equivalent if the payoffs in the non-anonymous setting depend on the average responses taken over particular coalitions of players, rather than on society as a whole; see the remark in Schmeidler's paper, and also a development in Chakrabarti-Khan (1991).

References

[1] R. J. Aumann, *Integrals of set valued functions,* Journal of Mathematical Analysis and Applications, **12** (1965), 1–12.

[2] R. J. Aumann, *An elementary proof that integration preserves uppersemicontinuity,* Journal of Mathematical Economics, **3** (1976), 15–18.

[3] S. K. Chakrabarti and M. Ali Khan, *Equilibria of large games with imperfect observability*, in C. D. Aliprantis, K. C. Border, W. A. J. Luxemburg (eds.), *Positive Operators, Reisz Spaces and Economics* (Berlin: Springer-Verlag), 1991.

[4] R. G. Douglas, *On extremal measures and subspace density*, Michigan Mathematics Journal, **11** (1964), 243–246.

[5] J. J. Gabszewicz, and J. F. Mertens, *An equivalence theorem for the core of an economy whose atoms are not "too" big*, Econometrica, **5** (1971), 713–721.

[6] R. E. Jamison, *A quick proof for a one-dimensional version of Liapunov's theorem*, American Mathematical Monthly, **81** (1974), 507–508.

[7] M. Ali Khan, *Equilibrium points of nonatomic games over a Banach space*, Transactions of the American Mathematical Society, **293** (1986), 737–749.

[8] M. Ali Khan, *On Cournot-Nash equilibrium distributions for games with a nonmetrizable action space and upper semicontinuous payoffs*, Transactions of the American Mathematical Society, **315** (1989), 127–146.

[9] M. Ali Khan and Y. Sun, *On symmetric Cournot-Nash equilibrium distributions in a finite-action atomless game*, B. E. B. R. Faculty Working Paper No.1327, February 1987. Reprinted in M. Ali Khan and N. C. Yannelis (eds.), *Equilibrium Theory in Infinite Dimensional Spaces* (Berlin: Springer-Verlag), 1991.

[10] M. Ali Khan and Y. Sun, *Extremal structures and symmetric equilibria with countable actions*, Johns Hopkins Working Paper No. 316, 1993, forthcoming in Journal of Mathematical Economics.

[11] J. Lindenstrauss, *A remark on extreme doubly stochastic measures*,American Mathematical Monthly, **72** (1965), 379–382.

[12] A. Mas-Colell, *An axiomatic approach to the efficiency of non-cooperative equilibrium in economies with a continuum of traders*, Stanford IMSSS Technical Report 2744., 1978. Reprinted in M. Ali Khan and N. C. Yannelis (eds.), *Equilibrium Theory in Infinite Dimensional Spaces* (Berlin: Springer-Verlag), 1991.

[13] A. Mas-Colell, *On a theorem of Schmeidler*, Journal of Mathematical Economics **13**, (1984), 206–210.

[14] K. Rath, *Representation of finite action large games, mimeo*, University of Notre Dame, 1991. International Journal of Game Theory, forthcoming.

[15] K. Rath, *A direct proof of the existence of pure strategy equilibria in games with a continuum of players*, Economic Theory, **2** (1992), 427–433.

[16] D. Schmeidler, *Equilibrium points of non-atomic games*,Journal of Statistical Physics, **7** (1973), 295–300.

[17] S. Yamashige, *Large games and large economies with incomplete information*, unpublished Ph. D dissertation, The Johns Hopkins University (1992).

[18] M. Yor, *Sous-espaces denses dans L^1 on H^1 et représantation des martingales,* Séminaire de probabilités XII, Lecture Notes in Mathematics 649, (New York: Springer-Verlag.), 1978.

On a Method Constructing Morse Flows

Norio Kikuchi
Department of Mathematics, Faculty of Science and Technology
Keio University, Hiyoshi, Yokohama 223, JAPAN

Abstract

By enumerating a treatment of a variational problem of harmonic type in illustration, we shall propose a method constructing Morse flows of variational problems.

Introduction

Around 1980, there appeared a nice result [1] due to M.Giaquinta and E.Giusti, which has succeeded in dealing with minimizers of variational problems of harmonic map type. The paper has derived Caccioppoli estimates for the minimizers and accomplished the higher integrability, in the sense of Gehring ([2,3,4]), of the gradients of the minimizers, which enables one to take one step toward studying the regularlity of the minimizersof the variational problem.

If one wants to make a trial to construct Morse flows - gradient flows - to the variational problem, there will turn up a difficulty to be surmounted, because it is not apriori estimates of all critical points but estimates only for the minimizers that the paper has successfully handled. For harmonic maps and related heat flows, refer to the papers [5,6,7,8,9,10,11,12,13,14,15].

To overcome the difficulty, we make a proposal of the scheme([16]): Beginning with an initial datum, we introduce a family of variational functionals associated to the variatioal problem, where the Euler-Lagrange equations of the introduced variational funtionals are of the Rothe's approximate equations([17]) of parabolic partial differential equations generating Morse flows for the variational problem.

By making use of a minimizer of each of the functionals, we construct an approximate, discrete, Morse flow, for which we establish local estimates, say, Caccioppoli estimates, thanks to the property of the minimizer.

Following the scheme thus presented and trying to construct Morse flows necessitate the establishment of the regularity theory of solutions to difference partial differential equations satisfied by the minimizers. There seems to exist an uncultivated field, on the regurarity of solutions to such discrete equations, to be refined in the future.

Some contrives would enable us to make the method, described in this paper, applicable to other variational problems. The authors are preparing for completing the papers [18,19] which treat a construction of Morse flows to such a variational problem on one-sided condition as treated in the papers [1,5].

The Rothe's approximates to the wave equations have a family of the canonical variational functionals. However, the author has no idea whether the minimality will play an essential role for the regularity problem of solutions to the wave equations. Rothe's approach works in the search of regular solutions to wave equations associated to the variational problem to be considered in this paper ([20]).

By way of the scheme adopted in this paper, there have been made some trials to pursue the evolution of the regularity or the singularlity of intial data along discrete Morse flows ([21]).

1. A scheme of constructing Morse flows and main results

Let Ω be a bounded domain in $\mathbf{R}^m, m \geq 2$, with C^2-boundary $\partial\Omega$. Maps mean the one from Ω to $\mathbf{R}^M, M \geq 1$. In the Sobolev space $H^1(\Omega) = H^1(\Omega, R^M)$, we consider the variational problem (P) with the functional

$$F(u) = \int_\Omega A_{ij}^{\alpha\beta}(x) D_\alpha u^i(x) D_\beta u^j(x) dx, \tag{P}$$

where $u = (u^i), D_\alpha u^i = \partial u^i / \partial x^\alpha (i = 1, \ldots, M; \alpha = 1, \ldots, m)$. The summation convention is used, the Greek indices running from 1 to m and the Latin ones from 1 to M.

The coefficients $A_{ij}^{\alpha\beta}$ are assumed to be measurable and bounded in Ω with a positive L:

$$|A_{ij}^{\alpha\beta}(x)| \leq L \quad \text{for } x \in \Omega$$

and to satisfy an ellipticity condition with a positive λ:

$$A_{ij}^{\alpha\beta}(x) \xi_\alpha^i \xi_\beta^j \geq \lambda \mid \xi \mid^2 \quad \text{for } \xi = (\xi_\alpha^i) \in \mathbf{R}^{mM} \quad \text{and } x \in \Omega.$$

'Morse flows' of variational functional F are defined as 'solutions' of parabolic partial differential equations

$$\frac{\partial u^i}{\partial t} = \frac{1}{2} D_\alpha ((A_{ij}^{\alpha\beta} + A_{ji}^{\beta\alpha}) D_\beta u^j) \quad (i = 1, \ldots, M).$$

By a Morse flow to variational problem (P), we mean a mapping $u \in L^\infty((0,T), H^1(\Omega)) \cap H^1((0,T), L^2(\Omega))$ which satisfies

$$\int_Q \frac{\partial u^i}{\partial t}\varphi^i dz + \frac{1}{2}\int_Q (A_{ij}^{\alpha\beta}(x) + A_{ji}^{\beta\alpha}(x)) D_\alpha u^j D_\beta \varphi^i dz = 0$$
$$\text{for every } \varphi \in L^2((0,T), H_0^1(\Omega)).$$

Let u_0 be a given map belonging to $H^1(\Omega)$ and T a positive number. We take a positive integer N, $N > T$, arbitrarily and put

$$h = T/N \text{ and } t_n = nh \ (n = 0, 1, \ldots, N).$$

We use a function space

$$H^1_{u_0}(\Omega) = \{u \in H^1(\Omega); u - u_0 \in H_0^1(\Omega)\},$$

$H_0^1(\Omega)$ being the space obtained by taking the closure of $C_0^\infty(\Omega)$ in the space $H^1(\Omega)$.

Given u_0, we inductively construct two sets of maps u_n and functionals F_n $(n = 1, \ldots, N)$ by the scheme: For each n $(n = 1, \ldots, N)$, we introduce a functional

$$F_n(u) = F(u) + \int_\Omega \frac{1}{h} \mid u - u_{n-1} \mid^2 dx$$

and define u_n as a minimizer of F_n in $H^1_{u_0}(\Omega)$, the existence of which is assured by the lower semi-continuity of F_n with respect to weak convergence and by the coercivity of F_n in the space $H^1(\Omega)$ ([10]), $H^1_{u_0}(\Omega)$ being convex.

We remark that the Euler-Lagrange equations of F_n in $H^1_{u_0}(\Omega)$ are of the form: For n $(n = 1, \ldots, N)$,

$$\frac{u_n^i - u_{n-1}^i}{h} = \frac{1}{2} D_\alpha((A_{ij}^{\alpha\beta} + A_{ji}^{\beta\alpha}) D_\beta u_n^j) \ \ (i = 1, \ldots, M),$$

which are approximate equations of Rothe type for Euler-Lagrange equations of functional F ([17]).

Upon comparing u_{n-1} with a minimizer u_n of F_n, we infer

$$F(u_n) + \int_\Omega \frac{1}{h} \mid u_n - u_{n-1} \mid^2 dx \le F(u_{n-1})$$

and thus have the following result. Such estimates do work in dealing with other variational functionals F.

Theorem 0([16]). *Let u_n be a minimizer of F_n $(n = 1, 2, \ldots, N)$ in the space $H^1_{u_0}(\Omega)$. Then there hold the estimates*

$$\int_\Omega A_{ij}^{\alpha\beta}(x) D_\alpha u_n^i(x) D_\beta u_n^j(x) dx \le \int_\Omega A_{ij}^{\alpha\beta}(x) D_\alpha u_0^i(x) D_\beta u_0^j(x) dx$$

for any $n(n = 1, \dots, N)$ and

$$h\sum_{n=1}^{N}\int_{\Omega}\left|\frac{u_n - u_{n-1}}{h}\right|^2 dx \leq \int_{\Omega} A(x)(Du_0, Du_0)dx.$$

By an approximate Morse flow in $H^1(\Omega, R^M)$ to variational problem (P), we will mean a mapping $u_h(t) \in H^1_{u_0}(\Omega, R^M), 0 \leq t \leq T$, connected by the relation:

$$u_h(t) = \begin{cases} u_n & \text{for } t_{n-1} < t \leq t_n \ (n = 1, \dots, N), \\ u_0 & \text{for } -h \leq t \leq 0, \end{cases}$$

where $\{u_n\}(n = 1, 2, \dots, N)$ is a family of minimizerss of functionals F_n. Put

$$\partial_t u_h(t) = \frac{1}{h}(u_n - u_{n-1}) \quad \text{for } t_{n-1} < t \leq t_n \ (n = 1, \dots, N)$$

and

$$\tilde{u}_h(t) = u_h(t - h) \quad \text{for } 0 \leq t \leq T.$$

To state main results, we shall prepare the natations: $Q = (0, T) \times \Omega$. For $z_0 = (t_0, x_0) \in Q$ and positive number s, put

$$B_s(x_0) = \{x \in \Omega; \mid x - x_0 \mid < s\}, \ \ Q_s(z_0) = \{t \in (0, T); \ t_0 - s^2 < t < t_0\} \times B_s(x_0).$$

For $u \in L^1(Q)$ and a measurable set $A \subset Q$, put

$$\fint_A u(z)dz = \frac{1}{|A|}\int u(z)dz, \ |A| = \text{ Lebesgue measure of } A.$$

For a real number a, $[a]$ denotes the largest integer not exceeding the number a, viz., Gauss's symbol.

The following is an estimate, in local domains, for an approximate Morse flow satisfying difference partial differential equations of elliptic-parabolic type.

Theorem 1. *Let u_h be an approximate Morse flow to variational problem (P). Then there exist positive ε and C depending on m, M, λ and L such that*

$$\Big(\fint_{Q_{r/4}(z_0)} \mid Du_h \mid^{2+\varepsilon} dz\Big)^{1/(2+\varepsilon)} \leq C\Big(\fint_{Q_r(z_0)} \mid Du_h \mid^2 dz\Big)^{1/2}$$

$$+Ch^{(\bar{p}-1)/2}\Big(\fint_{Q_r(z_0)} \mid \partial_t u_h \mid^{(1+\varepsilon/2)\bar{p}} \mid u_h - \tilde{u}_h \mid^{(1+\varepsilon/2)(2-\bar{p})} dz\Big)^{1/(2+\varepsilon)}$$

holds for any $Q_r(z_0) \subset Q$, $r < 2$, and $1 < \bar{p} < 2$.

Noting that the estimates in Theorem 0 and 1 are valid uniformly in h, we have an existence theorem of Morse flows, with the higher integrability of those gradients, to (P).

Theorem 2. *For any $u_0 \in H^1(\Omega)$, there exists a Morse flow u to variational problem (P) satisfying an initial and boundary condition in the sense:*

$$u(t) \in H^1_{u_0}(\Omega) \quad \textit{for almost every } t \in (0,T)$$

and

$$\lim_{t \downarrow 0} u(t) = u_0 \quad \textit{in } L^2(\Omega).$$

The flow u satisfies the estimate:

$$\left(\fint_{Q_{r/2}(z_0)} | Du |^{2+\varepsilon} dz \right)^{1/(2+\varepsilon)} \leq C \left(\fint_{Q_r(z_0)} | Du |^2 dz \right)^{1/2}$$

for any $Q_r(z_0) \subset Q$, where C and ε are positive numbers depending on m, M, λ and L. Furthermore, if $A^{\alpha\beta}_{ij}$ are continuous in Ω, the solution u is of $C^\alpha(Q)$ with any exponent $0 < \alpha < 1$.

2. Fundamental estimates for approximate Morse flows

We introduce ([22]) such a family of functions $\{\zeta_n\}(n = 1, 2, \ldots, N)$, which is available as a cut-off function for such difference-differential equations as treated in this paper [18]: For fixed n_0 ($n_0 = 1, \ldots, N$) and positive numbers k, l satisfying $k < l$, we define ζ_n ($n = 1, 2, \ldots, N$) as follows:

$$\zeta_n = \zeta_{n,(k,l)} = \begin{cases} 1 & n_0 - [k^2/h] \leq n, \\ \dfrac{n - n_0 + [l^2/h] - 1}{[l^2/h] - 2 - [k^2/h]} & n_0 - [l^2/h] + 1 \leq n \leq n_0 - [k^2/h] - 1, \\ 0 & n \leq n_0 - [l^2/h]. \end{cases}$$

Clearly,

$$0 \leq \zeta_n - \zeta_{n-1} \leq 4h(l-k)^{-2} \ (n = 1, \ldots, N) \quad \text{for } 0 < k < l \ \text{ with } (l-k)^2 > 4h.$$

Let $x_0 \in \Omega$ and r be a positive number satisfying $B_r(x_0) \subset \Omega$ and $\eta \in C^\infty_0(B_r(x_0))$ a usual cut-off function, $0 \leq \eta \leq 1$, such that $\eta \equiv 1$ in $B_{r/2}(x_0)$. For mapping $u \in L^1(\Omega)$ and positive $s < r$, we put

$$\overline{u}_{s,x_0} = \int_{B_s} \eta u dx \Big/ \int_{B_s} \eta dx.$$

We recall the fundamental result due to Gehring ([2,3,4]). We, however, need generalize it so as to be applicable to difference-partial differential equations. See [19] in this connection.

Let $Q_R = Q_R(\overline{z}_0)$, $\overline{z}_0 \in Q$, be a local parabolic cylinder arbitrarily taken in Q.

Proposition 0([18]). *Let g be a nonnegative L^q-integrable function, $q > 1$, and f a nonnegative $L^{\tilde{q}}$-integrable function, $\tilde{q} > q$, defined in Q_R. If g satisfies the inequality, with positive b, c, R_0 and δ,*

$$\fint_{Q_r(z_0)} g^q dz \leq b(\fint_{Q_{4r}(z_0)} g dz)^q dz + c \fint_{Q_{4r}(z_0)} f^q dz + \delta \fint_{Q_{4r}(z_0)} g^q dz$$

for any $z_0 = (t_{n_0}, x_0) \in Q_R (n_0 = 1, \cdots, N)$ and $0 < r < \min\{dist(z_0, \partial Q_R)/4, R_0\}$, then there exist positive numbers β, γ, δ_0 and ε, $0 < \varepsilon \leq \tilde{q} - q$, depending only on q, $\tilde{q}$, b, c, δ, and m, such that, if $\delta < \delta_0$, then $g \in L^p_{loc}(Q_R)$ for $p \in [q, q + \varepsilon)$, and there holds, for any parabolic cylinder $Q_{4r}(z_0) \subset Q_R$, $z_0 = (t_{n_0}, x_0)$ $(n_0 = 1, \cdots, N)$, with $r < R_0$,

$$(\fint_{Q_r(z_0)} g^p dz)^{1/p} \leq \beta(\fint_{Q_{4r}(z_0)} g^q dz)^{1/q} + \gamma(\fint_{Q_{4r}(z_0)} f^p dz)^{1/p}.$$

For approximate Morse flows, the following are fundamental estimates.

Lemma 1 (Caccioppoli type estimate). *Let u_h be an approximate Morse flow to variational problem (P). Then there exists a positive constant C depending on λ and L such that*

$$\begin{aligned} &\textstyle\int_{Q_r(z_0)} | Du_h |^2 dz \\ &\leq Cr^{-2} \textstyle\int_{Q_{2r}(z_0)} | u_h - \overline{u}_{h,2r} |^2 dz + Ch^{\overline{p}-1} \textstyle\int_{Q_{2r}(z_0)} | \partial_t u_h |^{\overline{p}} | u_h - \tilde{u}_h |^{2-\overline{p}} dz \end{aligned}$$

holds for any $Q_{2r}(z_0) \subset Q$, $0 < r < 1/2$, with $z_0 = (t_{n_0}, x_0)$ $(n_0 = 1, \ldots, N)$ and for any $1 < \overline{p} < 2$. Furthermore, for each $1 < \overline{p} < 2$, $|\partial_t u_h|^{\overline{p}} |u_h - \tilde{u}_h|^{2-\overline{p}}$ belongs to $L^p(Q)$ for any $1 < p \leq m/(m - 2 + \overline{p})$.

Proof. Let k and l be positive numbers satisfying $r < k < l < 2r$. Let $\eta \in C_0^\infty(B_l(x_0))$ be a cut-off function such that

$$\eta \equiv 1 \text{ in } B_k(x_0) \text{ and } 0 \leq \eta \leq 1, | D\eta | \leq 2(l - k)^{-1}.$$

As a comparison map in functional F_n $(n = 1, \ldots, N)$, we adopt v_n defined by

$$v_n = u_n - h\eta(u_n - \overline{u}_{n,l}).$$

Since u_n is a minimizer of F_n in $H^1_{u_0}(\Omega)$, some calculations bring us in the estimates:

$$h \int_{B_k} | Du_n |^2 dx \leq Ch \int_{B_l \setminus B_k} | Du_n |^2 dx + C(l - k)^{-2} h \int_{B_l} | u_n - \overline{u}_{n,l} |^2 dx$$

$$+C\int_{B_l} | u_n - \overline{u}_{n,l} |^2 \, dx + C\int_{B_l} | u_n - u_{n-1} |^2 \, dx \tag{A}$$

and

$$h\int_{B_k} | Du_n |^2 \, dx \le Ch\int_{B_l\setminus B_k} | Du_n |^2 \, dx + C(l-k)^{-2}h\int_{B_l} | u_n - \overline{u}_{n,l} |^2 \, dx$$
$$-C\Big(\int_{B_l} \eta \, | u_n - \overline{u}_{n,l} |^2 \, dx - \int_{B_l} \eta \, | u_{n-1} - \overline{u}_{n-1,l} |^2 \, dx\Big). \tag{B}$$

To complete the estimation, we take the way of making a classification between $(l-k)^2$ and h ([22]): $(l-k)^2 \le 4h$ or $(l-k)^2 > 4h$.
We first treat the former case, in which case (A) leads to

$$\int_{Q_k} | Du_h |^2 \, dz \le C\int_{Q_l\setminus Q_k} | Du_h |^2 \, dz + C(l-k)^{-2}\int_{Q_l} | u_h - \overline{u}_{h,l} |^2 \, dz$$
$$+Ch^{\overline{p}-1}\int_{Q_l} | \partial_t u_h |^{\overline{p}} | u_h - \tilde{u}_h |^{2-\overline{p}} \, dz,$$

where $\overline{u}_l = \overline{u}_l(t)$ is a map defined by

$$\overline{u}_l(t) = \overline{u}_{n,l} \quad \text{for } t_{n-1} < t \le t_n \ (n = 1, \ldots, N).$$

We next deal with the latter case. Multiplying (B) by ζ_n, a cut-off function defined at the beginning of Section 2, and making summation over n from $n_0 - [l^2/h] + 1$ to n_0, we infer

$$\int_{Q_k} | Du_h |^2 \, dz \le C\int_{Q_l\setminus Q_k} | Du_h |^2 \, dz + C(l-k)^{-2}\int_{Q_{2r}} | u_h - \overline{u}_{h,2r} |^2 \, dz.$$

Gathering two estimates, achieved for each case of the classifications made as above, implies that

$$\int_{Q_k} | Du_h |^2 \, dz \le C\int_{Q_l\setminus Q_k} | Du_h |^2 \, dz + C(l-k)^{-2}\int_{Q_{2r}} | u_h - \overline{u}_{h,2r} |^2 \, dz$$
$$+Ch^{\overline{p}-1}\int_{Q_{2r}} | \partial_t u_h |^{\overline{p}} | u_h - \tilde{u}_h |^{2-\overline{p}} \, dz$$

holds for any $1 < \overline{p} < 2$ and $r < k < l < 2r$, $\tilde{u}$ being the translated mapping of u. The estimate, with the help of Hölder inequality and Sobolev's imbedding theorem, results in

$$\int_{Q_r} | \partial_t u_h |^{p\overline{p}} | u_h - \tilde{u}_h |^{p(2-\overline{p})} \, dz$$

$$\le Cr^{2(1-1/s)}\Big(\sup_{t\in(t_0-r^2-h,t_0)}\int_{\Omega} | Du_h |^2 \, dx\Big)^{p(2-\overline{p})/2}\Big(\int_{Q_r} | \partial_t u_h |^{p\overline{p}s} \, dz\Big)^{1/s}.$$

We take a positive p satisfying $1 < p \leq m/(m-2+\overline{p})$, which implies, with $s,t > 1, 1/s + 1/t = 1$ and $p(2-\overline{p}) = 2^*$:Sobolev index, that the inequality $p\overline{p}s \leq 2$ holds. Therefore, it follows from the estimate just established and Theorem 0 that $| \partial_t u_h |^{\overline{p}} | u_h - \tilde{u}_h |^{2-\overline{p}}$ belongs to $L^p(Q)$ with $1 < p \leq m/(m-2+\overline{p})$.

Lemma 2. *Let u_h be an approximate Morse flow to variational problem (P). Then there exists a positive number C depending on λ and L such that*

$$\sup_{n_0-[r^2/h]\leq n\leq n_0} \int_{B_r(x_0)} | u_n(x) - \overline{u}_{n,r,x_0} |^2 \, dx \leq C \int_{Q_{2r}(z_0)} | Du_h |^2 \, dz$$

holds for any $Q_{2r}(z_0) \subset Q$ with $z_0 = (t_{n_0}, x_0)$ $(n_0 = 1, \ldots, N)$.

For the proof, there are distinguished two possible cases: $4h \geq r^2$ or $4h \leq r^2$.
In the former case, we resort to Poincaré inequality and in the latter case we adopt a map: $v_n = u_n - h\eta(u_n - \overline{u}_{n,2r})$, η being a cut-off function defined at the first part of the proof of Theorem 1, as a comparison map with the minimizer u_n in functional F_n, so that the desired estimate is established.

3. Proof of theorems

In this section we give the proof of Theorem 1 and 2. Let u_h be an approximate solution to variational problem (P).

Proof of Theorem 1. Following the same calculation as in the proof of [7, Theorem 2.1], we obtain from Lemma 1 and 2 that

$$\fint_{Q_r} | Du_h |^2 \, dz \leq \theta \fint_{Q_{4r}} | Du_h |^2 \, dz + C(\theta)(\fint_{Q_{4r}} | Du_h |^q \, dz)^{2/q}$$
$$+ Ch^{\overline{p}-1} \fint_{Q_{4r}} | \partial_t u_h |^{\overline{p}} | u_h - \tilde{u}_h |^{2-\overline{p}} \, dz$$

holds for any $\theta > 0$ and $1 < \overline{p} < 2$, where $q, 1 < q < 2$, is the conjugate exponent to Sobolev index 2^*. By applying Proposition 0 to the estimates thus obtaind. we have the result of Theorem 1.

Proof of Theorem 2. The estimates, presented in Theorem 0, imply that there exists a Morse flow of (P) satisfying the desired initial and boundary condition. If $A_{ij}^{\alpha\beta}$ are continuous in Ω, we follow the proof of [9, Theorem 3.1] to obtain the regularity result, there being assured $\partial u/\partial t \in L^2(Q)$ by dint of estimate in Theorem 0. This concludes the proof of Theorem 2.

Remark. The author was recently informed of the paper [23] due to K. Rektorys, which has utilized the method, similar to the one developed in this

paper, for the purpose of constructing solutions to a higher order parabolic differential scalar equation, the term $|u - u_{n-1}|^2/h$ being replaced by $(u^2 - 2u_{n-1}u)/h$. In [23], there has been achieved such a 'global' estimate of the type as stated in Theorem 0, which has enabled the author to construct weak solutions of the linear equation.
The details related to this paper have been stated in the paper [24].

Acknowledgement

The author enjoys a blessing to refer to the paper [25] due to Professor Masuo Hukuhara, which has been one of the origins to the author's study. He also would like to express his gratitude to Professor Tohru Maruyama, Keio University, for his considerate encouragement.

References

[1] M. Giaquinta and E.Giusti, *On the regularity of the minima of variational integrals*, Acta Math. **148** (1982) 31–46.

[2] F. W. Gehring, *The L^p-integrability of the partial derivatives of a quasi conformal mapping*, Acta Math. **130** (1973) 265–277.

[3] M. Giaquinta and G. Modica, *Regularity results for some classes of higher order non linear elliptic systems*, J. Reine Angew. Math. **311/312** (1979) 145–169.

[4] M. Giaquinta, *Multiple integrals in the calculus of variations and non linear elliptic systems*, Vorlesungsreihe des Sonderforschungsbereiches **72,6** Universität Bonn 1981

[5] S. Hildebrandt, *Nonlinear elliptic systems and harmonic mapping*, Proe. Beijing Symp. Diff. Geom. Diff. Eq., Beijing (1980).

[6] S. Campanato, *Equazioni paraboliche del secondo ordine e spazi $L^{2,\theta}(\Omega, \delta)$*. Ann. Mat. Pura Appl. **73** (1966) 55–102.

[7] S. Campanato, *L^p Regularity for Weak Solutions of Parabolic Systems*, Ann. Scuola Norm. Sup. Pisa Cl. Sci. Ser. **4,7** (1980) 65–85.

[8] G. Da Prato, *Spazi $L^{(p,\theta)}(\Omega, \delta)$ e loro proprietá*, Ann. Mat. Pura Appl. **69** (1965) 383–392.

[9] M. Giaquinta and Giusti, E., *Partial regularity for the solutions to nonlinear parabolic systems*, Ann. Mat. Pura Appl. **47** (1973) 253–266.

[10] M. Giaquinta and M. Struwe, *On the partial regularity of weak solutions of non-linear parabolic systems*, Math. Zeit. **142** (1982) 473–451.

[11] J. Eells and J. H. Sampson, *Harmonic mappings of Riemannian manifolds*. Amer. J. Math. **86** (1964) 109–160.

[12] M. Struwe, *On the Hölder contunuity of bounded weak solutions of quasilinear parabolic systems*, manusc. math. **35** (1981) 125–145.

[13] J. Jost, *Ein Existenzbeweis für harmonische Abbildungen, die ein Dirichletproblem lösen, Mittels der Methode des Wärmeflusses*, manusc. math. **34** (1981) 17–25.

[14] M. Struwe, *On the evolution of harmonic maps in higher dimensions*, J. Diff. Geom. **28** (1988) 485–502.

[15] Y. Chen and M. Struwe, *Existence and partial regularity results for the heat flow for harmonic maps*, Math. Z. **20** (1989) 83–103.

[16] N. Kikuchi, *An approach to the construction of Morse flows for variational functionals*, in Nematics-Mathematical and Physical Aspects, ed. J-M.Coron, J-M.Ghidaglia and F.Hélein (NATO Adv. Sci. Inst. Ser.C: Math. Phys. Sci. **332** Kluwer Acad. Publ., Dordrecht-Boston-London) (1991) 195–198.

[17] E. Rothe, *Wärmeleitungsgleichung mit nichtkostanten Koeffizienten*, Math. Ann. **104** (1931) 340–362.

[18] J. Haga and N. Kikuchi, *On the higher integrability of the gradients of the silutions to difference partial differential systems of elliptic-parabolic typepreprint.*

[19] J. Haga and N. Kikuchi, *Costruction of Morse flows to variational problems of harmonic map type. preprint.*

[20] K. Hoshino and N. Kikuchi, *On a construction of weak solutions to linear hyperbolic partial differential equationspreprint.*

[21] M. Kashiwagi and N. Kikuchi, *Numerical calculations of Morse flows for variational functionspreprint.*

[22] N. Kikuchi, *Hölder estimates of solutions for difference-differntial equations of elliptic-parabolic type*, Reseach Report, Keio Univ. (1992) 1–23; to appear in J.Geom.Anal.

[23] K. Rektorys, *On application of direct variational methods to the solution of parabolic boundary value problems of arbitrary order in the space variables*, Czech. Math. J. 21 **96** (1971) 318–339.

[24] N. Kikuchi, *A method of constructing Morse flows to variational functionals*, to appear in Nonlinear World.

[25] M. Hukuhara, *Existence proof of solutions by Cauchy polygons*, F. E. **14** (1961) 265–272 (Japanese).

On Covering Theorems of a Simplex and Their Generalizations

Hidetoshi Komiya
Faculty of Business and Commerce, Keio University
Kouhoku-ku, Yokohama 223, JAPAN

Abstract

Since Scarf established non-emptiness of the core of a balanced nontransferable utility game in 1967, others have attempted to prove it in a simple way. Most of them reduced the problem to that of a covering of a simplex in the Euclidean space. Thus, we have obtained several covering theorems of a simplex in the course of the endeavor to prove the Scarf's theorem. We shall give a theorem unifying these covering theorems.

1. Introduction

Let $N = \{1, \ldots, n\}$ and let $\mathcal{N}$ be the family of all non-empty subsets of N. Let R^n be the n-dimensional Euclidean space with the inner product $\cdot$. Let $\{e^i : i \in N\}$ be the standard basis of R^n, that is, e^i is an n-vector whose i-th coordinate is 1 and 0 otherwise. Let Δ^N be the simplex $\text{co}\{e^i : i \in N\}$ and Δ^S the face of Δ^N spanned by the set $\{e^i : i \in S\}$, that is, $\Delta^S = \text{co}\{e^i : i \in S\}$, for any $S \in \mathcal{N}$.

A subfamily $\mathcal{B}$ of $\mathcal{N}$ is said to be *balanced* if there are nonnegative weights λ^S, $S \in \mathcal{B}$, such that $\sum_{S \in \mathcal{B}} \lambda^S e^S = e^N$, where e^S denotes the n-vector whose ith coordinate is 1 if $i \in S$ and 0 otherwise. Note that any partition of N is balanced. It is easily seen that $\mathcal{B}$ is balanced if and only if $m^N \in \text{co}\{m^S : S \in \mathcal{B}\}$, where m^S denotes the center of gravity of the face Δ^S, that is, $m^S = \sum_{i \in S} e^i / |S|$.

After Scarf[8] showed that every balanced nontransferable utility game has the non-empty core, several authors proved the Scarf's result with the aid of covering theorems of a simplex. Shapley[9] developed the technique first and proved the following covering theorem that extended the classical result of Knaster, Kuratowski and Mazurkiewicz; it is now called the K-K-M-S theorem.

Theorem A *Let $\{C^S\}_{S\in\mathcal{N}}$ be a closed covering of Δ^N such that $\Delta^T \subset \bigcup_{S\subset T} C^S$ holds for each $T \in \mathcal{N}$. Then there is a balanced family $\mathcal{B}$ for which $\bigcap_{S\in\mathcal{B}} C^S \neq \emptyset$.*

Ichiishi[3] obtained another covering theorem:

Theorem B *Let $\{C^S\}_{S\in\mathcal{N}}$ be a closed covering of Δ^N such that $\Delta^T \subset \bigcup_{S\supset N\setminus T} C^S$ holds for each $T \in \mathcal{N}$. Then there is a balanced family $\mathcal{B}$ for which $\bigcap_{S\in\mathcal{B}} C^S \neq \emptyset$.*

As pointed out in [3], Theorem A and Theorem B can be regarded as the duals each other if we look at the fact that a subfamily $\mathcal{B}$ of non-empty and proper subsets of the set N is balanced if and only if the family $\mathcal{B}' = \{N \setminus S : S \in \mathcal{B}\}$ is balanced. With the aid of Theorem B, Ichiishi proved the Scarf's result adopting the idea of Keiding and Thorlund-Petersen[6] or that of Vohra[10].

Recently, Zhou[11] proved the Scarf's result using his open covering theorem:

Theorem C *Let $\{G^S\}_{S\in\mathcal{N}}$ be an open covering of Δ^N such that $\Delta^{N\setminus\{i\}} \subset G^{\{i\}}$ holds for each $i \in N$. Then there is a balanced family $\mathcal{B}$ for which $\bigcap_{S\in\mathcal{B}} G^S \neq \emptyset$.*

As pointed out in [11], Theorem C is easily derived from the following closed covering theorem due to Scarf[8]:

Theorem D *Let $\{C^S\}_{S\in\mathcal{N}}$ be a closed covering of Δ^N such that $\Delta^{N\setminus\{i\}} \subset C^{\{i\}}$ holds for each $i \in N$. Then there is a balanced family $\mathcal{B}$ for which $\bigcap_{S\in\mathcal{B}} C^S \neq \emptyset$.*

Theorem A attracted attention of several authors and they provided its interesting generalizations, for example, see [1], [2], [4] and [5]. The purpose of this note is to give a theorem unifying Theorem A, Theorem B and Theorem D. Moreover, our result covers coverings of a compact convex polyhedron instead of those of a simplex.

2. Results

Throughout this section, let $P = \mathrm{co}\{a^1, \ldots, a^m\}$ be a compact convex polyhedron in R^n, where $a^1, \ldots, a^m$ are the extreme points of P. Let $M = \{1, \ldots, m\}$, $\mathcal{M}$ the family of all non-empty subsets of M and $\mathcal{F}$ the family of all non-empty subsets T of M such that $F^T = \mathrm{co}\{a^i : i \in T\}$ is a face of P. For any $S \in \mathcal{M}$, let b^S be the center of gravity of the polyhedron $\mathrm{co}\{a^i : i \in S\}$, that is, $b^S = \sum_{i\in S} a^i/|S|$.

Theorem 1 *Let $\{C^T : T \in \mathcal{M}\}$ be a closed covering of P such that, for each $T \in \mathcal{F} \setminus \{M\}$,*

$$F^T \subset \bigcup_{\mathcal{S}} \bigcap_{S \in \mathcal{S}} C^S,$$

where $\mathcal{S}$ runs over all disjoint subfamilies of $\mathcal{M}$ such that $\bigcup_{S \in \mathcal{S}} S \supset M \setminus T$. Then there is a subfamily $\mathcal{B}$ of $\mathcal{M}$ such that

$$\bigcap_{T \in \mathcal{B}} C^T \neq \emptyset \quad \text{and} \quad b^M \in \operatorname{co}\{b^S : S \in \mathcal{B}\}.$$

Proof. We borrow the method of the proof in [7]. Let $I(x) = \{T \in \mathcal{M} : C^T \ni x\}$ for each $x \in P$. Since the family $\{I(x) : x \in P\}$ is finite, denote it by $\{\mathcal{M}_1, \ldots, \mathcal{M}_p\}$. Suppose the conclusion is false. Define the continuous function $\beta_k : P \to R_+$ by

$$\beta_k(x) = \begin{cases} d(x, \bigcup_{T \notin \mathcal{M}_k} C^T) & \text{for } x \in P \text{ if } \bigcup_{T \notin \mathcal{M}_k} C^T \neq \emptyset, \\ 1 & \text{for } x \in P \text{ otherwise,} \end{cases}$$

for every $k = 1, \ldots, p$, where $d(x, A)$ denotes the distance between a point x and a set A. Since $b^P \notin \operatorname{co}\{b^T : T \in \mathcal{M}_k\}$ for $k = 1, \ldots, p$, for any k, there is $n_k \in R^n$ such that $b^M \cdot n_k < b^T \cdot n_k$ for any $T \in \mathcal{M}_k$ by the separation theorem for convex sets. Define the continuous function $f : P \to R^n$ by

$$f(x) = \sum_{k=1}^{p} \beta_k(x) n_k, \quad x \in P$$

and a set-valued function $G : P \to 2^P$ by

$$G(x) = \{y \in P : y \cdot f(x) = \max\{z \cdot f(x) : z \in P\}\}, \quad x \in P.$$

It is easily seen that G is upper hemicontinuous, and closed and convex-valued. By the Kakutani's fixed point theorem, there is $x_0 \in P$ with $x_0 \in G(x_0)$. Since $G(x_0)$ is a non-empty face of P, we can write $G(x_0) = F^L$ for some $L \in \mathcal{F}$. It is obvious that L is non-empty; moreover, L is not equal to M. If $L = M$, then $b^M \cdot f(x_0) = a^i \cdot f(x_0)$ for all $i \in M$. But note that $b^M \cdot f(x_0) < b^T \cdot f(x_0)$ for any $T \in I(x_0)$ by the definition of f. This is a contradiction. Hence $L \neq M$ and there is a disjoint subfamily $\{S_1, \ldots, S_r\}$ of $\mathcal{M}$ such that $L \cup S_1 \cup \cdots \cup S_r = M$ and $\bigcap_{i=1}^r C^{S_i} \ni x_0$ by the hypothesis. Let $S = S_1 \cup \ldots \cup S_r$ and $W = M \setminus S$. From the definition of G, we have $b^M \cdot f(x_0) \leq a^i \cdot f(x_0)$ for all $i \in W$. On the other hand, since $x_0 \in C^{S_j}$ for $j = 1, \ldots, r$, we have $S_j \in I(x_0)$ for $j = 1, \ldots, r$. Hence $b^M \cdot f(x_0) < b^{S_j} \cdot f(x_0)$ for all $j = 1, \ldots, r$. Since

$$b^M = \frac{|W|}{m} b^W + \sum_{j=1}^{r} \frac{|S_j|}{m} b^{S_j},$$

we have

$$\begin{aligned} b^M \cdot f(x_0) &= \frac{|W|}{m} b^W \cdot f(x_0) + \sum_{j=1}^{r} \frac{|S_j|}{m} b^{S_j} \cdot f(x_0) \\ &> \frac{|W|}{m} b^M \cdot f(x_0) + \sum_{j=1}^{r} \frac{|S_j|}{m} b^M \cdot f(x_0) \\ &= b^M \cdot f(x_0), \end{aligned}$$

which is a contradiction. □

The following is a simplex version of Theorem 1.

Corollary 1 *Let $\{C^T : T \in \mathcal{N}\}$ be a closed covering of Δ^N such that, for each $T \in \mathcal{N} \setminus \{N\}$,*

$$\Delta^T \subset \bigcup_{\mathcal{S}} \bigcap_{S \in \mathcal{S}} C^S,$$

where $\mathcal{S}$ runs over all disjoint subfamilies of $\mathcal{N}$ such that $\bigcup_{S\in\mathcal{S}} S \supset N \setminus T$. Then there is a balanced subfamily $\mathcal{B}$ of $\mathcal{N}$ such that

$$\bigcap_{T\in\mathcal{B}} C^T \neq \emptyset.$$

The following is a polyhedron version of Theorem B.

Corollary 2 *Let $\{C^T : T \in \mathcal{M}\}$ be a closed covering of P such that*

$$F^T \subset \bigcup_{S \supset M\setminus T,\ S\in\mathcal{M}} C^S$$

holds for each $T \in \mathcal{F} \setminus \{M\}$. Then there is a subfamily $\mathcal{B}$ of $\mathcal{M}$ such that

$$\bigcap_{T\in\mathcal{B}} C^T \neq \emptyset \quad \text{and} \quad b^M \in \text{co}\{b^S : S \in \mathcal{B}\}.$$

The following is a polyhedron version of Theorem D.

Corollary 3 *Let $\{C^T : T \in \mathcal{M}\}$ be a closed covering of P such that*

$$\bigcup \{F^T : T \in \mathcal{F}, i \notin T\} \subset C^{\{i\}}$$

for each $i = 1, \ldots, m$. Then there is a subfamily $\mathcal{B}$ of $\mathcal{M}$ such that

$$\bigcap_{T\in\mathcal{B}} C^T \neq \emptyset \quad \text{and} \quad b^M \in \text{co}\{b^S : S \in \mathcal{B}\}.$$

The following is a polyhedron version of Theorem A.

Corollary 4 *Let $\{C^T : T \in \mathcal{M}\}$ be a closed covering of P such that*

$$F^T \subset \bigcup_{S \subset T,\ S \in \mathcal{M}} C^S$$

holds for any $T \in \mathcal{F}$. Then there is a subfamily $\mathcal{B}$ of $\mathcal{M}$ such that

$$\bigcap_{T \in \mathcal{B}} C^T \neq \emptyset \quad \text{and} \quad b^M \in \operatorname{co}\{b^S : S \in \mathcal{B}\}.$$

We can easily derive Corollary 4 from Corollary 2 by the same argument given in [3] when we have the following lemma.

Lemma 1 *Let $\mathcal{B}$ be a subfamily of $\mathcal{M} \setminus \{M\}$ and $\mathcal{B}' = \{M \setminus T : T \in \mathcal{B}\}$. Then, $b^M \in \operatorname{co}\{b^T : T \in \mathcal{B}\}$ if and only if $b^M \in \operatorname{co}\{b^S : S \in \mathcal{B}'\}$*

Proof. By symmetry, it is sufficient to prove the 'only if' part. Let $b^M = \sum_{T \in \mathcal{B}} \lambda^T b^T$ for some $\lambda^T \geq 0$ with $\sum_{T \in \mathcal{B}} \lambda^T = 1$. Then we have $\sum_{T \in \mathcal{B}} \lambda^T (b^M - b^T) = 0$. On the other hand, we have $b^M = (|T|/m) b^T + (|M \setminus T|/m) b^{M \setminus T}$; hence, $b^M - b^T = -(|M \setminus T|/|T|)(b^M - b^{M \setminus T})$. Therefore, we have

$$\sum_{T \in \mathcal{B}} \frac{\lambda^T |M \setminus T|}{|T|} (b^M - b^{M \setminus T}) = 0.$$

If we put $\mu = \sum_{T \in \mathcal{B}} \lambda^T |M \setminus T|/|T|$ and $\mu^T = \lambda^T |M \setminus T|/\mu|T| \geq 0$, then we have $b^M = \sum_{T \in \mathcal{B}} \mu^T b^{M \setminus T}$ and $\sum_{T \in \mathcal{B}} \mu^T = 1$. □

Note that Theorem A, Theorem B and Theorem D are simplex versions of Corollary 4, Corollary 2 and Corollary 3, respectively.

References

[1] K. Fan, A further generalization of Shapley's generalization of the Knaster-Kuratowski-Mazurkiewicz theorem, in *Game Theory and Mathematical Economics* (O. Moeschlin and D. Pallaschke eds.), 275–279, North-Holland, 1981.

[2] K. Fan, *Some properties of convex sets related to fixed point theorems,* Math. Ann. **266** (1984), 519–537.

[3] T. Ichiishi, *Alternative version of Shapley's theorem on closed coverings of a simplex,* Proc. Amer. Math. Soc. **104** (1988), 759–763.

[4] T. Ichiishi and A. Idzik, *Theorems on closed coverings of a simplex and their applications to cooperative game theory,* J. Math. Anal. Appl. **146** (1990), 259–270.

[5] T. Ichiishi and A. Idzik, *Closed covers of compact convex polyhedra,* International J. Game Theory **20** (1991), 161–169.

[6] H. Keiding and L. Thorlund-Petersen, *The core of cooperative game without side payments,* mimeo, 1985.

[7] H. Komiya, *A simple proof of K-K-M-S theorem,* to appear in Economic Theory.

[8] H. Scarf, *The core of an N person game,* Econometrica **35** (1967), 50–69.

[9] L. S. Shapley, On balanced games without side payments, in *Mathematical Programming*(T.C. Hu and S.M. Robinson, eds.), 261–290, Academic Press, 1973.

[10] R. Vohra, *On Scarf's theorem on the non-emptiness of the core: A direct proof through Kakutani's fixed-point theorem,* Working Paper No. **87-2**, Dept. of Economics, Brown Univ., 1987.

[11] L. Zhou, *A theorem on open coverings of a simplex and Scarf's core existence theorem through Brouwer's fixed point theorem,* mimeo, Yale University, 1993

Positive Nonlinear Systems in Economics

Ulrich Krause
Department of Mathematics, University of Bremen,
28359 Bremen, GERMANY

1. Introduction

In many cases an economic system can be modelled by a mapping transforming a state of the economic system at a certain period of time into the state of the system at the next period. If the transformation under consideration can be assumed to be linear then the well-established theory of linear operators can be applied; thereby spectral theory, including Perron–Frobenius theory for positive matrices and positive linear operators, is of particular importance. Very often, however, linearity is not an appropriate idealization, in which case a rigorous analysis may become very difficult or even impossible. It is this state of affairs which brings positive nonlinear systems into play, this not only in economics. Positivity and related mathematical properties are quite natural assumptions in economics. The state space is often given, e.g., if states are described by quantities or prices, by the positive orthant (or some more general convex cone) in Euclidean space. The transformation of such a state space may possess additional properties related to positivity as various forms of monotonicity. This is the case for the two economic problems considered in this paper: Balanced growth in a nonlinear multisectoral framework and price setting among several production units which depend on each other by technology. Given the transformation T mapping the state space K, a convex cone, into itself, the following questions will be addressed: Does there exist a unique equilibrium, that is does the fixed point equation $Tx^\star = x^\star$ possess a unique solution $x^\star \in K$ (up to a positive scalar)? If so, is the equilibrium $x^\star$ globally stable, i.e., does the sequence $(x_n)_{n\geq 0}$ generated by the discrete dynamical system $x_{n+1} = Tx_n$ converge to $x^\star$ for all starting points $0 \neq x_0 \in K$? Admitting for growth, an even more interesting question is if the nonlinear eigenvalue problem $Tx^\star = \lambda^\star x^\star$ has a unique solution $x^\star \in K$ (up to a positive scarlar), $\lambda^\star \geq 0$? If so, does the sequence of the normalized states $\frac{x_n}{\|x_n\|}$ ($\|\cdot\|$ some vector space norm) converge to $x^\star$? Answers to these questions will be presented in sections 2 and 3 for the problem of balanced growth. In sections 4 and 5 those questions arise again in the context of price setting, but this time in a more involved manner. This is due to the fact that the discrete dynamical system is no longer autonomous, i.e. the transformation T itself depends on time. Thus, instead of the asymptotic behavior of the iterates T^n the inhomogeneous iterates $T_n \circ T_{n-1} \circ \ldots \circ T_1$ need to be analyzed for $n \longrightarrow \infty$. As it turns out, stability for states cannot longer be expected but every path shows some stable behavior under certain conditions. When dealing with nonlinear systems the term "nonlinear" is used in this article to include linearity as a special case. In the special case of linear and finite dimensional systems the results presented in the following reduce to what could be called a "dynamic" Perron–Frobenius theory.

2. Balanced growth: relative stability

Consider a closed production model which models production as a transformation $T: K \longrightarrow K$ transforming an input vector $x_n \in K$ in period n into an output vector $x_{n+1} \in K$ in period $n+1$, where $n \in \mathbb{N} = \{0, 1, 2, \cdots\}$. For finitely many goods K is the positive orthant $\mathbb{R}^d_+$ in finite–dimensional Euclidean space $\mathbb{R}^d$, for infinitely many goods K may be taken the positive orthant $(l_1)_+$ in the space of sequences l_1. A *balanced growth path* is described by $x^\star_n = \lambda^{\star n} x^\star$ for $n \in \mathbb{N}$ where initial input $x^\star_0 = x^\star \in K$ and growth factor $\lambda^\star \geq 0$ are given by the nonlinear eigenvalue problem $Tx^\star = \lambda^\star x^\star$. The balanced growth path is *relatively stable* if for any non–zero initial input $x_0 = x$ the path given by $x_{n+1} = Tx_n, n \in \mathbb{N}$, behaves finally as the balanced growth path, more precisely, if

$$\lim_{n \to \infty} \frac{T^n x}{\lambda^{\star n}} = c(x) x^\star$$

where $x^\star \in K$ with $\| x^\star \| = 1$, $\lambda^\star > 0$ is understood to be a unique solution of $Tx^\star = \lambda^\star x^\star$; $c(x)$ is a nonnegative constant which may depend on the initial input. Relative stability in a nonlinear framework of balanced growth has been first analyzed by R.M. Solow and P.A. Samuelson in [34] followed by M. Morishiman [24] and H. Nikaido [27]. Subsequent investigations include [4, 5, 7, 8, 9, 13, 15, 19, 25, 26, 28, 30, 32, 36].

Before presenting results about relative stability we need to introduce some terminology, especially certain monotonicity conditions. Let $E \neq \{0\}$ be a real Banach space with norm $\| \cdot \|$ and let K be a closed convex cone contained in E. K induces an ordering relation on K by $x \leq y$ iff $y - x \in K$; $x \lneq y$ means $x \leq y$ but $x \neq y$; $x < y$ means that $y - x$ is contained in the interior $\dot{K}$ of K. We assume K to be normal and choose the norm such that $\| x \| \leq \| y \|$ whenever $0 \leq x \leq y$. A selfmapping $T: K \longrightarrow K$ is *increasing*, if

$$0 \leq x \leq y \text{ implies } 0 \leq Tx \leq Ty;$$

positively homogeneous, if

$$T(\lambda x) = \lambda Tx \text{ for all } x \in K, \lambda \geq 0;$$

ray preserving, if

to $\lambda \geq 0$ and $x \in K$ there exists some $\lambda(x) \geq 0$ such that $T(\lambda x) = \lambda(x)Tx$;

concave, if

$$T(\lambda x + (1-\lambda)y) \geq \lambda Tx + (1-\lambda)Ty \text{ for all } 0 \leq \lambda \leq 1 \text{ and } x, y \in K;$$

primitive at x, if there exists $s(x) \geq 1$ such that $x \lneq y$ implies $T^{s(x)}x < T^{s(x)}y$ for all $y \in K$;

ascending on $A \subset K$, if there exists a continuous selfmapping φ of the unit interval $[0,1]$ with $\lambda < \varphi(\lambda)$ for $0 < \lambda < 1$ and such that for all $x, y \in A$ and all $\lambda \in [0,1]$ the relation $\lambda x \leq y$ implies that $\varphi(\lambda)Tx \leq Ty$.

Theorem 1. ([16])
If T is ray preserving, ascending on $\{x \in K \mid \| x \| = 1\}$ and such that for $x, y \in K \backslash \{0\}$ there exists $a = a(x,y) > 0$ with $aTx \leq Ty$ then the eigenvalue problem $Tx = \lambda x$ has a unique solution $x^\star \in K, \| x^\star \| = 1, \lambda^\star > 0$ for which it holds that

$$\lim_{n \to \infty} \frac{T^n x}{\| T^n x \|} = x^\star \text{ for all } x \in K \text{ with } Tx \neq 0.$$

Theorem 2. ([19])
If T satisfies the assumptions of Theorem 1 and, additionally, T is positively homogeneous then there holds relative stability with a function $x \longmapsto c(x)$ which is increasing, positively homogeneous and strictly positive for $x \neq 0$. Furthermore, there holds the estimate

$$\| \frac{T^n x}{\lambda^{\star n}} - c(x)x^\star \| \leq \frac{\tau^{n-1}(\alpha(x))}{\beta(x)} \text{ all } n \geq 1, \text{ all } 0 \neq x \in K,$$

where τ is a selfmapping of $\mathbb{R}_+$ with $\tau(r) < r$ for $r > 0$ and $\alpha(x) > 0, \beta(x) > 0$.

Remarks.

1. It suffices to require T^r to be ascending for some $r \geq 1$. To make sure the existence of a balanced growth solution, that is $x^\star$ and $\lambda^\star$, no compactness condition is employed in the theorems above.

2. In [9, 13] results similar to Theorem 1 are proved for T *weakly homogeneous*, i.e. $T(\lambda x) = f(\lambda)Tx$ for all $\lambda \geq 0$, all $x \in K$ and f a selfmapping of $\mathbb{R}_+$ such that $f(0) = 0, \frac{1}{\lambda}f(\lambda)$ is nonincreasing for $\lambda > 0$. Results related to Theorem 1 can be found also in [9, 37].

The following two corollaries supply criteria for relative stability which are easier to handle.

Corollary 1. [19]
Let T be continuous, concave, positively homogeneous and such that for some numbers $a > 0, b > 0$ and $0 \neq e \in K$

$$ae \leq Tx \leq be \text{ for all } x \in K, \| x \| = 1.$$

Then there holds relative stability with an estimate

$$\| \frac{T^n x}{\lambda^{*n}} - c(x)x^* \| \leq \text{ constant } \cdot (1 - \frac{a}{b})^{n-1} \| x \|$$

$$\text{for all } n \geq 1 \text{ and } 0 \neq x \in K.$$

Corollary 2. ([24, 25, 31])
Let $E = \mathbb{R}^d$ (equipped with some increasing norm $\| \cdot \|$), $K = \mathbb{R}^d_+$ and let T be continuous, positively homogeneous, increasing and primitive at all points. Then there holds relative stability with an estimate

$$\| \frac{T^{ns} x}{\lambda^{*ns}} - c(x)x^* \| \leq \beta\tau^{n-1}(\alpha) \| x \|$$

for all $n \geq 1$ and $0 \neq x \in K$; s is the primitivity index of T; $\alpha > 0, \beta > 0$ are constants.

The following example illustrates Corollary 1 in infinitely many dimensions.

Example. Let
$E = l_1 = \{x = (x_1, x_2 \ldots) \mid x_i \in \mathbb{R}, \sum_{i=1}^{\infty} \mid x_i \mid < \infty\}$
$K = (l_1)_+ = \{x \in l_1 \mid x_i \geq 0 \text{ for } i = 1, 2, \ldots\}$
T the pointwise minimum of finitely many mappings of the following type
$S: l_1 \longrightarrow l_1, (Sx)_i = \sum_{j=1}^{\infty} a_{ij}x_j$ with $a_{ij} \geq 0$ and
$0 < u_i = \inf\{a_{ij} \mid j \geq 1\}, v_i = \sup\{a_{ij} \mid j \geq 1\} < \infty, \sum_{i=1}^{\infty} u_i < \infty, u_i \leq kv_i$ for some k and all $i \geq 1$. A special case of the Corollaries 1 and 2 is provided by $E = \mathbb{R}^d, K = \mathbb{R}^d_+$ and $Tx = A \cdot x$, where A is a matrix with strictly positive entries.

3. Balanced growth: absolute stability

For obtaining relative stability the transformation T was assumed to be positively homogeneous meaning that constant returns to scale apply. It is known that for decreasing returns to scale one may have ***absolute stability***, that is

$$\lim_{n\to\infty} T^n x = x^\star \text{ for all } 0 \neq x \in K.$$

Consider for example in the finite dimensional case, $E = \mathbb{R}^d$ (with some increasing norm) and $K = \mathbb{R}^d_+$, a selfmapping T of K which is continuous, increasing, indecomposable, primitive at 0, and *positively homogeneous of order* m with $0 < m < 1$, that is

$$T(\lambda x) = \lambda^m Tx \text{ for all } \lambda \geq 0,\ x \in K.$$

Then T possesses a unique fixed point $x^\star > 0$ for which absolute stability holds ([28]).
This result is generalized by the following theorem which holds even in the Banach space setting of the previous section.

Theorem 3. ([16])
If some iterate $T^r, r \geq 1$, of T is ascending on the whole cone K and maps $K\backslash\{0\}$ into the interior $\dot{K}$ of K and if T is continuous with $Tx \neq 0$ for $x \neq 0$ then the fixed point equation $Tx = x$ has a unique solution $x^\star \in \dot{K}$ and

$$\lim_{n\to\infty} T^n = x^\star \text{ for all } x \in K\backslash\{0\}.$$

Theorem 3 generalizes even in finite dimensions the above mentioned result of [28] since a positively homogeneous mapping of order m with $0 < m < 1$ is ascending on K with $\varphi(\lambda) = \lambda^m$.

Remark. The assumptions of Theorem 3 can be weakened if the existence of a unique fixed point $x^\star$ of T is already known. There holds, e.g., the following result in the finite dimensional case ([23, 31]):
If T is continuous, increasing, $T(\lambda x) \geq \lambda Tx$ for $x \in K, \lambda \in [0,1]$, and if T has a unique fixed point $x^\star > 0$ then $\lim_{n\to\infty} T^n x = x^\star$ for all $x > 0$.

If we drop the assumption of positive homogeneity of order m the existence of a fixed point $x^\star$ is not longer guaranteed, even if all the other assumptions in the introductory example are retained. What can be said, however, is that a *limit set trichotomy* for orbits $\{T^n x \mid n \geq 0\}$, $x \neq 0$, applies in the following sense:

Either

(i) all orbits are unbounded,
or

(ii) all orbits converge to 0 and 0 is the unique fixed point of T,
or

(iii) all orbits converge to a point of $\dot{K}$ which is the unique fixed point $\neq 0$ of T.

Obviously, *if* there holds relative stability then there also holds a limit set trichotomy according to the three possibilities $\lambda^* > 1$, $\lambda^* < 1$, $\lambda^* = 1$. The following result yields limit set trichotomy for a situation where positive homogeneity does not apply but some condition is satisfied which generalizes that of positive homogeneity of order m for $0 \leq m < 1$.

Theorem 4. ([22])
If $E = \mathbb{R}^d, K = \mathbb{R}^d_+$ and T is continuous, increasing, primitive at 0 and if for some $k \geq 1$

$$T^k(\lambda x) > \lambda T^k x \quad \text{for all} \quad x > 0, \text{ all } 0 < \lambda < 1$$

then there holds limit set trichotomy.

Remarks. In Theorem 4 all three cases of a limit set trichotomy can occur as is shown by the following little examples for $d = 2, x = (x_1, x_2)$:

case (i) for $Tx = (\sqrt{x_1} + 3x_2,\ 2x_1)$

case (ii) for $Tx = (\frac{x_1+x_2}{2+x_1+x_2}, \frac{x_1+x_2}{2+x_1+x_2})$

case (iii) for $Tx = (\sqrt{x_1} + \sqrt{x_2}, \sqrt{x_1})$

Theorem 4 can be used to show absolute stability by proving limit set trichotomy and excluding cases (i), (ii) thereafter as in the following application to the nonlinear resolvent problem.

Example. *Nonlinear resolvent problem*
Let $A: \mathbb{R}^d_+ \longrightarrow \mathbb{R}^d_+$ be continuous, increasing, $A(\lambda x) \geq \lambda Ax$ for $x \geq 0$, $\lambda \in [0, 1]$ and suppose that $A(x_0) < x_0$ for some $x_0 \geq 0$. For $c \in \mathbb{R}^d_+, c > 0, Tx = Ax + c$ satisfies the assumptions of Theorem 4 for $k = 1$. (T is not positively homogeneous of order m.) In the limit set trichotomy case (i) is impossible since the orbit starting in x_0 is bounded and case (ii) is impossible because of $T0 \geq c > 0$.

Theorem 4 can be sharpened and extended to the Banach space setting of the previous section. For example, it is shown in [23] that limit set trichotomy holds if in $(E, \|\cdot\|)$ all bounded orbits have compact closure and if T is continuous, maps $\dot{K} \neq \emptyset$ into itself and there exists $k \geq 1$ such that T^k maps $K\backslash\{0\}$ into $\dot{K}$ and T^k is a contraction for the part metric. Here, the part metric is defined by $p(x, y) = -\log \inf\{\lambda > 0 \mid \frac{1}{\lambda}x \leq y \leq \lambda x\}$. The contraction property for T^k generalizes the property $T^k(\lambda x) > \lambda T^k x$ of Theorem 4 which generalized positive homogeneity of order $0 < m < 1$.

4. Interdependent price setting: path stability / weak ergodicity

Consider d producers who are interdependent in the sense that producer $j, j \in \{1, \ldots, d\}$, produces the specific commodity j by means of the commodities produced by the other producers and by employing (homogeneous) labor. Denote by $A_j(u,t)$ the *production possibility set of producer* j, that is $A_j(u,t)$ consists of all inputs (a,l) where $a \in \mathbb{R}_+^d$ is a vector of material inputs and $l \in \mathbb{R}_+$ is a labor input, by which producer j is able to produce at time $t \in \{0,1,\ldots\}$ u units of good j. Given a vector $p \in \mathbb{R}_+^d$ of commodity prices and wage $w \in \mathbb{R}_+$, *the minimum cost for producer* j to produce u units of good j at time t is

$$C_j(p,w,u,t) = \inf\{pa + wl) \mid (a,l) \in A_j(u,t)\},$$

where pa denotes the inner product $\sum_{i=1}^{d} p_i a_i$ for $p = (p_1, \ldots, p_d)$, $a = (a_1, \ldots, a_d)$. If output u, time t, good j are fixed, the function $(p,w) \longmapsto C_j(p,w,u,t)$ is concave and positively homogeneous. In considering how to set a price for good j based on his cost of production, producer j may derive a *relevant cost function* $c_j(p,w,u,t)$ from minimal cost C_j. The relevant cost function could be given, for example, by marginal cost $\frac{\partial C_j(p,w,u,t)}{\partial u}$ or average cost $\frac{C_j(p,w,u,t)}{u}$ (which coincide, by Shephard's lemma, for the case of constant returns) or by something else. We shall assume that the relevant cost function, too, is concave and positively homogeneous in (p,w). This is the case if relevant cost is given by marginal or by average cost.

To illustrate, consider interdependencies given by a time–independent (linear) Leontief technology and by a Cobb–Douglas technology, respectively. For a Leontief technology average (marginal) cost is given by $c_j(p,w,u,t) = a_j p + wl_j$, where $a_j \in \mathbb{R}_+^d$ collects the amounts of goods used to produce one unit of j and l_j is the amount of labor used; the matrix with columns $a_j, 1 \leq j \leq d$, is the Leontief input–output matrix. In this very special case the relevant cost function for j is even linear and does not depend on output u. This is different for a Cobb–Douglas technology. In this case average (marginal) cost is given by

$$c_j(p,w,u,t) = k_j \prod_{i=1}^{d} p_i^{\alpha_{ij}/\beta_j} u^{\frac{1}{\beta_j}-1} + wl_j$$

with

$$0 \leq \alpha_{ij} \leq 1, \ \sum_{i=1}^{d} \alpha_{ij} = \beta_j > 0.$$

The relevant cost function for j is not linear but is concave and positively homogeneous in (p,w) and it does depend on output u except for $\beta_j = 1$.

In what follows we reduce the independent variables by assuming the wage to depend on prices, more precisely, $w = w(p)$ and $p \longmapsto w(p)$ concave, positively homogeneous. (For example, $w(p) = pc, c \in \mathbb{R}^d_+$ a given real wage; in a more subtle way one could also allow utiliy maximization, cf. [10].) Hence the relevant cost function for j will be denoted also by $c_j(p, u, t)$ and is concave and positively homogeneous with respect to $p \in \mathbb{R}^d_+$.

Concerning the process of setting the prices for the next period we employ the classical rule of price setting according to which (relative) prices are driven by (relative) cost (cf. [1]). That is

$$\frac{p_i(t+1)}{p_j(t+1)} = \frac{c_i(p(t), y_i(t), t)}{c_j(p(t), y_j(t), t)} \text{ for all } i, j \in \{1, \ldots, d\} \text{ and all } t \in \{0, 1, \ldots, \},$$

or, equivalently,

$$p(t+1) = \lambda(t)\ c\ (p(t), y(t), t) \qquad (\star).$$

Here $p(t) \in \mathbb{R}^d_+$ is the vector of prices $p_i(t)$ at time $t, \lambda(t) > 0$ is a factor of proportionality and $c(p(t), y(t), t)$ is the vector of relevant costs at t the i–th component of which is $c_i(p(t), y_i(t), t)$ with $y_i(t)$ the output of good i at time t. The above model of price setting $(\star)$ has been developed in [18]. For models of price setting in a similar framework see [2, 5, 6, 10, 13, 14, 15, 17, 35, 38]. Define the *cost operator* $T(t)$ by

$$T(t)p = c(p, y(t), t).$$

$T(t)\colon \mathbb{R}^d_+ \longrightarrow \mathbb{R}^d_+$ is a nonlinear mapping which is concave and positively homogeneous. Equation $(\star)$ then becomes

$$p(t+1) = \lambda(t)T(t)\ p\ (t) \qquad (\star\star).$$

By the trick of introducing the time–dependent cost operator we don't need to study $t \longmapsto y(t)$ in detail. The dependence of relevant costs and prices set on output does appear via t now as a parametric change in the law of motion $T(t)$. If we would assume in $(\star\star)$ $\lambda(t) = 1$ for all t (cf. [38]) then $(\star\star)$ could be treated as a fixed point problem for a sequence of mappings $(T(t))_{t \geq 0}$. It seems, however, that admitting for $\lambda(t) \neq 1$ seems to be appropriate when dealing with relative prices. Normalizing prices we eliminate $\lambda(t)$ as follows:

$$\text{Let } \| p \| = \sum_{i=1}^{d} | p_i |,\ \tilde{p}(t) = \frac{p(t)}{\| p(t) \|}, \tilde{T}p = \frac{Tp}{\| Tp \|}$$

(where defined). Then $(\star\star)$ becomes $\tilde{p}(t+1) = \tilde{T}(t)\tilde{p}(t)$, or, denoting relative prices $\tilde{p}(t)$ simply by $p(t)$ we arrive at the following nonlinear and non–autonomous discrete dynamical system

$$\left.\begin{array}{rl} p(t+1) = & \tilde{T}(t)p(t),\ p(t) \in S = \{p \in \mathbb{R}^d_+ \mid \| p \| = 1\} \\ \text{or} & \\ p(t+1) = & \tilde{T}(t) \circ \tilde{T}(t-1) \circ \ldots \circ \tilde{T}(1) \circ \tilde{T}(0)p(0) \end{array}\right\} \quad (\star\star\star).$$

This dynamical system is, by the normalization, nonlinear even for the standard linear Leontief model. It is non–autonomous even if relevant costs do not depend explicitly on time, because of the dependence of relevant costs on output. In the following we shall assume that relevant costs do not explicitly depend on time. (Technical change will be considered in the next section.)

A non–autonomous discrete dynamical system given by a sequence of mappings $T_n: K \longrightarrow K$ for $n \geq 0, K$ being a closed convex cone in Banach space $(E, \| \cdot \|)$, is said to possess *path stability* if for arbitrary starting points $x_0, y_0 \in K \setminus \{0\}$ the paths defined for $n \geq 0$ by $x_{n+1} = T_n x_n$ and $y_{n+1} = T_n y_n$, respectively approach finally each other, i.e.

$$\lim_{n \to \infty} \| x_n - y_n \| = 0.$$

The system possesses *weak ergodicity* if for arbitrary $x_0, y_0 \in K \setminus \{0\}$ the above paths stay in $K \setminus \{0\}$ and

$$\lim_{n \to \infty} \left\| \frac{x_n}{\| x_n \|} - \frac{y_n}{\| y_n \|} \right\| = 0.$$

Path stability (weak ergodicity) means that all paths (normalized paths) show finally the same behaviour which, however, may be rather irregular. One might also say that each path is stable in the sense that when being distorted by an exogeneous shock the path comes back after a while to its original form.

To prove path stability/weak ergodicity for the system $(\star\star\star)$, that is for

$$p(t+1) = \tilde{T}(t)p(t), p(t) \in S = \{p \in \mathbb{R}^d_+ \mid \| p \| = 1\}$$

$$\text{where } \tilde{T}(t)p = \frac{T(t)p}{\| T(t)p \|} \text{ and } T(t)p = c(p, w(p), y(t))$$

we shall make the following assumptions. (By positive homogeneity of the mappings $T(t)$, weak ergodicity for the system given by $(T(t))_{t \geq 0}$ amounts to the same as path stability for the system given by $(\tilde{T}(t))_{t \geq 0}$.)

Assumptions of positivity and monotonicity:

(i) Costs are positive, i.e.
there exists some i such that $c_j(e_i, w(e_i), y_j) > 0$ for all j with $y_j > 0$(e_i the i–th unit vector in $\mathbb{R}^d$).

(ii) Costs are irreducible, i.e.
for $\emptyset \subsetneqq I \subsetneqq \{1, \ldots d\}$ there exist $i \in I, j \notin I$ such that $c_j(e_i, w(e_i), y_j) > 0$.

(iii) Costs increase with output, i.e.
$y_j \longmapsto c_j(p, w, y_j)$ is increasing.

(iv) Outputs remain finally bounded and positive, i.e.
there exist numbers $u_j, v_j > 0$ and $t_0 \geq 0$ such that $u_j \leq y_j(t) \leq v_j$ for all j, all $t \geq t_0$.

Theorem 6. ([18])
Under the above assumptions (i) – (iv) the process of price setting is weakly ergodic, that is

$$\lim_{t \to \infty} \| p(t) - q(t) \| = 0.$$

for all normalized price paths $t \longmapsto p(t) \in S, t \longmapsto q(t) \in S$ with arbitrary initial prices $p(0), p(0) \in S$.

Remarks.

(1) The proof of Theorem 6 uses a result from [11].

(2) For path stability/weak ergodicity see also [11, 17, 18, 20, 21, 29].

(3) If relevant costs are independent of output but depend explicitly on time that is for $c_j(p, w(p), t)$, a result similar to Theorem 6 can be proved. Cf. also the discussion of technological change in the next section.

Coming back to the examples of a Leontief technology and a Cobb–Douglas technology mentioned above, the assumptions in Theorem 6 can be satisfied as follows. Assuming simply that all labor inputs are (strictly) positive and $w(p) > 0$ for $p \gneqq 0$ assumptions (i) and (ii) are met in both examples. For a Leontief technology relevant costs are independent of output wherefore (iii) is trivially satisfied and (iv) is an unnecessary assumption. For a Cobb–Douglas technology (iii) is satisfied for $0 \leq \beta_j \leq 1$ for all j. If $\beta_j = 1$ then relevant costs do not depend on output and (iv) is an unnecessary assumption; for $\beta_j < 1$ (iv) is an assumption related to the outside of the system, e.g. to demand.

5. Interdependent price setting: strong ergodicity

A non–autonomous discrete dynamical system given by a sequence of mappings $T_n: K \longrightarrow K$ for $n \geq 0, K$ being a closed convex cone in Banach space

$(E, \|\cdot\|)$, is said to possess *strong ergodicity* if for arbitrary $x_0 \in K \setminus \{0\}$ the path defined by $x_{n+1} = T_n x_n$ for $n \geq 0$ stays in $K \setminus \{0\}$ and there exists some $x^\star \in K$ such that

$$\lim_{n\to\infty} \frac{x_n}{\| x_n \|} = x^\star.$$

The following result provides a criterium for strong ergodicity in the above Banach space setting .

Theorem 7. ([11, 12])
Consider a non–autonomous discrete dynamical system with mappings $T_n: K \longrightarrow K$ which are ray preserving with $T_n x = 0$ iff $x = 0$. Assume the mappings T_n converge uniformly on $S = \{x \in K \mid \| x \|= 1\}$ to a selfmapping T of K. For T assume that T is ascending and uniformly continuous on S and such that for some numbers $a, b > 0$ and $0 \neq e \in K$ it holds that $ae \leq Tx \leq be$ for all $x \in S$.
Then $Tx = \lambda x$ has a unique solution $x^\star \in K, \| x^\star \|= 1, \lambda^\star > 0$ and

$$\lim_{n\to\infty} \frac{x_n}{\| x_n \|} = x^\star \text{ uniformly for } x_0 \in K \setminus \{0\}.$$

Using Theorem 7 one can prove the following result about strong ergodicity for the price setting process considered in the previous section.

Theorem 8. ([18])
Consider a price setting process which satisfies beside assumptions (i) and (ii) in Theorem 6 the following two assumptions:

(v) $c_j(p, w(p), y_j)$ is continuous in p and uniformly continuous in y_j on $\{p \in \mathbb{R}^d_+ \mid \| p \|= 1\}$.

(vi) Outputs converge, i.e. $\lim_{t\to\infty} y(t) = y^\star \in \mathbb{R}^d_+$.

Then for $T: \mathbb{R}^d_+ \longrightarrow \mathbb{R}^d_+$ defined by components $(Tp)_j = c_j(p, w(p), y^\star_j)$ the equation $Tp = \lambda p$ has a unique solution $p^\star \geq 0, \| p^\star \|= 1, \lambda^\star > 0$ and for the relative prices $p(t) \in S$ it holds that $\lim_{n\to\infty} p(t) = p^\star$.

As already mentioned we may admit also technological change during the process of price setting, that is relevant costs which depend explicitly on time. To simplify we suppose relevant costs independent of output as , e.g., for Leontief technologies with input–output matrices depending on time. In general, strong ergodicity cannot be expected as shown by the following simple example which exhibits path stability.

Example. ([17, 20])
Consider an economy producing a single good, corn, by this good and labor

alone. Suppose that to produce one unit of corn as material input $\frac{1}{3}$ units of corn are needed and that in summer (t even) no labor input is needed whereas in the winter time (t odd) $\frac{2}{3}$ hours of labor are needed per unit of output. This gives

a production possibility set $A(u,t) = \begin{cases} (\frac{1}{3}u, o), & t \text{ even} \\ (\frac{1}{3}u, \frac{2}{3}u), & t \text{ odd} \end{cases}$

and, measuring prices in labor commanded, i.e. $w = 1$,

a minimum cost function $C(p,w,u,t) = \begin{cases} \frac{1}{3}pu, & t \text{ even} \\ \frac{1}{3}pu + \frac{2}{3}u, & t \text{ odd.} \end{cases}$

The cost operator $T(t)\colon \mathbb{R}_+ \longrightarrow \mathbb{R}_+$ is an affine–linear mapping given by

$$T(t)p = \begin{cases} \frac{1}{3}p, & t \text{ even} \\ \frac{1}{3}p + \frac{2}{3}, & t \text{ odd.} \end{cases}$$

For the prices given by the non–autonomous, affine, discrete dynamical system
$p(t+1) = T(t)p(t)$ for $t \geq 0$ it follows that irrespective of the initial price $p(0) > 0$
$\lim_{t\to\infty} p(2t) = \frac{3}{4}$ and $\lim_{t\to\infty} p(2t+1) = \frac{1}{4}$.
Therefore, $\lim_{t\to\infty} p(t)$ does not exist but $\lim_{t\to\infty} \mid p(t) - q(t) \mid = 0$ for any two paths, i.e. path stability holds.

By the way, fixing the initial price $p(0)$ but allowing arbitrary infinite compositions of the mappings $T(0)$ and $T(1)$ (not just $\ldots \circ T(1) \circ T(0) \circ T(1) \circ T(0)$) then the set of accumulation points forms a fractal set, namely the Cantor set. Path stability means that this fractal set is independent of $p(0)$.

The following result provides conditions under which price setting in the case of technical change is strongly ergodic.

Theorem 9. ([10])
Consider a price setting process with technical change which satisfies the following assumptions:

(a) $A_j(u,t) = uA_j(1,t)$ for all $j \in \{1,\ldots,d\}$, all $u \geq 0$, all $t = 0,1,2,\ldots$.

(b) $A_j(1,t) \subset A_j(1,t+1)$ for all j, all t.

(c) $\inf\{l \mid (a,l) \in A_j(1,t)\} \geq k > 0$ for all j, all t.

Then for $T\colon \mathbb{R}^d_+ \longrightarrow \mathbb{R}^d_+$ defined by components $(Tp)_j = \inf\{pa + w(p)l \mid (a,l) \in A_j(1,t), t \geq 0\}$ ($p \longmapsto w(p)$ concave and positively homogeneous) the equation $Tp = \lambda p$ has a unique solution $p^\star \geq 0, \| p^\star \| = 1, \lambda^\star > 0$ and for the relative prices $p(t) \in S$ it holds that

$$\lim_{t\to\infty} p(t) = p^\star.$$

Remarks.

(1) In Theorem 9, assumption (b) means that during technological change older techniques will not be forgotten; (c) means that there is not perfect automation.

(2) In Theorem 9, as a special case we may put $A_j(1,t) = A_j$ for all j, all t. The assumptions in Theorem 9 are satisfied if we take, e.g., A_j to be a finite set of Leontief techniques with positive labor inputs. In other words, Theorem 9 covers a linear Leontief model where a choice of techniques is admitted.

References

[1] K. Arrow and D. Starrett, *Cost–theoretical and demand–theoretical approaches to the theory of price determination*, in: *Carl Menger and the Austrian School of Economics*, ed. J. R. Hicks and W. Weber, Oxford Clarendon Press, Oxford (1973), 129–148.

[2] L. Boggio, *Stability of production prices in a model of general interdependence*, in [33], 83–114.

[3] M. Caminati and F. Petri (eds.), *Convergence to long–period positions*, Special issue of Political Economy, **6** (1990).

[4] P. Chander, *The nonlinear input–output model*, J. Econ. Theory, **30** (1983), 219–229.

[5] R.A. Dana, M. Florenzano, C. Le Van and D. Lévy, *Production prices and general equilibirium prices*, J. Math. Econ., **18** (1989), 263–280.

[6] G. Duménil and D. Lévy, *The classicals and the neoclassicals: a rejoinder to Frank Hahn*, Cambridge J. Econ., **9** (1985), 327–345.

[7] T. Fujimoto, *Non–linear generalization of the Frobenius theorem*, J. Math. Econ., **6** (1979), 17–21.

[8] ——, *Non–linear Leontief models in abstract spaces*, J. Math. Econ., **15** (1986),
151–156.

[9] T. Fujimoto and U. Krause, *Strong ergodicity for strictly increasing nonlinear operators*, Lin. Alg. Appl., **71** (1985), 101–112.

[10] ——, *Ergodic price setting with technical progress*, in [33], 115–124.

[11] ——, *Asymptotic properties for inhomogeneous iterations of nonlinear operators*, SIAM J. Math. Anal., **19** (1988), 841–853.

[12] ——, *Stable inhomogeneous iterations of nonlinear positive operators on Banach spaces*, to appear in SIAM J. Math. Anal.

[13] E. Kohlberg, *The Perron–Frobenius theorem without additivity*, J. Math. Econ., **10** (1982), 299–303.

[14] U. Krause, *Minimal cost pricing leads to prices of production*, Cahiers de la R. C. P. Systemes de Prix de Production 2, 3 ("La Gravitation"), Paris 1984.

[15] ——, *Perron's stability theorem for nonlinear mappings*, J. Math. Econ., **15** (1986), 275–282.

[16] ——, *A nonlinear extension of the Birkhoff–Jentzsch theorem*, J. Math. Anal. Appl., **114** (1986), 552–568.

[17] ——, *Gravitation processes and technical change: convergence to fractal patterns and path stability*, in [3], 317–327.

[18] ——, *Path stability of prices in a nonlinear Leontief model*, Ann. Oper. Res., **37** (1992), 141–148.

[19] ——, *Relative stability for ascending and positively homogeneous operators on Banach spaces*, to appear in J. Math. Anal. Appl.

[20] ——, *Path stability in positive discrete dynamical systems*, to appear.

[21] ——, *Positive nonlinear systems: Some results and applications*, to appear in the Proceedings of the First World Congress of Nonlinear Analysts, Tampa 1992.

[22] U. Krause and P. Ranft, *A limit set trichotomy for monotone nonlinear dynamical systems*, Nonlin. Anal. TMA, **19** (1992), 375–392.

[23] U. Krause and R. Nussbaum, *A limit set trichotomy for self-mappings of normal cones in Banach spaces*, Nonlin. Anal. TMA, **20** (1993), 855–870.

[24] M. Morishima, *Generalizations of the Frobenius-Wielandt theorems for non-negative square matrices*, J. London Math. Soc., **36** (1961), 211–220.

[25] ——, *Equilibrium, Stability, and Growth*, Oxford University Press, London (1964).

[26] M. Morishima and T. Fujimoto, *The Frobenius theorem, its Solow–Samuelson extension and the Kuhn–Tucker theorem*, J. Math. Econ., 1 (1974), 199–205.

[27] H. Nikaido, *Balanced growth in multi-sectoral income propagation under autonomous expenditure schemes*, Rev. Econ. Studies, **31** (1964), 25–42.

[28] ——, *Convex Structures and Economic Theory*, Academic Press, New York (1968).

[29] R. Nussbaum, *Some nonlinear weak ergodic theorems*, SIAM J. Math. Anal., **21** (1990), 436–460.

[30] Y. Oshime, *An extension of Morishima's nonlinear Perron–Frobenius theorem*, J. Math. Kyoto Univ., **23** (1983), 803–830.

[31] ——, *Perron–Frobenius problem for weakly sublinear maps in a Euclidean positive orthant*, Japan J. Ind. Appl. Math., **9** (1992), 313–350.

[32] T. T. Read, *Balanced growth without constant returns to scale*, J. Math. Econ., **15** (1986), 171–178.

[33] W. Semmler (ed.), *Competition, Instability, and Nonlinear Cycles*, Springer–Verlag, Berlin etc. (1986).

[34] R. M. Solow and P. A. Samuelson, *Balanced growth under constant returns to scale*, Econometrica, **21** (1953), 412–424.

[35] I. Steedman, *Natural prices, differential profit rates and the classical competitive process*, The Manchester School, June ed. (1984), 123–140.

[36] B. P. Stigum, *Balanced growth under uncertainty*, J. Econ. Th., **5** (1972), 42–68.

[37] D. Weller, *Hilbert's metric, part metric, and self-mappings of a cone*, Ph. D. diss., Universität Bremen (1987).

[38] C. C. von Weizsäcker, *Steady State Capital Theory*, Springer–Verlag, Berlin etc. (1971).

Option Replication Cost with Transaction Costs

Shigeo Kusuoka
Department of Mathematical Sciences, University of Tokyo,
7-3-1 Hongo, Tokyo, 113 JAPAN

1. Introduction

Recently there appeared several papers which discussed the option replication in the case that transaction costs exist (e.g. Leland(1985), Merton(1990), Boyle-Vorst(1992)). In the present paper, we introduce the notion of option replication cost and see the effect of the existence of the transaction costs.

We consider only two securities, the security 0: bond, and the security 1: stock. We assume that the security 0 is transaction cost free. In Section 1, we think of a continuous time model, Black-Scholes model. In Section 2, we think of a discrete time model, Cox-Ross-Rubinstein model and think of the asymptotic behaviour of the option replication costs as the unit time interval is getting to zero.

2. Continuous Time Model

Let $\Omega = C([0,\infty); R)$ and $\tilde{\mathcal{F}}_t = \sigma\{\omega(s); s \in [0,t]\}$, $t \in [0,\infty)$, and let $\mathcal{F}_t = \cap_{s>t}\tilde{\mathcal{F}}_s$, $t \in [0,\infty)$. Then $\{\mathcal{F}_t\}_{t\in[0,T]}$ is a family of right continuous increasing σ-algebras on Ω. Let P be a standard Wiener measure on Ω. Let $r \geq 0, \sigma > 0, \mu \in R$ and let

$$P^0(t;\omega) \ = \ \exp(rt),$$

and

$$P^1(t;\omega) \ = \ \exp(\sigma\omega(t) + \mu t)$$

for each $t \in [0,\infty)$ and $\omega \in \Omega$. We think that the price of security i at time t is $P^i(t)$, $i = 0,1$. We also think that the rate of the transaction cost on sales of security 1 is c_0 and the rate of the transaction cost on purchases of security 1 is c_1, $c_0 \in [0,1)$ and $c_1 \geq 0$.

Now let $\mathcal{I}$ be the family of the $\mathcal{F}_t$-adapted left-continuous increasing process $\{I(t);\ t \in [0,\infty)\}$ such that $I(0) = 0$ and ess.sup $I(t) < \infty$ for all $t > 0$.

We call an element $\{I_j;\ j = 0, 1\} \in \mathcal{I}^2$ a strategy. Here we think of the trading strategy such that in a period $(s, t]$, we buy the security 1 by the amount $I_1(t) - I_1(s)$ and we sell the security 1 by the amount of $I_0(t) - I_0(s)$.

Let $\tilde{P}(t;\omega)$ denote $P^0(t;\omega)^{-1}P^1(t;\omega)$, $t \in [0,\infty)$, $\omega \in \Omega$. For each $x = (x^0, x^1) \in R^2$ and $I_* = \{I_j;\ j = 0, 1\} \in \mathcal{I}^2$, let $X(t; x, I_*) = (X^0(t; x, I_*), X^1(t; x, I_*))$, $t \in [0,\infty)$, be the right-continuous adapted R^2-valued process given by

$$X^0(t; x, I_*) = x^0 + \int_{[0,t+]} (1 - c_0)\tilde{P}(s)dI_0(s) - \int_{[0,t+]} (1 + c_1)\tilde{P}(s)dI_1(s)$$

and

$$X^1(t; x, I_*) = x^1 + I_1(t+) - I_0(t+).$$

Then it is obvious that $X(t; x, I_*)$ is the posterior portfolio at time t, if the initial portfolio is given by x and the investigation strategy is given by I_*.

Let $f : R \to R$ be the function given by

$$f(x) = \begin{cases} (1 - c_0)x, & x \geq 0 \\ (1 + c_1)x, & x < 0 \end{cases}$$

Then one can see that $X^0P^0(t) + f(X^1)P^1(t)$ is the amount of money one gets at time t, if he has the portfolio $X = (X^0, X^1)$ at time t and liquidates all securities.

Definition 1 We say that the pair $(x, I_*) \in R^2 \times \mathcal{I}^2$ *is admissible,* if

$$\text{ess.inf } \inf\{X^0(t; x, I_*)P^0(t) + f(X^1(t; x, I_*))P^1(t);\ t \in [0,T]\} > -\infty$$

for all $T > 0$.

Definition 2 For each $Y = (Y^0, Y^1) \in L^\infty(\Omega; R^2, \mathcal{F}_T, dP)$, *the replication cost* $\pi_T(Y; c_0, c_1)$ *for a contingent claim* Y *at time* T is defined by

$$\begin{aligned} &\pi_T(Y; c_0, c_1) \\ = \ &\inf\{x^0P^0(0);\ \exists \textit{admissible strategy } I_* \textit{ s.t.} \\ &\qquad X^0(T; (x^0, 0), I_*) \geq Y^0,\ X^1(T; (x^0, 0), I_*) \geq Y^1\}. \end{aligned}$$

Let $\mathcal{P}_m(r)$ be the set of probability measures μ on $W = C([0,T]; R)$ such that $\{e^{-rt}w(t);\ t \in [0,T]\}$ be a non-negative martingale under $\mu(dw)$ and $E^\mu[\sup\{w(t);\ t \in [0,T]\}^p] < \infty$ for all $p \in (1,\infty)$.

Then we have the following.

Theorem 1 *Let* $g^i : C([0,T]; R) \to R$, $i = 0, 1$, *be continuous functions such that*

$$|\, g^i(w) \,| \leq C(1 + \sup_{t\in[0,T]} |\, w(t) \,|)^p, \quad w \in W$$

for some $C > 0$ *and* $p \in (1,\infty)$. *Let* $Y(\omega) = (g^0(P^1(\cdot;\omega)), g^1(P^1(\cdot;\omega)), \Omega \in \Omega$. *Then we have*

$$\begin{aligned} &\pi_T(Y; c_0, c_1) \\ \geq\ &\sup\{E^{\mu}[g^0(w) + e^{-rT}w(T)\xi(w)g^1(w)]; \\ &\qquad \mu \in \mathcal{P}_m(r),\ \xi : C([0,T];R) \to [1-c_0, 1+c_1] \text{ measurable}\} \end{aligned}$$

for any $c_0 \in (0,1)$ *and* $c_1 \in (0,\infty)$.

3. Discrete Time Model

Let $\Omega = \{-1,1\}^N$ and $\mathcal{F}$ be the Borel σ -algebra on Ω. Let $Z_k : \Omega \to R$, $k \in N$, be given by $Z_k(\omega) = \omega_k$, $\omega = (\omega_1, \omega_2, \ldots) \in \Omega$. Let P be the Bernoulli measure on Ω such that $P(Z_k = -1) = P(Z_k = 1) = \frac{1}{2}$, $k \in N$, and $Z_k, k = 1,2,\ldots$ are independent under P. Let $\mathcal{F}_0 = \{\phi, \Omega\}$ and $\mathcal{F}_k = \sigma\{Z_1,\ldots,Z_k\}, k = 1,2,\ldots$.

Now let $r, \sigma \in (0,\infty)$, $\mu \in R$, $T > 0$ and $c_0, c_1 \in [0,\infty)$, and fix them throughout this section. Also let $r_n, \sigma_n, \mu_n, c_{1,n} \in (0,\infty)$ and $c_{0,n} \in (0,1)$, $n = 1,2,\ldots$, be such that

$$\mu_n - \sigma_n < r_n < \mu_n + \sigma_n, \quad n \in N,$$

$$\lim_{n\to\infty}(n^{-1}T)^{-1}r_n = r,$$

$$\lim_{n\to\infty}(n^{-1}T)^{-1}\mu_n = \mu,$$

$$\lim_{n\to\infty}(n^{-1}T)^{-1/2}\sigma_n = \sigma,$$

$$\lim_{n\to\infty}(n^{-1}T)^{-1/2}c_{0,n} = c_0,$$

and

$$\lim_{n\to\infty}(n^{-1}T)^{-1/2}c_{1,n} = c_1.$$

For each $n \geq 1$, let

$$P_n^0(k;\omega) = \exp(r_n k),$$

and

$$P_n^1(k;\omega) = \exp(\sigma_n \sum_{\ell=1}^{k} Z(\ell) + \mu_n k), \quad k = 0,\ldots,n, \omega \in \Omega.$$

We consider the model such that the unit time interval is $n^{-1}T$, the price of the security i, $i = 0,1$, at time $(k/n)T$ is given by $P_n^i(k;\omega)$ and the maturity is T. This is equivalent to the n-step multiplicative binomial lattice model employed by Cox, Ross, and Rubinstein. We assume that the rate of the transaction cost on sales of security 1 is $c_{0,n}$, and the rate of the transaction cost on purchases of security 1 is $c_{1,n}$. Let $f_n : R \to R$ be given by

$$f_n(x) = \begin{cases} (1 + c_{1,n})x, & x \geq 0 \\ (1 - c_{0,n})x, & x < 0 \end{cases}$$

Then we see that we get money by the amount $-f_n(-z)P^1(k)$ if we sell the security 1 by the amount z, $z \geq 0$, and that we need money by the amount $f_n(z)P^1(k)$ to buy the security 1 by the amount z, $z > 0$, at time kT/n.

Let $\mathcal{I}$ be the family of adapted stochastic processes $\{I(k)\}_{k\geq 0}$. We call an element $I \in \mathcal{I}$ a strategy again. Here we think of the situation that at time kT/n, we buy the security 1 by the amount $I(k)$ if $I(k) \geq 0$ and we sell the security 1 by the amount of $-I(k)$ if $I(k) < 0$.

Let $\tilde{P}_n(k;\omega)$ denote $P_n^0(k;\omega)^{-1}P_n^1(k;\omega)$, $k \geq 0, \omega \in \Omega$. For each $x = (x^0, x^1) \in R^2$ and $I \in \mathcal{I}$, let $X_n(k;x,I) = (X_n^0(k;x,I), X_n^1(k;x,I))$, $k \geq 0$, be the adapted R^2-valued process defined by

$$X_n^0(k;x,I) = x^0 - \sum_{\ell=0}^{k} f_n(I(\ell))P_n(\ell)$$

and

$$X_n^1(k;x,I) = x^1 + \sum_{\ell=0}^{k} I(\ell).$$

Then $X_n(k;x,I)$ is the posterior portfolio at time kT/n, if the initial portfolio is given by x and the investigation strategy is given by I.

Definition 3 *For each* $Y \in L^\infty(\Omega; R^2, \mathcal{F}_n, dP)$, *the replication cost* $\pi_n^*(Y) = \pi_n^*(Y; c_{0,n}, c_{1,n})$ *of* Y is defined by

$$\begin{aligned} &\pi_n^*(Y) \\ = \; &\inf\{x^0 P_n^0(0);\ x^0 \in R,\ \exists I \in \mathcal{I}\ s.t. \\ &\qquad Y^0 \leq X_n^0(n;(x^0,0),I),\ Y^1 \leq X_n^1(n;(x^0,0),I)\ P-a.s.\}. \end{aligned}$$

Let $W_n : R^{1+n} \to C([0,T];R)$ be a linear interpolation operator given by

$$\begin{aligned} &W_n(\{z(k)\}_{k=0}^n)(t) \\ = \; &([nt/T] + 1 - nt/T)z([nt/T]) + (nt/T - [nt/T])z([nt/T]+1) \end{aligned}$$

for any $\{z(k)\}_{k=0}^n \in R^{1+n}$ and $t \in [0,T]$.

Let $\gamma = c_0 + c_1$, and let $\mathcal{P}_M(\sigma,\gamma,r)$ be the set of probability measures Q on $C([0,T];R)$ such that $\{e^{-rt}w(t);\ t \in [0,T]\}$ is a positive martingale under Q, $Q(w(0) = 1) = 1$, and that the quadratic variation $< \log w >_t$ of $\{\log w(t);\ t \in [0,T]\}$ satisfies that

$$\sigma(\sigma - \gamma)dt \leq d < \log w >_t \leq \sigma(\sigma + \gamma)dt, \quad t \in [0,T],\ Q-a.s.w.$$

Then we have the following.

Theorem 2 *Let $F : C([0,T];R) \to R^2$ be a continuous function such that there are $C \in (0,\infty)$ and $p \in (1,\infty)$ for which*

$$| F(w) | \leq C(1 + \max_{t\in[0,T]} | w(t) |)^p, \quad w \in C([0,T];R).$$

Then we have

$$\begin{aligned} &\lim_{n\to\infty} \pi_n^*(F(W_n(\{P_n^1(k;\cdot)\}_{k=0}^n))) \\ = \ &\sup\{E^Q[F_0(w) + e^{-rT}w(T)F_1(w)];\ Q \in \mathcal{P}_M(\sigma,\gamma,r)\}, \end{aligned}$$

where $F(w) = (F_0(w), F_1(w))$, $w \in C([0,T];R)$.

Corollary 1 *Let $F : C([0,T];R) \to R^2$ be a continuous function such that there are $C \in (0,\infty)$ and $p \in (1,\infty)$ such that*

$$| F(w) | \leq C(1 + \max_{t\in[0,T]} | w(t) |)^p.$$

Let $G : C([0,T];R) \to R$ be given by $G(w) = F_0(w) + e^{-rT}w(T)F_1(w)$, $w \in C([0,T];R)$.
(1) If $G : C([0,T];R) \to R$ is concave, then

$$\lim_{n\to\infty} \pi_n^*(F(W_n(\{P_n^1(k;\cdot)\}_{k=0}^n))) = E^Q[G(w)].$$

Here Q is the probability law of $\{\exp(\sigma B(t) + (r - \frac{\sigma^2}{2})t);\ t \in [0,T]\}$, $\sigma = (\sigma(\sigma+\gamma))^{1/2}$ and $\{B(t); t \in [0,T]\}$ is a standard Brownian motion.
(2) If there is a convex function $\tilde{g} : R \to R$ such that $G(w) = \tilde{g}(w(T))$, $w \in C([0,T];R)$, then

$$\begin{aligned} &\lim_{n\to\infty} \pi_n^*(F(W_n(\{P_n^1(k;\cdot)\}_{k=0}^n))) \\ = \ &\int_R (2\pi T)^{-1/2}\ \tilde{g}(\exp(\tilde{\sigma}x + rT - \frac{\sigma^2 T}{2})\exp(-\frac{x^2}{2T})dx. \end{aligned}$$

Here $\tilde{\sigma} = \{(\sigma(\sigma-\gamma)) \vee 0\}^{1/2}$.

Remark. In the case of a European call option with the exercise price α, we have $G(w) = 0 \vee (w(T) - \alpha)$. So we can apply the above Corollary. This result coincides with the consequence of Lemma 1 in Boyle-Vorst(1992) naturally.

References

[1] P.P. Boyle and T. Vorst, *Option Replication in Discrete Time with Transaction Costs*, Journal of Finance **47** (1992), 271-293.

[2] F. Delbaen and W. Schachermayer, *A general version of the fundamental theorem of asset pricing*, Preprint, 1992.

[3] M. Harrison and D. Kreps, *Martingales and arbitrage in multiperiod security markets*, J. Econ. Theory **20** (1979), 381-408.

[4] M. Harrison and S. Pliska, *Martingales and stochastic integrals in the theory of continuous trading*, Stoch. Processes and Their Appl. **11** (1981), 215-260.

[5] M. Harrison and S. Pliska, *A stochastic calculus model of continuous trading: complete market*, Stoch. Processes and Their Appl. **15** (1983), 313-316.

[6] S. Kusuoka, *Limit Theorem on Option Replication Costs with Transaction Costs*, Submitted, 1993.

[7] H. Leland, *Option Pricing and Replication with Transaction Costs*, Journal of Finance **40** (1985), 1283-1301.

[8] R.C. Merton, *Continuous Time Finance*, Basil Blackwell Ltd, Oxford, 1990.

Fixed Point and Finite Dimensional Invariant Subspace Properties for Semigroups and Amenability

Anthony To-Ming Lau*
Department of Mathematics, University of Alberta
Edmonton, Alberta, CANADA T6G 2G1

1. Introduction

A well-known theorem of Markov-Kakutani [5, p. 456] asserts that if S is a commutative semigroup, then S has the following fixed point property:

(1) whenever $\mathcal{S} = \{T_s;\ s \in S\}$ is a representation of S as affine continuous mappings from a non-empty compact convex subset K of a separated locally covex space (i.e. $T_s(\lambda x + (1-\lambda)y) = \lambda T_s(x) + (1-\lambda)T_s(y)$, $0 \leq \lambda \leq 1$, $x, y \in K$), then K contains a common fixed point for $\mathcal{S}$.

DeMarr [4] proved that if S is commutative, then S has the following fixed point property for non-expansive mappings

(2) whenever $\mathcal{S} = \{T_s;\ s \in S\}$ is a representation of S as non-expansive mappings from a non-empty compact convex subset K of a Banach space into K (i.e. $\|T_s x - T_s y\| \leq \|x - y\|$, $x, y \in K$), then K has a common fixed point for $\mathcal{S}$.

In [3], Day showed that semigroups S satisfying (1) is precisely the class of left amenable semigroups (i.e. $\ell^\infty(S)$ has a left invariant mean: positive norm one left translation invariant linear functional). Also, Takahashi [25] showed that any left amenable semigroup S has fixed point property (2). On the other hand, Fan [6] obtained the following finite dimensional subspace property for left amenable semigroups S :

$P(n)$: Let E be a separated locally convex space, and X be a subset of E containing an n-dimensional subspace. If $\mathcal{S} = \{T_s;\ s \in S\}$ is a representation of S as continuous linear transformations from E into E such that $T_s(L)$ is an n-dimensional subspace contained in X whenever L is an n-dimenaional

*This research is supported by an NSERC grant.

invariant subspace contained in X, and there exists a closed $\mathcal{S}$-invariant subspace H in E of codimension n with the property that $x + H \cap X$ is compact and convex for each $x \in H$, then there exists an n-dimensional subspace L_0 contained in X such that $T_s(L_0) = L_0$ for all $s \in S$.

It is the purpose of this paper to report on some results and problems concerning the relationship between the existence of a left invariant mean on $\ell^\infty(S)$ (or its subspaces) and fixed point properties of types (1) and (2), or the finite dimensional invariant subspace property $P(n)$ of Fan. It is an extended version of my talk given during the Conference on Nonlinear and Convex Analysis in Economic Theory organized by Tokyo Institute of Technology and Keio Economic Society. I would like to thank the organisers of the conference for their kind invitation to speak, the academically stimulating environment created during the conference and their very warm hospitality.

2. Preliminaries and Some Notations

Throughout this paper, S will denote a *semitopological semigroup* i.e. a semigroup with a Hausdorff topology such that for each $a \in S$, the mappings $s \to sa$, $s \to as$ from S into S are continuous. S is *left reversible* if S has finite intersection property for closed left ideals.

Let $\ell^\infty(S)$ denote the Banach algebra of bounded real-valued functions on S with the suprenum norm and pointwise multiplication. For each $a \in S$ and $f \in \ell^\infty(S)$, let $\ell_a f$ and $r_a f$ denote, respectively, the left and right translates of f by a i.e. $(\ell_a f)(s) = f(as)$ and $(r_a f)(S) = f(sa)$, $s \in S$. Let X be a closed subspace of $\ell^\infty(S)$ containing constants and invariant under translations. Then a linear functional $m \in X^*$ is called a *mean* if $\|m\| = m(1) = 1$; m is called a *left invariant mean,* denoted by LIM, if $m(\ell_a f) = m(f)$ for all $a \in S$, $f \in X$.

Let $C(S)$ denote the space of all bounded continuous real-valued functions on S and let $LUC(S)$ denote the space of *left uniformly continuous* real-valued functions on S, i.e. all $f \in C(S)$ such that the mappings $a \mapsto \ell_a f$ from S into $C(S)$ is continuous when $C(S)$ has the sup norm topology. Then $LUC(S)$ is a translation invariant closed subalgebra of $C(S)$ containing constants. Further informations on $LUC(S)$ and its dual Banach algebra may be found in [2]. Note that when S is a topological group, $LUC(S)$ is the space of right uniformly continuous functions defined in [9].

Let $A(S)$ denote the space of all $f \in C(S)$ such that $\mathcal{LO}(f) = \{\ell_a f;\ a \in S\}$ is relatively compact in the norm topology of $C(S)$ and $W(S)$ denote the space of all $f \in C(S)$ such that $\mathcal{LO}(f)$ is relatively compact in the weak topology of $C(S)$. Functions in $A(S)$ (resp. $W(S)$) are called *almost periodic* (resp. *weakly almost periodic*). In general, $A(S) \subseteq W(S) \cap LUC(S)$. Also if S is left reversible, then $A(S)$ has a LIM (see [13]). If S is normal and $C(S)$

has a LIM, then S is left reversible (see [10]). On the other hand, as shown by Mitchell (see [15, p. 328]), if S is the semigroup generated by $\{a, b, c, 1\}$, where 1 is the identity with the relation $ab = 1$ and $ac = 1$, then S is not left reversible, but $AP(S)$ has a left invariant mean. Also, Hsu [11] proved that if S is a discrete left reversible semigroup, then $W(S)$ has a LIM.

Let E be a separated locally convex space and Q be a family of continuous seminorms which determine the topology on E. A non-empty subset M of E is said to have *normal structure* (with respect to Q) if for any $p \in Q$ such that $\delta_p(M) = \sup\{p(x-y);\ x, y \in M\} > 0$, there exists an element $u \in \overline{co\, M}$ (closed convex hull of M) such that $\sup\{p(u-y) | y \in M\} < \delta_p(M)$. Non-empty compact subset of a E has normal structure (see [9; Lemma 3]). However, weakly compact convex subset of a Banach space $(E, \|\cdot\|)$ need not have normal structure with respect to $\|\cdot\|$ (see [1]).

3. Fixed Point Properties

Let X be a compact Hausdorff space, and $\mathcal{S} = \{T_s;\ s \in S\}$ be a continuous representation of S as continuous mappings from X into X. Then $\mathcal{S}$ may be regarded as a subsemigroup of the semigroup X^X (with composition) equipped with the product topology. Let $\overline{\mathcal{S}}$ denote the closure of $\mathcal{S}$ in X^X. Then $\overline{\mathcal{S}}$, with the relative topology, is compact, such that

(i) for each $U \in \overline{\mathcal{S}}$, the map $V \to V \circ U$ from $\overline{\mathcal{S}}$ into $\overline{\mathcal{S}}$ is continuous;
(ii) for each $s \in S$, the map $V \to T_s \circ V$ from $\overline{\mathcal{S}}$ into $\overline{\mathcal{S}}$ is continuous.

In particular $\overline{\mathcal{S}}$ is a compact right topological semigroup [2]. Hence $\overline{\mathcal{S}}$ has minimal left ideals and the minimal left ideals are closed, and pairwise homeomorphic. Now if $\mathcal{S}$ is equicontinuous, then $\overline{\mathcal{S}}$ is a compact topological semigroup (i.e. the multiplication in $\overline{\mathcal{S}}$ is jointly continuous); if $AP(S)$ has a LIM, then $C(\overline{\mathcal{S}})$ also has a LIM, and so the right ideals in $\overline{\mathcal{S}}$ has finite intersection property [23]. So $\overline{\mathcal{S}}$ has a unique minimal right ideal. Hence the minimal left ideals of $\overline{\mathcal{S}}$ are compact topological groups (see [2] for details).

Theorem 3.1 [12] *$AP(S)$ has LIM $\Longleftrightarrow$ S has the following fixed point property:*

(F_1) whenever $\mathcal{S} = \{T_s :\ s \in S\}$ is a continuous representation of S as affine equicontinuous mappings from a non-empty compact convex subset X of a separated locally convex space into X, then X contains a common fixed point for $\mathcal{S}$.

Proof Let L be a minimal left ideal in $\overline{\mathcal{S}}$. Then L is a compact group of affine continuous map from X into X. Let C be a minimal S-invariant compact convex subset of X. An application of the Krein Milman theorem shows that if x, $y \in C$, then $\sigma(x) = \sigma(y)$ for all $\sigma \in L$ (see [2, p. 57]). Hence if $x \in C$, $\theta \in L$ is fixed, then $\theta\sigma(x) = \theta(x)$ for all $\sigma \in L$. So $\sigma(x) = x$ for all $\sigma \in L$. Now if $\sigma \in \overline{\mathcal{S}}$, then $\sigma\theta \in L$. So $\sigma x = \sigma\theta x = x$ i.e. x is also a fixed point for $\mathcal{S}$.

The converse can be established by considering the action of S on the set $K = \{\phi \in AP(S)^*, \phi \geq 0, \|\phi\| = 1\}$ defined by $(s, \phi) \to \ell_s^*\phi$, with the locally convex topology τ defined by the seminorms $\{p_f;\ f \in AP(S)\}$, where $p_f(\phi) = \sup\ \{|\phi(\ell_s f|, |\phi(f)|; s \in S\}$. □

Let (E, Q) or simply E be the locally convex space where Q denotes a (fixed) family of continuous seminorms on E which determine the topology of E. Then the representation of $\mathcal{S} = \{T_s;\ s \in S\}$ on X is *Q-non-expansive* if $p(s \cdot x - s \cdot y) \leq p(x - y)$ for all $s \in S$, $x, y \in X$ and $p \in Q$.

Theorem 3.2 ([13]) *$AP(S)$ has a LIM if and only if S has the following fixed point property:*

(G_1) whenever $\mathcal{S} = \{T_s;\ s \in S\}$ is a continuous representation of S as Q-non-expansive mappings on a non-empty compact convex subset X of a separated locally convex space E, then X has a common fixed point for $\mathcal{S}$.

Proof As in the proof of Theorem 3.1, each minimal left ideal L of $\overline{\mathcal{S}}$ is a compact group and L acts non-expansively on X. By Zorn's lemma, there exists a non-empty compact convex subset K of X which is minimal with respect to being closed, convex and invariant under each element of L. A second application of Zorn's lemma shows that there exists a non-empty subset F of K which is minimal with respect to being closed and invariant under each element of L. Since L is a group, $TF = F$ for each $T \in L$. Also since X has normal structure [10, Lemma 3], it follows by an argument similar to [4, Lemma 2] that F must consist of a single point, which is a common fixed point for L (and hence $\mathcal{S}$). The converse can be established as in Theorem 3.1. □

Let $\overline{\mathcal{S}}^a$ denote the almost periodic compactification of S (see [2]). If S is left reversible, then the compact topological semigroup $\overline{\mathcal{S}}^a$ is left reversible. Consequently, it follows from [23] that $AP(S)$ has a LIM. In particular, S has fixed point properties (F_1) and (G_1). More generally, R. Hsu establishes the following for *discrete* left reversible semigroups:

Theorem 3.3 (Hsu [11]) *If S is a left reversible discrete semigroup, then S has the following fixed point properties:*

(F_2) Whenever $\mathcal{S}$ is weakly continuous representation of S as τ-equicontinuous affine self-maps on a non-empty weakly compact convex subset X of a separated locally convex space E with topology τ, then X has a common fixed point for $\mathcal{S}$.

(G_2) *Whenever $\mathcal{S}$ is a weakly continuous representation of S as Q-nonexpansive mappings on a non-empty compact convex set X of a separated locally convex space (E, Q), then X contains a common fixed point for $\mathcal{S}$.*

Theorem 3.4 ([16]) *$W(S)$ has a LIM if and only if S has fixed point property (F_2).*

Proof Let Φ denote the convex hull of $\mathcal{S}$ in the product space E^X and $\overline{\Phi}$, $\overline{\mathcal{S}}$ denote the closure of Φ and $\mathcal{S}$ in E^X. By τ-equicontinuity of $\mathcal{S}$, $\overline{\Phi}$ consists entirely of continuous affine maps from (X, τ) into (X, τ). Consequently, maps in $\overline{\Phi}$ are also continuous from (X, weak) into (X, weak). Since $W(S)$ has a LIM, it follows from [2, Theorem 2.11, p. 143] that $C(\overline{\mathcal{S}})$ also has a LIM. Consequently each minimal left ideal L of $\overline{\mathcal{S}}$ is a compact group of affine continuous mappings from X into X (see [2]). So an argument similar to that for Theorem 3.1 shows that $\mathcal{S}$ has a common fixed point in X. The converse can also be established as in Theorem 3.1. □

PROBLEM 1. It is not known if a semitopological semigroup S would have fixed point property (F_2) if S satisfies

(a) S is left reversible;

or

(b) $AP(S)$ has a LIM.

PROBLEM 2. It is also not known if a semitopological semigroup S would have fixed point property (G_2) if S satisfies one of the following properties:

(a) S is left reversible;
(b) $W(S)$ has a LIM;
(c) $AP(S)$ has a LIM;
(d) $LUC(S)$ (or even $C(S)$) has a LIM.

On the other hand, Lim proved:

Theorem 3.5 (Lim [20]) *Let S be a left reversible semitopological semigroup. Then S has the following fixed point property:*

(G_2') *If $\mathcal{S} = \{T_s;\ s \in S\}$ is a τ-continuous representation of S as Q-non -expansive mappings on a non-empty weakly compact convex subset X of a separated locally-convex space (E, τ) with the τ-topology determined by the family of seminorms Q, and X has Q-normal structure, then X contains a common fixed point for $\mathcal{S}$.*

PROBLEM 3. Does S have fixed point property (G_2') when $A(S)$ (or even $W(S)$,
$LUC(S)$ or $C(S)$) has a LIM?

Mitchell proved the following generalization of Day's fixed point theorem ([3]):

Theorem 3.6 (Mitchell [22]) *$LUC(S)$ has a LIM if and only if S has the following fixed point property:*

(F_3) Whenever $\mathcal{S}$ is a jointly continuous representation of S as affine mappings on a non-empty compact convex subset X of a separated locally convex space, then X contains a common fixed point for $\mathcal{S}$.

(*Note:* a representation $\mathcal{S}$ of S on X is *jointly continuous* means the map $S \times X \to X$, $(s, x) \to T_s x$ is continuous when $S \times X$ has the product topology.)

PROBLEM 4. Let S be a semitopological semigroup such that $LUC(S)$ has a LIM. We do not know if S has any of the following fixed point properties:

(F_3') Whenever $\mathcal{S}$ is a continuous representation of S as affine mappings on a non-empty compact convex subset X of a separated locally convex space, then X contains a common fixed point for $\mathcal{S}$.

(F_3'') Whenever $\mathcal{S} = \{T_s;\ s \in S\}$ is a weakly measurable representation of S as continuous affine mappings on a non-empty compact convex subset X of a locally convex space E (i.e. for each $\phi \in E^*$, $x \in E$, the function $s \to \phi(T_s(x))$ is Borel measurable on S), then X contains a common fixed point for $\mathcal{S}$.

It is easy to see that $(F_3'') \Longrightarrow (F_3') \Longrightarrow (F_3)$. Also, if S is a locally compact group then all the four conditions are equivalent (see, [18] and [24]). However, it is not known (even for S compact) whether any of the implications can be reversed.

Consider on a semitopological semigroup S the following fixed point property:

(G_3) Whenever $\mathcal{S}$ is a weak* jointly continuous representation of S as norm non-expansive mappings on a weak*-compact non-empty convex subset X of a dual Banach space, then X contains a common fixed point for $\mathcal{S}$.

It follows easily from [7, Theorem 2] that if S has property (G_3), then $LUC(S)$ has a LIM. However the converse is not known. Recently, Takahashi and I establish the following result for separable dual Banach space:

Theorem 3.7 (Lau and Takahashi [19]) *Let S be a semitopological semigroup. If $LUC(S)$ has a LIM, or if S is left reversible as a discrete semigroup, then whenever $\mathcal{S}$ is a weak* jointly continuous representation of S as norm non-expansive mappings on a non-empty weak*-compact convex subset X of a separable dual Banach space, then X contains a common fixed point for $\mathcal{S}$.*

We do not know if Theorem 3.7 remains true when S is left reversible topologically. On the other hand Lim proved:

Theorem 3.8 (Lim [21]) *If S is a left reversible semitopological semigroup, and $\mathcal{S} = \{T_s;\ s \in S\}$ is a norm continuous representation of S as norm non-expansive mappings on a non-empty weak*-compact convex subset X of ℓ_1, then X contains a common fixed point for $\mathcal{S}$.*

Alspach [1] has shown that there is a weakly compact convex subset X of $L'[0,1]$ and a non-expansive mapping $T : X \to X$ without a fixed point. Since $L'[0,1]$ may be embedded isometrically into $(\ell^\infty(\mathbb{N}), \|\cdot\|_\infty)$, K regarded as a subset of $\ell^\infty(\mathbb{N})$ is still weakly compact and convex. However K is weak*-compact. Let $S = \mathbb{N}$, the positive integer with addition (which is left reversible). Then $\mathcal{S} = \{T^n;\ n \in \mathbb{N}\}$ is a representation of S as non-expansive mappings from K into K without fixed point.

The following analogue of Theorem 3.6 was established by Mitchell for $n = 1$ [22] and the author [12] for general n :

Theorem 3.9 *Let n be a positive integer. Then $LUC(S)$ has a left invariant mean of the form $\frac{1}{n}\sum_{i=1}^{n}\phi_i$, where each ϕ_i is a multiplicative mean on $LUC(S)$ if and only if*
$M(n)$: Whenever $\{T_s;\ s \in S\}$ $\mathcal{S}$ is a continuous representation of S as jointly continuous self-maps on a compact Haudorff space X, there exists a non-empty finite subset $F \subseteq X$, $|F| \leq n$, $|F|$ divides n, such that $s \cdot F = F$ for all $s \in S$.

For discrete semigroups with fixed point property $M(1)$ can be characterized algebraically:

Theorem 3.10 (Granirer [8]) *Let S be a discrete semigroup. Then S has fixed point property $M(1)$ if and only for any two elements in S has a common right zero i.e. for any a, b in S, there exists c in S such that $ac = bc = c$.*

4. Finite Dimensional Invariant Subspaces

Let S be a semigroup and $\mathcal{S} = \{T_s;\ s \in S\}$ be a representation of S as continuous linear mappings on a separated locally convex space E. Let X be a subset of E, and $\mathcal{L}_n(X)$ denote all n-dimensional subspaces of E. We say that X is *n-consistent* if:

(a) $\mathcal{L}_n(X)$ is non-empty and $\mathcal{S}$-invariant (i.e. $T_s(L) \subseteq L$ for all $s \in S$); and

(b) there exists a closed $\mathcal{S}$-invariant subspace H of E with codimension n and $(x + H) \cap X$ is compact convex for each $x \in E$.

Theorem 4.1 *(a) If $LUC(S)$ has a LIM, then S satisfies $P(n)$ for each n :*
$P(n)$: Let E be a separated locally convex space and $\mathcal{S} = \{T_s;\ s \in S\}$ is a continuous representation of S as linear operators from E into E and jointly continuous on compact convex subsets of E. Let X be an n-consistent subset of E. Then there exists $L_0 \in \mathcal{L}_n(X)$ such that $T_s(L_0) = L_0$ for each $s \in S$.

(b) If $P(n)$ holds for some $n = 1, 2, \ldots$ then $LUC(S)$ has a LIM. In particular $P(m)$ and $P(n)$ are equivalent for all $m,\ n = 1, 2, \ldots$.

Remark 4.2 (a) Theorem 4.1(a) was first established by Ky Fan [6] for discrete S, and by Lau [14] for topological S (see [1], p. 555] for correction). To prove (b), we can modify the proof in [17, Theorem 4.2]: Let $H = LUC(S)^*$ with the weak*-topology and $E = H \times \mathbb{R}^n$ with the product topology. Let M be the set of means on $LUC(S)$ i.e. all $m \in LUC(S)^*$ such that $\|m\| = m(1) = 1$. For each $m \in M$, let $V_m =$ span $\{(m, b);\ b \in B\}$ where $B = \{e_i\} \subseteq \mathbb{R}^n$, $e_i = (0, \cdots 0, 1, 0 \cdots, 0)$ (with "1" in the ith place), and identity H with $H \times \{0\}$, $\mathbb{R}^n$ with $\{0\} \times \mathbb{R}^n$. Then $E = H \oplus V_m$ for each $m \in M$. Let $X = \cup\{V_m;\ m \in M\}$. Consider the continuous representation $\mathcal{S} = \{T_s;\ s \in S\}$ of S on E defined by $T_s(\phi, t) = (\ell_s^* \phi, t)$, $\phi \in H$, $t \in \mathbb{R}^n$. Then $\mathcal{S}$ is jointly continuous on compact convex subsets of E. Note that $T_s V_m = V_{\ell_s^* m}$ for each $s \in S$, $m \in M$. Also for each $x = (\phi, t) \in E$, $\phi \in LUC(S)^*$, $t \in \mathbb{R}^n$, and $t = \sum_{i=1}^{n} \lambda_i e_i$, then

$$(x + H) \cap X = (t + H) \cap X = t + (\sum_{i=1}^{n} \lambda_i) M$$

which is compact and convex. By $P(n)$, we can find $L_0 \in \mathcal{L}_n(X)$ such that $T_s(L_0) = L_0$ for each $s \in S$. Necessarily $L_0 = V_m$ for some $m \in M$. Clearly m is a LIM on $LUC(S)$.

(b) Consider on S the following finite dimensional invariant subspace properties:

$P'(n)$: Let E be a separated locally convex space and $\mathcal{S} = \{T_s;\ s \in S\}$ be a weakly continuous representation of S as linear operators from E into E. Let X be an n-consistent subset of E. Then there exists $L_0 \in \mathcal{L}_n(X)$ such that $T_s(L_0) = L_0$ for each $s \in S$.

$P''(n)$: Let E be a separated locally convex space and $\mathcal{S} = \{T_s;\ s \in S\}$ be a measurable representation of S as continuous linear maps from E into E. Let X be an n-consistent subset of E. Then there exists $L_0 \in \mathcal{L}_n(X)$ such that $T_s(L_0) = L_0$ for each $s \in S$.

Clearly $P''(n) \Longrightarrow P'(n) \Longrightarrow P(n)$, and $P''(n) \Longleftrightarrow (F_3'')$, $P'(n) \Longleftrightarrow (F_3')$, and $P(n) \Longleftrightarrow (F_3)$ using argument in (a) and [18, Theorem 3.3 and

Theorem 3.4]. However, we do not know if $P(n) \Longrightarrow P'(n)$ or $P'(n) \Longrightarrow P''(n)$ even when S is compact. When G is a locally compact group, then all these conditions are equivalent.

(c) Let $P_A(n)$ denote the same property as $P(n)$ with joint continuity replaced by equicontinuity on compact convex subsets and $P_W(n)$ denotes $P(n)$ with joint continuity replaced by quasi-equicontinuity on compact convex subsets K of E (i.e. the closure of $\mathcal{S}$ in E^K with the product topology consists entirely of continuous maps). Then an argument similar to that given for Theorem 4.1(b) in (a) above shows that for any positive integer n :

(i) $P_A(n)$ implies $AP(S)$ has a LIM

(ii) $P_W(n)$ implies $W(S)$ has a LIM.

However, it is not known whether the converse of (i) or (ii) also holds (see [18] for details).

Using Mitchell's fixed point theorem [22] and argument as given in Remark 4.2(a), we have (see [14, Theorem 4] and [18, p. 555]):

Theorem 4.3 *(a) If $LUC(S)$ has a multiplicative LIM, then S satisfies: $Q(n)$: Let E be a separated locally convex space and $\mathcal{S} = \{T_s;\ s \in S\}$ be a continuous representation of S as linear operators from E into E jointly continuous on compact subsets of E. Let X be a subset of E such that there exists a closed $\mathcal{S}$-invariant subspace H of E with co-dimension n, and $x + H \cap X$ is compact for each $x \in E$. If $\mathcal{L}_n(X)$ is non-empty and $\mathcal{S}$-invariant, then there exists $L_0 \in \mathcal{L}_n(X)$ such that $T_s(L_0) = L_0$ for each $s \in S$.*

(b) If S has $Q(n)$, for some $n = 1, 2, \ldots$, then $LUC(S)$ has a multiplication LIM.

References

[1] D. Alspach, *A fixed point free nonexpansive map,* Proceedings A.M.S., **82** (1981), 423–424.

[2] J.F. Berglund, H.D. Junghenn and P. Milnes, *Analysis on Semigroups,* John Wiley & Sons, New York (1989).

[3] M.M. Day, *Fixed point theorem for compact convex sets,* Illinois J. Math., **5** (1961), 585–590.

[4] R. DeMarr, *Common fixed points for commuting contracting mappings,* Pacific J. Math., **13** (1963), 1139–1141.

[5] N. Dunford and J.T. Schwartz, *Linear Operators, Part I,* Interscience Publishers, Inc., New York (1967).

[6] K. Fan, *Invariant subspaces for a semigroup of linear operators,* Indag. Math., **27** (1965), 447–451.

[7] E. Granirer and A.T. Lau, *Invariant means on locally compact groups,* Illinois J. Math., **15** (1971), 249–257.

[8] E. Granirer, *Extremely amenable semigroups*, Math. Scand., **17** (1965), 177–197.

[9] E. Hewitt and K. Ross, *Abstract Harmonic Analysis I,* Springer-Verlag, Berlin etc. (1963).

[10] R.D. Holmes and A.T. Lau, *Non-expansive actions of topological semigroups and fixed points,* Journal of London Math. Soc., **5** (1972), 330–336.

[11] R. Hsu, *Topics on weakly almost periodic functions*, Dissertation, SUNY at Buffalo (1985).

[12] A.T. Lau, *Topological semigroups with invariant means in the convex hull of multiplication means,* Transactions A.M.S., **148** (1970), 69–84.

[13] A.T. Lau, *Invariant means on almost periodic functions and fixed point properties,* Rocky Mountain J. Math., **3** (1973), 69–76.

[14] A.T. Lau, *Finite dimensional invariant subspaces for a semigroup of linear operators*, J. Math. Analysis and Applications, **97** (1983), 374–379.

[15] A.T. Lau, *Amenability of semigroups* in "The Analytic and Topological Theory of Semigroups" (K.H. Hofmann, J.D. Lawson and J.S. Pym, Eds.), pp. 313–334, De Gruyter Expositions in Mathematics 1, Berlin, New York (1990).

[16] A.T. Lau, *Some fixed point theorems and their applications to w^*-algebras*, in "Fixed Point Theory and its Applications," (S. Swaminathan, Ed.), pp. 121–129, Academic Press (1976).

[17] A.T. Lau, L.T. Paterson and J.C.S. Wong, *Invariant subspace theorems for amenable groups,* Proc. Edinburgh Math. Soc., **32** (1989), 415–430.

[18] A.T. Lau and J.C.S. Wong, *Finite dimensional invariant subspaces for measurable semigroups of linear operators,* J. Math. Analysis and Applications, **127** (1987), 548–558.

[19] A.T. Lau and W. Takahashi, *Invariant means and fixed point properties for non-expansive representations of topological semigroups,* Topological methods in nonlinear analysis (to appear).

[20] T.C. Lim, *Characterizations of normal structure,* Proceedings A.M.S., **43** (1974), 313–319.

[21] T.C. Lim, *Asymptotic centers and nonexpansive mappings in conjugate Banach spaces,* Pacific J. Math., **90** (1980), 135–143.

[22] T. Mitchell, *Topological semigroups and fixed points,* Illinois J. Math., **14** (1970), 630–641.

[23] W.G. Rosen, *On invariant means over compact semigroups,* Proc. A.M.S., **7** (1956), 1076–1082.

[24] B. Simon, *A remark on groups with fixed point property,* Proc. A.M.S., **32** (1972), 623–624.

[25] W. Takahashi, *Fixed point theorem for amenable semigroup of non-expansive mappings,* Kodai Math. Sem. Report, **21** (1969), 383–386.

Pricing of Bonds and their Derivatives with Multi-factor Stochastic Interest Rates : A Note

Miura Ryozo
Department of Commerce
Hitotsubashi University
Naka, Kunitachi-shi, Tokyo 186, JAPAN

Kishino Hirohisa*
Department of International and Social Relations
University of Tokyo Komaba, Meguro-ku, Tokyo 153, JAPAN

Abstract and Introduction

Vasicek(1977) derived a Bond Pricing formula for the interest rate following a single factor Ornstein-Uhlenbeck process. Along with this line, Jamshidian(1989) derived a bond option pricing formula, and Chen (1992) derived formulae for futures and futures option on bonds.

The data analysis for the Japanese Government Bonds traded in Tokyo, however, is said to show that the rate of return of JGB seems to have more than one factors, maybe three factors. This motivated me to work on the extension of the Vasicek-Jamshidian's pricing framework toward the multi-factor settings.

The extension of the framework had gone through, and it covers the pricing of bonds and their derivatives; bond options, bond forwards, bond futures and bond futures options. The exact pricing formulae for a specific case were obtained. While preparing for this conference, I had come across with the works of Duffie (1992) and Beaglehole and Tenney (1991) which contain implicitly and/or explicitly the mathematical set-up and the derived formulae in our work. So this note does not provide any new result essentially. However, for a minor supplement to these literature and also for a record of the talk in the conference (upon the request of its organizer), we write down these results as a note.

*The second author, Kishino, joined the work within the calculation of the pricing formulae.

A data-analysis of the interest-rates observed at swap trading in Tokyo market (Shibata et al.(1993)) is referred here. The result of the data-analysis will be briefly summarized, and based on which, the following two aspect, that may be important for modelling the stochastic interest rate process, will be discussed; the number of factors for the term structure of the interest rates and the two levels of stochastics involved in the behavior of the interest rates.

This note is organized in the following way. There are five sections. Section1 gives the definition of the multi-factor interest rate process. Section2 derives the partial differential equation which the price of any interest rate security in the market has to satisfy. section3 derives the exact pricing formulae for zero coupon bonds and European call options on the zero coupon bonds, and also derives forward prices and futures prices on the zero coupon bonds. Section4 summarizes briefly the results in Shibata et al.(1993). Section5 discusses a few important aspects for building a realistic model of interest rate processes.

1. Interest rates

Let $r_1(t)$ be specifically the instantaneous interest rate, and let $r_2(t)$, $r_3(t), ..., r_k(t)$ be stochastic processes related to other interest rates or let them be these yields observed at time t in the term structure for the term points $s_2, s_3, ..., s_k$.

It is talked about that the data of daily changes for Japanese Government Bond (coupon bonds) in Tokyo market show that bond price changes may be based on three (or so) stochastic factors, which are common to all the traded J.G. bonds.

They might suggest the following model of bond price changes;

$$\begin{bmatrix} dP_1/P_1 \\ \vdots \\ dP_n/P_n \end{bmatrix} = \begin{bmatrix} \mu_1 \\ \vdots \\ \mu_n \end{bmatrix} + \begin{bmatrix} b_{11} \\ \vdots \\ b_{n1} \end{bmatrix} f_1 + \begin{bmatrix} b_{12} \\ \vdots \\ b_{n2} \end{bmatrix} f_2 + \begin{bmatrix} b_{13} \\ \vdots \\ b_{n3} \end{bmatrix} f_3 + \begin{bmatrix} e_1 \\ \vdots \\ e_n \end{bmatrix}.$$

These suggestions for the existence of stochastic factors might be based on the analysis of bond price data using some specific statistical methods such as principal component analysis and/or factor analysis. Since the mathematical discussion does not require to specify the number of stochastic factors, we let k be the number of factors in the beginning three sections.

Let $z_1(t), z_2(t), \ldots, z_k(t)$ be the k independent standardized Wiener processes.

[A1] Assume that $r_1(t), r_2(t), \ldots, r_k(t)$ satisfy the following equations:

$$\begin{bmatrix} dr_1(t) \\ \vdots \\ dr_k(t) \end{bmatrix} =$$

$$\begin{bmatrix} f_1(t, r_1(t), r_2(t), \ldots, r_k(t)) \\ \vdots \\ f_k(t, r_1(t), r_2(t), \ldots, r_k(t)) \end{bmatrix} dt + \begin{bmatrix} g_{11}(t), & \ldots, & g_{1k}(t) \\ \vdots & \vdots & \vdots \\ g_{k1}(t), & \ldots, & g_{kk}(t) \end{bmatrix} \begin{bmatrix} dz_1(t) \\ \vdots \\ dz_k(t) \end{bmatrix}$$

where $dz_i(t) \cdot dz_j(t) = 0$, for $i, j = 1, 2, \ldots, k$, $(i < j)$.

$f_j(t, r_1(t), r_2(t), \ldots, r_k(t))$'s will be specified later, in section 3, to be a linear function of $r_1, r_2, \ldots, r_k$ for deriving explicit pricing formulae. Until then, they can be quite arbitrary (with necessary regularity).

Duffie(1992. page 137.) takes the same view as that in [A1] for modelling the behavior of the term structure of interest rates.

[A2] Assume that for any t under consideration , the matrix

$$\begin{bmatrix} g_{11}(t), & \ldots, & g_{1k}(t) \\ \cdots & \cdots & \cdots \\ g_{k1}(t), & \ldots, & g_{kk}(t) \end{bmatrix}$$

be non-singular.

Vasicek(1977), Jamshidian(1989) and Chen(1992) used the one factor model ($k = 1$) for the short term interest rate, and derived the exact pricing formulae for zero coupon bond, European bond option, bond futures and European bond futures option. Our setting [A1] and [A2] is a natural extension of their one factor model.

A meaningful extension to a multi-factor framework requires that the number of interest rate variables (r's) be more than one if there are k factors (z's), as long as we assume the interest rate security prices be functions of interest rate variables: if the prices are functions of r_1 only, then all the stochastic factors are included within dr_1 at Ito's stochastic calculus. Then, all the price movements are based on dr_1 and are perfectly correlated as is case of the one-factor model, and hence the multi-factors are not utilized in the analysis. To make full use of the k factors, the number of interest rate variables (r's) should be k. If we take more than k interest rate variables, the exceeds are waisted in the analysis.

2. The Partial Differential Equation for the Prices of Interest Rate Securities

Let $U_1, U_2, \ldots$ be the prices of interest rate securities.

[A3] Assume that the prices of interest rate securities be functions of t, and $r_1(t), r_2(t), \ldots, r_k(t)$. Write $U_i(t, r_1, r_2, \ldots, r_k), i = 1, 2, \cdots$ for the prices.

[A4] Assume that for any k-combination of securities $(i_1, i_2, \ldots, i_k)$, the matrix

$$\begin{bmatrix} \frac{1}{U_{i_1}}\frac{\partial U_{i_1}}{\partial r_1}, & \cdots, & \frac{1}{U_{i_k}}\frac{\partial U_{i_k}}{\partial r_1} \\ \vdots & & \vdots \\ \frac{1}{U_{i_1}}\frac{\partial U_{i_1}}{\partial r_k}, & \cdots, & \frac{1}{U_{i_k}}\frac{\partial U_{i_k}}{\partial r_k} \end{bmatrix}$$

be non-singular.

This assumption requires that the price of a security behave differently from those of the other securities in response to the movements of interest rate variables.

By Ito's calculus, we have for any $i, i = 1, 2, \cdots$

$$\frac{dU_i}{U_i} = \mu_i dt + \sigma_{i1}\, dz_1 + \cdots + \sigma_{ik}\, dz_k$$

where

$$\mu_i = \frac{1}{U_i}\left[\frac{\partial U_i}{\partial t} + \sum_{j=1}^{k} f_j \cdot \frac{\partial U_i}{\partial r_j} + \frac{1}{2}\sum_{j=1}^{k}\sum_{\ell=1}^{k}\frac{\partial^2 U_i}{\partial r_j \partial r_\ell}\cdot\left\{\sum_{m=1}^{k} g_{jm}\, g_{\ell m}\right\}\right]$$

$$\sigma_{i1} = \frac{1}{U_i}\left[\sum_{j=1}^{k} g_{j1}\cdot\frac{\partial U_i}{\partial r_j}\right], \cdots, \sigma_{ik} = \frac{1}{U_i}\left[\sum_{j=1}^{k} g_{jk}\cdot\frac{\partial U_i}{\partial r_k}\right].$$

2.1 Arbitrage Portfolio

[A5] Assume that there be no arbitrage in the market.

Take any $k+1$ securities and invest on them with portfolio weights $w_1, w_2, \ldots, w_{k+1}$. Then the rate of return of the portfolio W is written as follows.

$$\begin{aligned} \frac{dW}{W} &= \sum_{j=1}^{k+1}\frac{dU_j}{U_j}\cdot w_j, \qquad \left(\sum_{j=1}^{k+1} w_j = 1, \quad w_j \neq 0\right) \\ &= \left(\sum_{j=1}^{k+1} w_j\, \mu_j\right) dt + \sum_{\ell=1}^{k}\left(\sum_{j=1}^{k+1} w_j\, \sigma_{j\ell}\right) dz_\ell. \end{aligned}$$

For this portfolio to be riskless at time t, the weights $w_1, w_2, \ldots, w_{k+1}$ have to satisfy

$$\sum_{j=1}^{k+1} w_j\, \sigma_{j\ell} = 0, \qquad \ell = 1, 2, \ldots, k.$$

Let $w^{*\prime} = (w_1^*, w_2^*, \ldots, w_{k+1}^*)$ be the solution of this set of linear equation, and let W^* be the value of the portfolio with this weight. Then under the assumption [A5], we have

$$dW^*/W^* = r_1(t)\cdot dt.$$

This implies that

$$\sum_{j=1}^{k+1} w_j^* \mu_j = r_1(t).$$

Hence

$$\sum_{j=1}^{k+1} w_j^* (\mu_j - r_1) = 0.$$

2.2 Market Price of Risk

In summarizing the above, we see that the portfolio weight w^* satisfies the following linear equations.

$$\sum_{j=1}^{k+1} w_j^* = 1$$

$$\sum_{j=1}^{k+1} w_j^* \sigma_{j\ell} = 0, \qquad \ell = 1, 2, \ldots, k$$

$$\sum_{j=1}^{k+1} w_j^* (\mu_j - r_1) = 0$$

This set of equations implies that the $k+1$ vectors

$$\begin{bmatrix} \mu_1 - r_1 \\ \mu_2 - r_1 \\ \vdots \\ \mu_{k+1} - r_1 \end{bmatrix}, \begin{bmatrix} \sigma_{1,1} \\ \sigma_{2,1} \\ \vdots \\ \sigma_{k+1,1} \end{bmatrix}, \begin{bmatrix} \sigma_{1,2} \\ \sigma_{2,2} \\ \vdots \\ \sigma_{k+1,2} \end{bmatrix}, \ldots, \begin{bmatrix} \sigma_{1,k} \\ \sigma_{2,k} \\ \vdots \\ \sigma_{k+1,k} \end{bmatrix}$$

are linearly dependent. Therefore, there exist $\phi_1(t), \phi_2(t), \ldots, \phi_k(t)$ (dependent on the time t, the value of $r_1, r_2, \ldots, r_k$, and the choice of $k+1$ securities.) such that

$$\mu_j - r_1 = \phi_1 \sigma_{j1} + \phi_2 \sigma_{j2} + \cdots + \phi_k \sigma_{jk}, \quad j = 1, 2, \ldots, k+1.$$

It can be easily shown that $\phi_1, \ldots, \phi_k$ do not depend on the choice of the $k+1$ securities. ([A2] and [A4] are used for this proof.) Thus, $\phi_1, \ldots, \phi_k$ are common to all the securities in the market. Hence we have

$$\mu - r_1 = \phi_1 \sigma_1 + \phi_2 \sigma_2 + \cdots + \phi_k \sigma_k .$$

These values $\phi_1, \ldots, \phi_k$ are called "market price of risk".

Note that what we have done here is the same as that for the Exact multi-factor model of the APT in the mean-variance analysis. By substituting the definition of μ and σ_i 's , we have the following partial differential equation which all the interest rate securities in the market have to satisfy.

$$\frac{\partial U}{\partial t}+\sum_{j=1}^{k}\left\{f_j-\sum_{\ell=1}^{k}\phi_\ell\, g_{j\ell}\right\}\frac{\partial U}{\partial r_j}+\frac{1}{2}\sum_{j=1}^{k}\sum_{\ell=1}^{k}\left(\sum_{m=1}^{k} g_{jm}\, g_{\ell m}\right)\frac{\partial^2 U}{\partial r_j \partial r_\ell}-r_1 U=0$$

When dealing with the real market data, the following remarks may be helpful.

If we set r_j to be the yield of the bonds with maturity s_j , for $j = 2, 3, \ldots, k$, observed at time t, and if these bonds are traded in the market, then the partial differential equation will provide certain relations among the parameters and the variables. These relations will reduce the number of parameters; especially the market price of risk parameters. They will also replace the coefficients in the partial differential equation.

The relations thus obtained would be simpler if, instead, we set r_j to be the prices themselves of the traded bonds. It could be interesting for some engineering purposes, though theoretically uninteresting.

3. The Derivation of the Pricing Formulae

In this section we sketch the derivation procedure for the prices of zero coupon bonds, European options on zero coupon bonds, forward contracts and futures contracts on zero coupon bonds, and European options on the futures contracts on zero coupon bonds.

They can be obtained by specifying their values (payoffs) at the maturity of the contracts, and by solving, for each of them, the partial differential equation with that payoff as the initial condition.

3.1 Zero Coupon Bond

Solve the partial differential equation with the initial condition:

$$U(s, r_1(s), \ldots, r_k(s)) = 1$$

where s is the maturity date of the zero coupon bond.

The solution is given by the following:

$$U(t, r_1(t), \ldots, r_k(t)) = e\left[1 \cdot \exp\left\{-\int_t^s \tilde{r}_1(u)du\right\} \Big| \tilde{r}_1(t) = r_1(t), \ldots, \tilde{r}_k(t) = r_k(t)\right]$$

where $\tilde{\mathbf{r}}(\cdot) = (\tilde{r}_1(\cdot), \ldots, \tilde{r}_k(\cdot))^t$ satisfies the stochastic differential equation;

$$d\tilde{\mathbf{r}}_t = (\mathbf{f}_t - \mathbf{h}_t \varphi_t)dt + g^{1/2} d\tilde{\mathbf{z}}_t \,.$$

here **f** and **h** are

$$\mathbf{f} = (f_1, \ldots, f_k)^t \quad \varphi = (\phi_1, \ldots, \phi_k)^t \quad \mathbf{h} = \begin{pmatrix} g_{11} & \cdots & g_{1k} \\ \vdots & & \vdots \\ g_{k1} & \cdots & g_{kk} \end{pmatrix},$$

and $g^{1/2}$ is the positive square root of the matrix $\mathbf{g}$:

$$\mathbf{g} = \left(g_{j\ell} = \sum_{m=1}^{k} g_{jm}\, g_{\ell m} \qquad \begin{matrix} j = 1, 2, \ldots, k \\ \ell = 1, 2, \ldots, k \end{matrix} \right)$$

and $\tilde{z}_1, \ldots, \tilde{z}_k$ are some independent wiener processes.

assuming the linearity of f's (i.e. $\mathbf{f}_t = \mathbf{a}_t \tilde{\mathbf{r}}_t + \mathbf{b}_t$) and some regularity on g's , we have an explicit form for $\tilde{r}_1, \ldots, \tilde{r}_k$:. for $s > t$

$$\tilde{\mathbf{r}}_s = \psi(s,t) + \int_t^s \psi(s,u)(\mathbf{b}_u - \mathbf{h}_u \varphi_u)du + \int_t^s \psi(s,u)\, g^{1/2} d\tilde{\mathbf{z}}_u$$

$$\text{where } \psi(s,t) = \exp\left\{\int_t^s \mathbf{a}_u du\right\}.$$

from this, we get

$$\int_t^s \tilde{\mathbf{r}}_u du = \left\{\int_t^s \psi(u,t)du\right\} \tilde{\mathbf{r}}_t \; + \; \int_t^s \left\{\int_t^u \psi(u,\acute{u})(\mathbf{b}_{\acute{u}} - \mathbf{h}_{\acute{u}}\varphi_{\acute{u}})d\acute{u}\right\} du$$
$$+ \int_t^s \left\{\int_{\acute{u}}^s \psi(u,\acute{u})du\right\} g^{1/2} d\tilde{\mathbf{z}}_{\acute{u}} \,.$$

denoting $\quad \psi = \begin{bmatrix} \psi_1^t \\ \cdots \\ \psi_k^t, \end{bmatrix}, \qquad$ we have

$$p(t, \mathbf{r}(t), s) = \exp\Big[- \left\{\int_t^s \psi_1^{\mathrm{T}}(u,t)du\right\}$$
$$- \int_t^s \left\{\int_t^u \psi_1^{\mathrm{T}}(u,\acute{u})(\mathbf{b}_{\acute{u}} - \mathbf{H}_{\acute{u}}\varphi_{\acute{u}})d\acute{u}\right\} du$$
$$+ \frac{1}{2}\int_t^s \left\{\int_{\acute{u}}^s \psi_1^{\mathrm{T}}(u,\acute{u})du\right\} \mathbf{G}_{\acute{u}} \left\{\int_t^s \psi_1^{\mathrm{T}}(v,t)dv\right\} d\acute{u}\Big]$$

from the characteristic of the log normal distribution.

When the coeficients, $\mathbf{A}_t$, $\mathbf{b}_t$ and φ_t can be assumed to be constant as $\mathbf{A}$, $\mathbf{b}$ and φ respectively, we get a simple matrix expression. Decomposing the matrix $\mathbf{A}$ by its eigen vectors as

$$\mathbf{A} = \mathbf{C}\mathbf{D}\mathbf{C}^{-1}$$

where $\mathbf{D} = Diag(\lambda_1, \ldots, \lambda_k)$, the above equation becomes

$$\begin{aligned}P(t, \mathbf{r}(t), s) = \exp[\quad & -\mathbf{c}_1^T \mathbf{D}^{-1} \left\{ e^{(s-t)\mathbf{D}} - \mathbf{I} \right\} \mathbf{C}^{-1} \mathbf{r}(t) \\ & -\mathbf{c}_1^T \mathbf{D}^{-1} \left\{ \mathbf{D}^{-1} \left(e^{(s-t)\mathbf{D}} - \mathbf{I} \right) - (s-t)\, \mathbf{I} \right\} \mathbf{C}^{-1} (\mathbf{b} - \mathbf{H}\varphi) \\ & + \frac{1}{2} \mathbf{c}_1^T \mathbf{D}^{-1} \left\{ \left(e^{(s-t)\mathbf{D}} \boldsymbol{\Delta}\, e^{(s-t)\mathbf{D}} - \boldsymbol{\Delta} \right) - \tilde{\mathbf{G}}\, \mathbf{D}^{-1} \left(e^{(s-t)\mathbf{D}} - \mathbf{I} \right) \right. \\ & \left. \qquad - \mathbf{D}^{-1} \left(e^{(s-t)\mathbf{D}} - \mathbf{I} \right) \tilde{\mathbf{G}} + (s-t) \tilde{\mathbf{G}} \right\} \mathbf{D}^{-1} \mathbf{c}_1 \,].\end{aligned}$$

Here,

$$\begin{aligned}\tilde{\mathbf{G}} &= (\tilde{g}_{ij}) \equiv \mathbf{C}^{-1} \mathbf{G} \left(\mathbf{C}^{-1} \right)^{\mathrm{T}}, \\ (\mathbf{c}_1, \ldots, \mathbf{c}_k) &\equiv \mathbf{C}^{\mathrm{T}} \\ \boldsymbol{\Delta} &= (\Delta_{ij}) \equiv \left(\frac{\tilde{g}_{ij}}{\lambda_i + \lambda_j} \right).\end{aligned}$$

3.2 European Call Options On Zero Coupon Bond

```
-------+---------------------------+-------------+-----------> time
       t                           T             s
```

Solve the same partial differential equation with the initial condition;

$$U(T, \tilde{r}_1(T), \ldots, \tilde{r}_k(T)) = \max\{P(T, s, \tilde{r}_1(T), \ldots, \tilde{r}_k(T)) - K, 0\}$$

where $P(T, s, \ldots)$ is the price at time T of bond with maturity s. K is a constant and
T is the maturity date of the option.

The solution is given in the following form.

$$\begin{aligned}& C(t, r_1(t), \ldots, r_k(t) : T, K, s) \equiv U(t, r_1(t), \ldots, r_k(t)) \\ &= E\left[\max\{P(T, \tilde{r}_1(t), \ldots, \tilde{r}_k(t) : s) - K, 0\} \cdot \exp\left\{ -\int_t^T \tilde{r}_1(u) du \right\} \;\middle|\; \begin{matrix} \tilde{r}_1(t) = r_1(t) \\ \vdots \\ \tilde{r}_k(t) = r_k(t) \end{matrix} \right] \\ &= \exp\left\{ \mu_x + \mu_y + \frac{1}{2}({\sigma_x}^2 + 2\sigma_{xy} + {\sigma_y}^2) \right\} \left(1 - \Phi\left(\frac{\log K - \mu_x - {\sigma_x}^2 - \sigma_{xy}}{\sigma_x} \right) \right) \\ &\quad - K \exp\left(\mu_y + \frac{1}{2}{\sigma_y}^2 \right) \left(1 - \Phi\left(\frac{\log K - \mu_x - \sigma_{xy}}{\sigma_x} \right) \right)\end{aligned}$$

where

$$\mu_x = -\left\{ \int_T^s \varphi_1^{\mathrm{T}}(u, t) du \right\} \mathbf{r}(t) - \int_T^s \left\{ \int_t^u \varphi_1^{\mathrm{T}}(u, \acute{u}) \left(\mathbf{b}_{\acute{u}} - \mathbf{H}_{\acute{u}} \varphi_{\acute{u}} \right) d\acute{u} \right\} du$$

$$+\frac{1}{2}\int_T^s \left\{\int_{\acute{u}}^s \varphi_1^{\mathrm{T}}(u,\acute{u})du\right\} \mathbf{G}_{\acute{u}} \left\{\int_{\acute{u}}^s \varphi_1(v,\acute{u})dv\right\} du$$

$$\mu_y = -\left\{\int_t^T \varphi_1^{\mathrm{T}}(u,t)du\right\}\mathbf{r}(t) - \int_t^T \left\{\int_t^u \varphi^{\mathrm{T}}(u,\acute{u})\,(\mathbf{b}_{\acute{u}} - \mathbf{H}_{\acute{u}}\varphi_{\acute{u}})d\acute{u}\right\} du$$

$$\sigma_x{}^2 = \int_t^T \left\{\int_T^s \varphi_1^{\mathrm{T}}(u,\acute{u})du\right\} \mathbf{G}_{\acute{u}} \left\{\int_T^s \varphi_1(v,\acute{u})dv\right\} d\acute{u}$$

$$\sigma_y{}^2 = \int_t^T \left\{\int_{\acute{u}}^T \varphi_1^{\mathrm{T}}(u,\acute{u})du\right\} \mathbf{G}_{\acute{u}} \left\{\int_{\acute{u}}^T \varphi_1(v,\acute{u})dv\right\} d\acute{u}$$

$$\sigma_{xy} = \int_t^T \left\{\int_T^s \varphi_1^{\mathrm{T}}(u,\acute{u})du\right\} \mathbf{G}_{\acute{u}} \left\{\int_{\acute{u}}^T \varphi_1(v,\acute{u})dv\right\} d\acute{u}$$

3.3 Forward Prices $G(t)$

(This G is different from that in the subsection 3.1 . Sorry for the bad notation.)

The pay-off at time T of the forward contract is $V(T)-G(t)$ where $V(T)$is the value at time T of the underlying asset (in the present case, zero coupon bond), and $G(t)$ is the forward price in the forward contract at time t. (The contract is made at time t, and $G(t)$ is fixed in the contract .) Then $G(t)$ is obtained from the following general relation;

$$0 = E_t\left[\{V(T) - G(t)\}\cdot \exp\left\{-\int_t^T \tilde{r}_1(u)du\right\}\middle|\, \tilde{r}_1(t) = r_1(t),\cdots\right].$$

This relation means that the value of the contract at time t is zero. Therefore,

$$G(t) = \frac{E_t\left[V(T)\cdot \exp\left\{-\int_t^T \tilde{r}_1(u)du\right\}\middle|\, \tilde{r}_1(t) = r_1(t),\cdots\right]}{E_t\left[\exp\left\{-\int_t^T \tilde{r}_1(u)du\right\}\middle|\, \tilde{r}_1(t) = r_1(t),\cdots\right]}$$

$$\equiv \frac{E_t\left[P(T,\tilde{r}_1(T),\ldots,\tilde{r}_k(T),s)\cdot \exp\left\{-\int_t^T \tilde{r}_1(u)du\right\}\middle|\, \tilde{r}_1(t) = r_1(t),\cdots\right]}{P(t,r_1(t),\ldots,r_k(t),T)}.$$

The numerator is calculated as

$$\exp\left[-\left\{\int_t^s \boldsymbol{\psi}_1^{\mathrm{T}}(u,t)du\right\}\mathbf{r}(t)-\int_t^s\left\{\int_t^u \boldsymbol{\psi}_1^{\mathrm{T}}(u,\acute{u})\left(\mathbf{b}_{\acute{u}}-\mathbf{H}_{\acute{u}}\boldsymbol{\varphi}_{\acute{u}}\right)d\acute{u}\right\}du\right.$$
$$\left.+\frac{1}{2}\int_t^s\left\{\int_{\acute{u}}^s \boldsymbol{\psi}_1^{\mathrm{T}}(u,\acute{u})du\right\}\mathbf{G}_{\acute{u}}\left\{\int_{\acute{u}}^s \boldsymbol{\psi}_1(v,\acute{u})dv\right\}du\right]$$

and so we get

$$\begin{aligned} G(t) \;=\; \exp&\left[-\left\{\int_T^s \boldsymbol{\psi}_1^{\mathrm{T}}(u,t)du\right\}\mathbf{r}(t)-\int_T^s\left\{\int_t^u \boldsymbol{\psi}_1^{\mathrm{T}}(u,\acute{u})\left(\mathbf{b}_{\acute{u}}-\mathbf{H}_{\acute{u}}\boldsymbol{\varphi}_{\acute{u}}\right)d\acute{u}\right\}du\right.\\ &+\frac{1}{2}\int_T^s\left\{\int_{\acute{u}}^s \boldsymbol{\psi}_1^{\mathrm{T}}(u,\acute{u})du\right\}\mathbf{G}\acute{u}\left\{\int_{\acute{u}}^s \boldsymbol{\psi}_1(v,\acute{u})dv\right\}du\\ &\left.-\frac{1}{2}\int_t^T\left\{\int_{\acute{u}}^T \boldsymbol{\psi}_1^{\mathrm{T}}(u,\acute{u})du\right\}\mathbf{G}\acute{u}\left\{\int_{\acute{u}}^T \boldsymbol{\psi}_1(v,\acute{u})dv\right\}du\right]\end{aligned}$$

When the coefficients are constant, we get as before

$$\begin{aligned} G(t)=\exp\Big[\; &-\mathbf{c}_1^T\mathbf{D}^{-1}\left\{e^{(s-t)\mathbf{D}}-e^{(T-t)\mathbf{D}}\right\}\mathbf{C}^{-1}\mathbf{r}(t)\\ &-\mathbf{c}_1^T\mathbf{D}^{-1}\left\{\mathbf{D}^{-1}\left(e^{(s-t)\mathbf{D}}-e^{(T-t)\mathbf{D}}\right)-(s-T)\,\mathbf{I}\right\}\mathbf{C}^{-1}(\mathbf{b}-\mathbf{H}\boldsymbol{\varphi})\\ &+\frac{1}{2}\mathbf{c}_1^T\mathbf{D}^{-1}\Big\{\left(e^{(s-T)\mathbf{D}}\boldsymbol{\Delta}\,e^{(s-T)\mathbf{D}}-\boldsymbol{\Delta}\right)-\tilde{\mathbf{G}}\,\mathbf{D}^{-1}\left(e^{(s-T)\mathbf{D}}-\mathbf{I}\right)\\ &\qquad-\mathbf{D}^{-1}\left(e^{(s-T)\mathbf{D}}-\mathbf{I}\right)\,\tilde{\mathbf{G}}+(s-T)\tilde{\mathbf{G}}\Big\}\mathbf{D}^{-1}\mathbf{c}_1\\ &-\frac{1}{2}\mathbf{c}_1^T\mathbf{D}^{-1}\Big\{\left(e^{(s-t)\mathbf{D}}\,\boldsymbol{\Delta}\,e^{(s-t)\mathbf{D}}-e^{(s-T)\mathbf{D}}\,\boldsymbol{\Delta}\,e^{(s-T)\mathbf{D}}\right)\\ &\qquad-\left(e^{(s-t)\mathbf{D}}\,\boldsymbol{\Delta}\,e^{(T-t)\mathbf{D}}-e^{(s-T)\mathbf{D}}\,\boldsymbol{\Delta}\right)\\ &\qquad-\left(e^{(T-t)\mathbf{D}}\,\boldsymbol{\Delta}\,e^{(s-t)\mathbf{D}}-\boldsymbol{\Delta}\,e^{(s-T)\mathbf{D}}\right)\\ &\qquad+\left(e^{(T-t)\mathbf{D}}\,\boldsymbol{\Delta}\,e^{(T-t)\mathbf{D}}-\boldsymbol{\Delta}\right)\Big\}\mathbf{D}^{-1}\mathbf{c}_1\Big].\end{aligned}$$

3.4 Futures Price $H(t)$

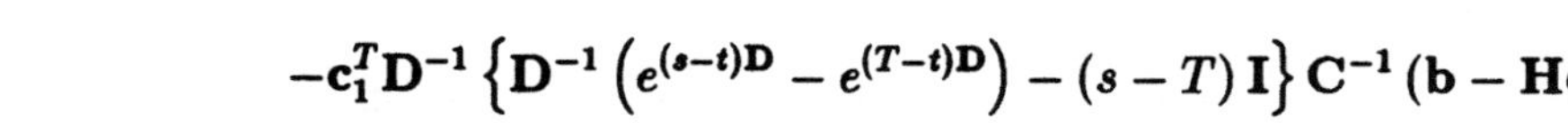

Futures contracts are very similar to forward contracts. The difference is in that futures contracts be "marked to market" every trading day. This "marking" in practice is done only once in each trading day; at a discrete time point on the continuous time line.

In the mathematical theory, however, this "marking" is treated as being done continuously. This is an approximation to the reality. The error of this approximation can be seen, for example, in the following form. (See Cox et al.(1981))

$$\begin{aligned} H(t) &= \exp\left\{\int_t^T r_1(u)du\right\} \\ &\quad + \sum_{j=0}^{n-1} \exp\left\{\int_{t_0}^{t_j} r_1(u)du\right\} \cdot [H(t_{j+1}) - H(t_j)] \exp\left\{\int_{t_{j+1}}^T r_1(u)du\right\} \\ &= V(t)\cdot \exp\left\{\int_t^T r_1(u)du\right\} \\ &\quad - \exp\left\{\int_t^T r_1(u)du\right\} \sum_{j=0}^{n-1} \left\{1 - \exp\left(-\int_{t_j}^{t_{j+1}} r_1(u)du\right)\right\} \\ &\qquad [H(t_{j+1}) - H(t_j)] \end{aligned}$$

This is the pay-off of a certain dynamical portfolio of futures contracts which reinvest at each "marking" time point. This portfolio buys one unit of future at time t_0, and $\exp\left\{\int_{t_0}^{t_j} r_1(u)du\right\}$ unit of future at time t_j , for $j = 1, 2, \ldots, n-1$. The error term (the second term) will be smaller if the investor predicts at time t_j the instantaneous interest rate of the next day (time t_{j+1}). By neglecting the second term, we have the pay-off of futures contract at time T (maturity date of the futures contract) that is $V(T)\cdot\exp\left\{\int_t^T \tilde{r}_1(u)du\right\}$ where $V(T)$ is the value at time T of the underlying asset (in our case, zero coupon bond).

Thus we have

$$\begin{aligned} &H(t) \\ &= E_t\left[V(t)\cdot\exp\left\{\int_t^T \tilde{r}_1(u)du\right\}\cdot\exp\left\{-\int_t^T \tilde{r}_1(u)du\right\}\Big|\ldots\right] \\ &= E_t[V(t)|\cdots] \\ &= E_t[P(T, \tilde{r}_1(T), \ldots, \tilde{r}_k(T), s) \mid \tilde{r}_1(t) = r_1(t), \cdots] \\ &= \exp\Bigg[-\left\{\int_T^s \boldsymbol{\psi}_1^{\mathrm{T}}(u,t)du\right\}\mathbf{r}(t) - \int_T^s\left\{\int_t^u \boldsymbol{\psi}_1^{\mathrm{T}}(u,\acute{u})\,(\mathbf{b}_{\acute{u}} - \mathbf{H}_{\acute{u}}\boldsymbol{\varphi}_{\acute{u}})d\acute{u}\right\}du \\ &\qquad + \frac{1}{2}\int_T^s\left\{\int_{\acute{u}}^s \boldsymbol{\psi}_1^{\mathrm{T}}(u,\acute{u})du\right\}\mathbf{G}\acute{u}\left\{\int_{\acute{u}}^s \boldsymbol{\psi}_1(v,\acute{u})dv\right\}du \\ &\qquad + \frac{1}{2}\int_t^T\left\{\int_T^s \boldsymbol{\psi}_1^{\mathrm{T}}(u,\acute{u})du\right\}\mathbf{G}\acute{u}\left\{\int_T^s \boldsymbol{\psi}_1(v,\acute{u})dv\right\}du\Bigg] \end{aligned}$$

If the coefficients are all constants,

$$H(t) = \exp[\quad -\mathbf{c}_1^T\mathbf{D}^{-1}\left\{e^{(s-t)\mathbf{D}} - e^{(T-t)\mathbf{D}}\right\}\mathbf{C}^{-1}\mathbf{r}(t)$$

$$
\begin{aligned}
&-\mathbf{c}_1^T\mathbf{D}^{-1}\left\{\mathbf{D}^{-1}\left(e^{(s-t)\mathbf{D}}-e^{(T-t)\mathbf{D}}\right)-(s-T)\mathbf{I}\right\}\mathbf{C}^{-1}(\mathbf{b}-\mathbf{H}\boldsymbol{\varphi})\\
&+\frac{1}{2}\mathbf{c}_1^T\mathbf{D}^{-1}\Big\{\left(e^{(s-t)\mathbf{D}}\boldsymbol{\Delta}\,e^{(s-T)\mathbf{D}}-\boldsymbol{\Delta}\right)-\tilde{\mathbf{G}}\,\mathbf{D}^{-1}\left(e^{(s-T)\mathbf{D}}-\mathbf{I}\right)\\
&\qquad+\mathbf{D}^{-1}\left(e^{(s-T)\mathbf{D}}-\mathbf{I}\right)\tilde{\mathbf{G}}+(s-T)\tilde{\mathbf{G}}\Big\}\mathbf{D}^{-1}\mathbf{c}_1\\
&-\frac{1}{2}\mathbf{c}_1^T\mathbf{D}^{-1}\Big\{\left(e^{(s-t)\mathbf{D}}\,\boldsymbol{\Delta}\,e^{(s-t)\mathbf{D}}-e^{(s-T)\mathbf{D}}\,\boldsymbol{\Delta}\,e^{(s-T)\mathbf{D}}\right)\\
&\qquad-\left(e^{(s-t)\mathbf{D}}\,\boldsymbol{\Delta}\,e^{(T-t)\mathbf{D}}-e^{(s-T)\mathbf{D}}\,\boldsymbol{\Delta}\right)\\
&\qquad-\left(e^{(T-t)\mathbf{D}}\,\boldsymbol{\Delta}\,e^{(s-t)\mathbf{D}}-\boldsymbol{\Delta}\,e^{(s-T)\mathbf{D}}\right)\\
&\qquad+\left(e^{(T-t)\mathbf{D}}\,\boldsymbol{\Delta}\,e^{(T-t)\mathbf{D}}-\boldsymbol{\Delta}\right)\Big\}\mathbf{D}^{-1}\mathbf{c}_1\Big].
\end{aligned}
$$

3.5 Options On Futures Contracts

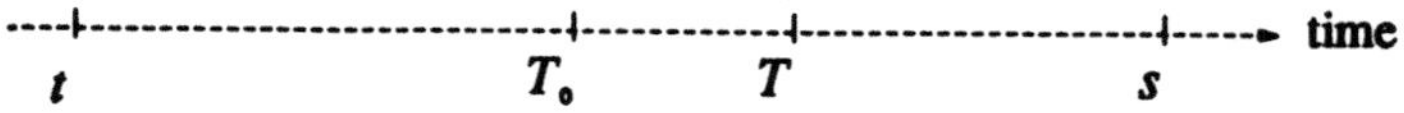

Once the functional form of the future price, $H(t, r_1, \ldots, r_k : T, s)$, is obtained, the pay-off at time T_0 (maturity date of the option contract) of the option on future is given. That for the call option is

$$\max\{H(T_0, r_1(T_0), \ldots r_k(T_0) : T, s) - K, 0\}$$

We can proceed in the same way as before, to obtain the pricing formula of the European call option on the bond future future. The formulae is not provided here.

4. A Data Analysis Of Swap Interest Rate

Shibata at al.(1993) analyzed the daily data of Euro-yen interest rates (3-months,6-months and 1-year) and Yen-to-yen Swap interest rates (3-years, 5-years, 7-years and 10-years). The data period starts from December 1,1986 and ends on September 16, 1992. The results of their data analysis depicts several important aspects we should be aware in building the model of stochastic interest rates. Specifically, the stochastics or uncertainties we deal with may be decomposed into those of different levels; e.g. the ones in daily moves and the others in monthly levels.

Let's summarize their analysis briefly. They decomposed each of the seven time series data into the sum of three time series: "Long term trend", "Short term trend" and "Irregular". The first two time series are obtained by a smoothing procedure (using LOWESS in the S-PLUS); long term trend at a day t is calculated by local regression (LOWESS) with data of 365 continuing days centering at the day t. Then using the residuals (original number - long term trend), short term trends are calculated with data of 30 days for each in the same manner. Heuristically one might regard that these trends are very sofisticated version of 365 days and 30 days moving averages. Irregulars are the left-overs (original number - long term trend - short term trend).

One might take seasonal series in place of the short term trend. They found in that case,however, that the irregular series is quite non-stationary.

Also they thought that it would be too much if one assume strict seasonalities. These are the reasons why they used the short term trend.

(i) Long term trend

For all of the seven time series, we see (by eye-sighting) that the move of the long term trend resembles to each other and that of the prime rate (plus some constant). The level of the interest rates are given by this trend. The daily increment of this trend is roughly 0.005.

(ii) Short term trend

The first three (3m,6m,1year) rates move similarly to each other. The 6-month series is leading the other two. The last four series move similarly to each other, but a little bit differently from the first three. All the seven series starts moving up around June, and move down toward the end of the calendar year.

These specific features might be due to (demand and supply caused by) traditional business activities (bonus,...), regulations, and some short term changes in the macro economic structure. — We should deal with uncertainties involved herein explicitly in the model. The theory of Economics should help explaining the move of the short term trend.

The daily increment of this trend is roughly 0.005, the same as that of long term trend.

(iii) Irregular

They are stationary. The seven Irregular series are fitted to multivariate autoregressive process of order two. And the residual vector are uncorrelated between different days. But the components are correlated to each other and they seem to have two principal components. Each component distributes non-normally, but rather very much like Cauchy.

The daily increment of this irregular series is roughly 0.05.

The irregular part of the decomposition seems to fit very well to the model described in the section 1 of the present note, except the order in the autoregressive behavior. However, (this was commented by Prof.Duffie .) their data-analysis utilized the future-time data at each time-point. So we should not yet claim that the model in section 1 be suitable for the Japanese Yen swap rates until a thorough data-analysis without using the future-time data is done.

It is possible to squeeze out other trends in addition, for example 3-months trend and/or two weeks trend.

5. Model and Uncertainty

The data analysis in the section 4 gives some hints for the analysis and modelling of stochastics and uncertainties, that those be related to the levels of time period lengths : the stochastics in daily moves and the uncertainties in monthly moves (conceptually this may not be stochastics to many statisticians, at least to me.)

"Level of time period length",or a partition of the time axis, is the important concept now. The above method of data analysis makes it possible to squeeze out the movement of interest rates related to each level of time period length.

(1) If one is thinking of the daily moves (of bonds prices and their derivative assets' prices) for a very short time period like five days, then the long and short term trends are rather constant are in the irregular part.

(2) If one is thinking of the monthly moves of prices of assets, then the irregular part may be ignored and therefore the one is to model the moves of the short term trend. The principle is to disregard the daily moves which are scattering around zero in the decomposition of the time series data, and to pick up the monthly moves and relate them to the structure of the economics and business activities. (It is easy to say this. But there is no theoretical support for this yet.)

Under the decomposition of the time series data, the daily moves and minute-to-minute moves may relate to or due to demand and supply at dealers minds' level. The monthly moves may be beyond the dealer's psychology,but rather due to the circumstances (macro/micro economic structures) of the market. The informations arriving at the market and give effects to the movement of asset prices, should have certain natures that relate to the levels of time period lengths.

One might think that it is better ,after taking off the long term trend from the original series, to proceed, in stead of taking short term trend off, directly to take multivariate autoregressive part off and leave the residuals (left-overs) for principal component analysis (or heteroskedasticity analysis). We regard this rather rough in their mixing the monthly moves with the daily moves.

References

[1] R. Chen, ***Exact Solution for Futures and European Futures Options on Pure Discount Bond***, Journal of Financial and Quantitative Analysis **27**

(1992), 97–107.

[2] J. C. Cox, J. E. Jr. Ingersoll and S. A. Ross, *The Relation between Forward Prices and Futures Prices*, Journal of Financial Economics **9** (1981), 321–346.

[3] D. Duffie, *Dynamic Asset Pricing Theory*, Prinston University Press, Chapter 7, 1992.

[4] F. Jamshidian, *An Exact Bond Option Formula*, Journal of Finance **44** (1989), 205–209.

[5] O. Vasicek, *An Equilibrium Characterization of the Term Structure*, Journal of Financial Economics **5** (1977), 177–188.

[6] D. R. Beaglehole and M. S. Tenney, *General Solutions of Some Interest Rate-Contingent Claim Pricing Equations*, Journal of Fixed Income, September (1991), 69–83.

[7] R. Shibata, R. Miura and K. Uchida, *Decomposition of Swap Interest Rates -Time series Data by Local Regression* (In Japanese), Keio Science and Technology Research Report. No.

Non-Linearity and Business Cycles in a Two-Sector Equilibrium Model: An Example with Cobb-Douglas Production Functions

Kazuo Nishimura
Institute of Economic Research, Kyoto University, Kyoto, JAPAN

Makoto Yano
Department of Economics, Yokohama National University, Yokohama, JAPAN

Abstract

This study presents a two-sector optimal growth model with Cobb-Douglas production functions in which optimal dynamics exhibits sharp non-linearity giving rise to cyclical optimal paths. This result demonstrates that such optimal paths may appear for any value of the discount factor of future utilities. Moreover, once a cyclical optimal path appears for a particular value, it appears for any value that discounts the future utilities stronger than that particular value.

1. Introduction

It has long been known in the static framework that the existence of factor intensity reversals upsets many of the standard theorems of international trade theory and causes various pathological phenomena. The recent studies on equilibrium dynamics have revealed a seemingly similar phenomenon.[1] That is to say, if factor intensity reversals are allowed for, an optimal path of capital accumulation may follow complex non-linear dynamics and, as a result, behave irregularly or cyclically.[2] This finding sharply contrasts with the conventional wisdom that an optimal path tends to converge to a stationary state over time.[3] These roles of factor intensity reversals raise an

[1]See Benhabib and Nishimura (1979) and (1985), Deneckere and Pelikan (1986), Boldrin and Montrucchio (1986) and Boldrin and Deneckere (1990).

[2]While our study is concerned with an optimal growth model, the work of Benhabib and Day (1982) and Grandmont (1985) demonstrates the possibility of complex non-linear dynamics in overlapping generations models.

[3]See McKenzie (1986).

interesting question. Namely, is the complex non-linear dynamics in the optimal growth model as pathological a phenomenon as those that trade theorists have attributed to factor intensity reversals?

Our result demonstrates that the answer to this question is "no." In fact, we construct a tent-shaped (non-linear) optimal transition function without factor intensity reversals; an optimal transition function is a function that maps each beginning-of-a-period stock level to the optimal level of end-of-a-period stock. This complex non-linearity in an optimal transition function is attributable to the finiteness of the output levels from given input levels and the existence of the consumption good sector that is always capital intensive (no factor intensity reversal).

In order to demonstrate these facts, we construct a simple example of the two-sector model with Cobb-Douglas production functions (a typical case in which a factor intensity reversal is impossible). We derive the condition on parameter values of the production functions under which cyclical optimal paths of period 2 appear. This condition reveals that such optimal paths may appear for any value of the discount factor of future utilities. Moreover, once a cyclical optimal path of period 2 appears for a particular value of the discount factor, it appears for any value that corresponds to stronger discounting. These findings demonstrate that the simplest setting of the standard two-sector model, with Cobb-Douglas production functions or without factor intensive reversals, suffices to bring out complex non-linear dynamics in optimal paths of capital accumulation. In this sense, complex non-linear dynamics in the optimal growth model is not as unusual a phenomenon as the pathological phenomena attributed to factor intensity reversals in trade theory.[4]

In what follows, Section 2 presents our model in a general framework and discusses some basic properties of the model. Section 3 demonstrates that the existence of the feasibility condition may, in a natural manner, give rise to non-linear optimal dynamics. Section 4 provides a sufficient condition of cyclical optimal paths of period 2. Section 6 presents a specific example with Cobb-Douglas production functions and a linear utility function in which cyclical optimal paths of period 2 appear. Section 6 discusses on the case of a general utility function. Section 7 makes concluding remarks.

[4] Using different types of production functions, Nishimura and Yano (1993c,1993d,1993e) demonstrate the possibility of optimal chaos resulting from the lack of factor intensity reversal.

2. The Two-Sector Model and Its Basic Properties

2.1 Model

We first set up the model in the form of an optimal growth model and, then, present a way in which an optimal growth path may be thought of a competitive equilibrium in an intertemporal market economy. The model we consider is a discrete time version of Uzawa's two-sector optimal growth model. There are two goods: the pure consumption good, C, and the pure capital good, K. Each sector uses both capital K and labor L as input. Capital input must be made one period prior to the period in which output is produced. Labor input is made in the same period as output is produced. Denote by $c = F_C(K_C, L_C)$ and $y = F_K(K_K, L_K)$ the production functions of sectors C and K, respectively. The production functions are assumed to have all the standard properties of a neoclassical production function.

Denote by c_t and y_t the outputs of sectors C and K, respectively, in period t. Moreover, denote by K_{Ct-1} and L_{Ct} the capital and labor inputs that sector C uses in order to produce c_t and by K_{Kt-1} and L_{Kt} those that sector K uses in order to produce y_t; *i.e.*,

$$c_t = F_C(K_{Ct-1}, L_{Ct}); \tag{1}$$

$$y_t = F_K(K_{Kt-1}, L_{Kt}). \tag{2}$$

Denote by k_{t-1} the aggregate capital input, *i.e.*,

$$K_{Ct-1} + K_{Kt-1} = k_{t-1}. \tag{3}$$

The capital good input for a particular period's production activity cannot exceed the capital good output from the previous period's activity, *i.e.*,

$$k_t \leq y_t. \tag{4}$$

The labor endowment of the economy is constant and time-independent. Its amount is set to be unity by normalization, *i.e.*,

$$L_{Ct} + L_{Kt} = 1. \tag{5}$$

Denote by $u(c)$ the representative consumer's utility from his consumption in a period. With these preparations, the optimal growth model is described by the following maximization problem,

$$\begin{array}{ll} \max & \sum_{t=1}^{\infty} \rho^{t-1} u(c_t) \\ s.t. & k_0 \leq k \, and \, constraints \, (1) \, through \, (5). \end{array} \tag{6}$$

where $0 < \rho < 1$.

In order to analyze the dynamics of the above model, it is convenient to express, for each given amount of capital input, the trade-off between outputs by $c = \gamma(y; k)$, *i.e.*,

$$\gamma(y; k) = \max F_C(K_C, L_C) \quad s.t. \, F_K(K_K, L_K) = y, \; L_C + L_K = 1 \, and \, K_C + K_K = k. \tag{7}$$

Note that, given an amount of capital input, k, $c = \gamma(y; k)$ captures the production possibility frontier in the output plane (see Figure 1). With this expression, the optimal growth model described by (6) can be transformed into the following form.

$$V(k) := \max \sum_{t=1}^{\infty} \rho^{t-1} v(k_{t-1}, k_t) \, s.t. \, k_0 = k, \tag{8}$$

where $v(k, y) := u(\gamma(y; k))$. Functions $v(k, y)$ and $V(k)$ are often called, respectively, reduced form utility and value functions (McKenzie,1986).

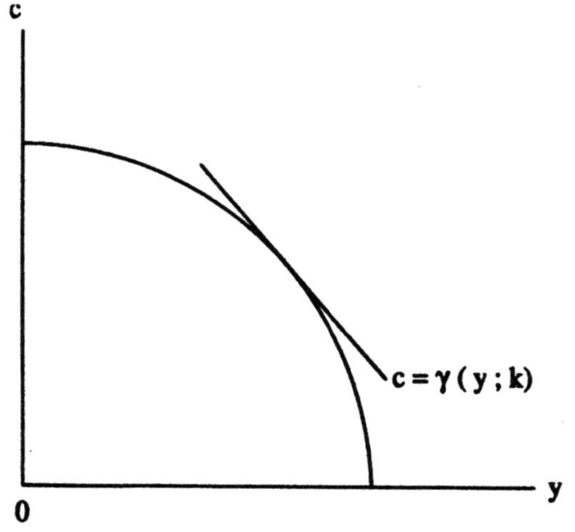

Figure 1

These functions allow to express the infinite time horizon optimization problem as a period-by-period optimization problem

$$V(k) = \max[v(k, y) + \rho V(y)] \tag{9}$$

(Bellman's principle of optimality).

As is well-known, an optimal path of capital accumulation may be thought of as a competitive equilibrium (a perfect foresight equilibrium), which is supported by a path of prices. An illustrative presentation of this fact can be constructed by using demand and supply curves of capital output. For the sake of simplicity, for the moment, consider the case in which value function $V(y)$ is differentiable. In this case, (9) gives rise to the following first order condition of optimization.

$$\rho V'(y) = -v_2(k, y). \tag{10}$$

The left-hand side of (10) captures the present value of the increase in the discounted utility sum over the future periods (cum the present period) that a marginal increase in capital output in the present period can bring about that is, the marginal value of capital output (MV). The right-hand side of (10) is the utility loss from the reductions in the output of the consumption good in the present period that an increase in capital output causes, that is the marginal cost of capital output $(MC_{|k})$.

In Figure 2, curve D depicts the marginal value of capital output. Since given a price of the capital good, p, the optimal level of capital input for the next period production is given by $\rho V'(y) = p$, curve D may be thought of as the capital demand curve. Curve $S_{|k}$ depicts the marginal cost of capital output, for given k. Since, given p, the optimal level of capital output in this period is given by $-v_2(k, y) = p$ for each given level of initial capital k, curve $S_{|k}$ may be thought of as the capital supply curve.

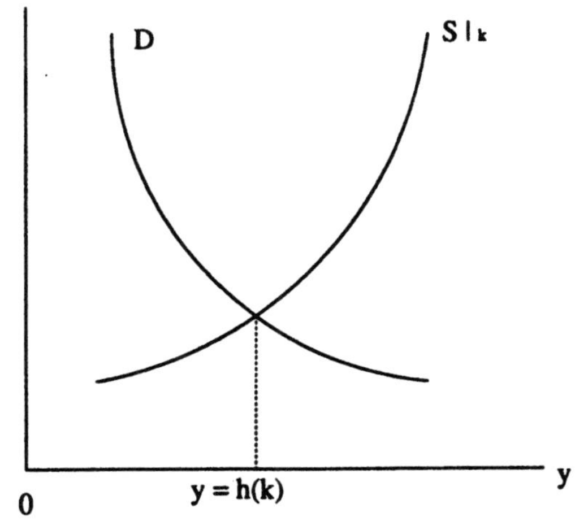

Figure 2

The optimal capital output is determined at the intersection of the demand and supply curves, E. Since the position of the demand curve depends on initial capital k, so does the optimal output. Denote by $y = h(k)$ the relationship between initial capital k and the optimal output. In other words, if a path of capital accumulation, k_t, is optimal (*i.e.*, solves optimization problem (8)), then it satisfies

$$k_t = h(k_{t-1}). \tag{11}$$

In this sense, h is often called an optimal transition function.

2.2 Monotonicity

The shape of an optimal transition function depends on the way in which supply curve $S_{|k}$ shifts as initial stock k changes. In the case in which the economy is incompletely specialized (*i.e.*, produces both outputs), as Figure 2 indicates, optimal transition function h is monotone increasing if an increase in k shifts supply curve $S_{|k}$ downwards and monotone decreasing if it shifts supply curve $S_{|k}$ upwards. The shape of optimal transition function h determines the pattern of capital accumulation along an optimal path. That is to say, optimal capital accumulation is monotone if h is monotone increasing and fluctuant if it is monotone decreasing; see Figures 3 and 4.

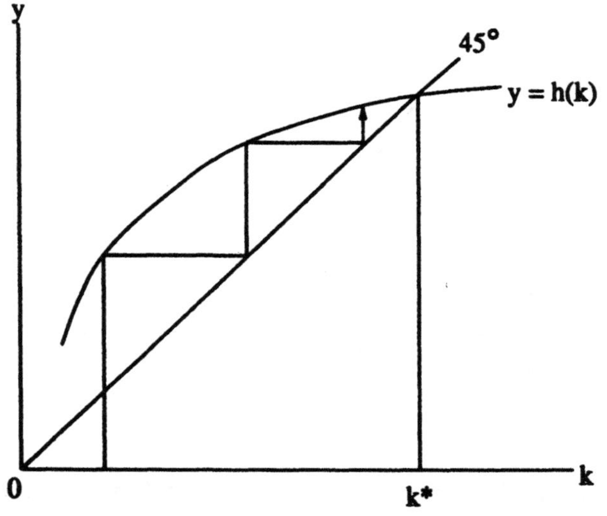

Figure 3

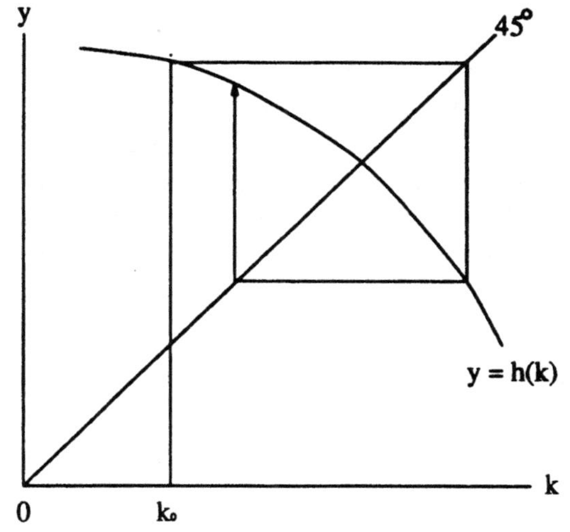

Figure 4

The way in which supply curve $S_{|k}$ shifts as k changes is characterized by the cross partial derivative of $v(k, y)$. This is because, as is noted above, $S_{|k}$ depicts the relationship in $p = -v_2(k, y)$ between p and y, for given k. That the economy produces both outputs means that (k, y) is in the interior of the domain of $v(k, y)$. Moreover, $v_{21} > 0$ means the downward shift of $S_{|k}$ whereas $v_{21} < 0$ means its upward shift, as k increases. These facts explain the following result due to Benhabib and Nishimura (1985). Denote by D° the interior of the domain of v.

Lemma 1 *A: Suppose that $v_{21} < 0$ on D°. Optimal transition function $h(k)$ is monotone decreasing at k such that $(k, h(k)) \in D^\circ$.*

B: Suppose that $v_{21} > 0$ on D°. Optimal transition function $h(k)$ is monotone increasing at k such that $(k, h(k)) \in D^\circ$.

It is noteworthy that the above argument makes an "almost complete" proof of these lemmas, except for the fact that the argument is based on the differentiability of value function V. The proof, however, dispenses with the differentiability. Because $V(k)$ is concave, it has a subgradient everywhere. Because the graph of the subgradient is a monotone decreasing curve on the $k - p$ plane, the capital demand can be expressed by a downward-sloping demand curve, which could have flat and vertical portions. With this fact, the above lemmas can be proved by an argument similar to the above.

3. Non-Linear Equilibrium Dynamics

The fact that $v_{21} < 0$ in the interior of the domain of v, D°, implies a stark non-linearity of optimal transition function $h(k)$. This non-linearity does not reflect any pathological relationship but is a natural consequence of the feasibility condition of an economy.

For the sake of explanation, see Figure 5. Curve K illustrates the graph of production function $y = F_K(k, 1)$, which corresponds to the case in which the capital good sector employs all the production factors available in the economy, *i.e.*, initial capital k and one unit of labor. In short, curve K constitutes the boundary of the set of feasible pairs of capital input and output.

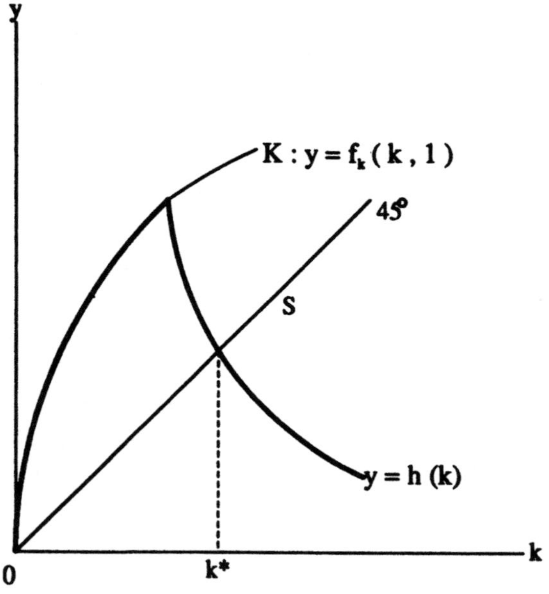

Figure 5

Suppose $v_{21} < 0$. Then, optimal transition function $h(k)$ is upward sloping. As k reduces, the value of optimal transition function $h(k)$ increases. However, since the boundary of the feasible set, K, is upward sloping, the graph of $h(k)$ must hit boundary K and, as k reduces further, must fall along the boundary; this is a result of the continuity of optimal transition function, which holds under the standard assumption (see McKenzie, 1986). This demonstrates that if $v_{21} < 0$, optimal transition function h is unimodal; a unimodal function is a function the graph of which consists of a monotone increasing part below a certain cut-off value of the argument and a monotone decreasing part above the cut-off value. The above argument can be summarized as follows.

Proposition 1 *A: If $v_{12} < 0$ on the interior of the domain of v, the optimal transition function, h, is unimodal.*

B: If $v_{12} > 0$ on the interior of the domain of v, the optimal transition function, h, is monotone increasing.

The discussion so far has all been based on the properties of reduced form utility function $v(k, y)$. It is, however, a natural question how those properties are related to those of the fundamental structure of the model, specified by production and utility functions.

For this purpose, we first analyze technological factors by focusing on the case in which the utility function, u, is linear in consumption, *i.e.*, $u(c) = c$; this is the case that most of the existing examples of non-linear optimal dynamics have focused on. In a later section, we will examine the role of the consumption side by generalizing a general utility function.

By the linearity of the utility function, $v(k, y) = \gamma(y; k)$. Therefore, the marginal cost of capital output $-v_2$ is equal to the slope of a production possibility frontier (PPF); recall that $c = \gamma(y; k)$ captures the PPF for each given level of capital input k. As capital input k increases, the PPF shifts outwards. Suppose that the consumption good sector is capital intensive. [5] According to the Rybczynski theorem, as k increases, the output of the consumption good sector increases while that of the capital good sector decreases, given a relative price between the two goods. This implies that for a given capital output y, an increase in capital input k steepens the slope of the PPF, *i.e.*, shifts capital supply curve $S_{|k}$ upwards. Since, as is seen above, the upward shift of $S_{|k}$ implies monotone decreasing $h(k)$, the following result holds.

Proposition 2 *Suppose that the utility function, $u(c)$, is linear in consumption c. If the consumption good sector is capital intensive, the optimal transition function, $h(k)$, is unimodal.*

The above discussion demonstrates that, on the production side, one factor contributing to non-linear optimal dynamics is the existence of a capital intensive consumption good sector.

4. Business Cycles

4.1 Period 2 Cycles

In this section, assume $F_K(0, 1) = 0$ (necessity of capital input). Suppose that, as is shown in Figure 5, optimal transition function h has a unique interior stationary state, say S, that is locally unstable. Proposition 1-A implies that, on the one hand, any optimal path starting from a point near S, oscillating during periods directly following the initial period, moves away from the stationary state. On the other hand, any optimal path starting

[5] This is the case considered by Uzawa (1961).

from a point near the origin must monotonically increase for a certain span of periods subsequent to the initial period. This establishes the existence of a cyclical optimal path of period 2 or, intuitively speaking, persisting business cycles.

The above discussion demonstrates that in order to guarantee the existence of persisting business cycles, it suffices to guarantee that of a unique interior stationary state that is locally unstable. To this end, take the Euler equation,

$$v_2(k_{t-1}, k_t) + \rho v_1(k_t, k_{t+1}) = 0. \tag{12}$$

We call k a stationary solution of (12) if it satisfies (12) as well as $k = k_{t-1} = k_t = k_{t+1}$. Let k^ρ be such a stationary solution. By locally expanding (12) at $(k_{t-1}, k_t, k_{t+1}) = (k^\rho, k^\rho, k^\rho)$,

$$v_{12}^\rho z_{t-1} + (v_{22}^\rho + \rho v_{11}^\rho) z_t + \rho v_{12}^\rho z_{t+1} = 0, \tag{13}$$

where $z_t = k_t - k^\rho$ and where v_{ij}^ρ is the second partial derivative of $v(k, y)$ evaluated at $(k, y) = (k^\rho, k^\rho)$. The characteristic roots of recursive equation (13) are

$$z = \frac{-(v_{22}^\rho + \rho v_{11}^\rho) \pm \sqrt{(v_{22}^\rho + \rho v_{11}^\rho)^2 - 4(v_{12}^\rho)^2}}{2v_{12}^\rho}. \tag{14}$$

Any optimal path from a point in a sufficiently small neighborhood of k^ρ must satisfy the Euler equation, (12), for a certain span of periods subsequent to the initial period. This implies that the slope of optimal transition function $h(k)$ at $k = k^\rho$ is equal to one of the characteristic roots given (14).

The above discussion may be summarized as follows:

Proposition 3 *Suppose $F_K(0,1) = 0$ (necessity of capital input) and that* (12) *has a unique stationary solution. If, at the stationary solution, both of the characteristic roots of* (13) *given by* (14) *are smaller than -1, there is a cyclical optimal path of period 2.*

5. An Example with Cobb-Douglas Production Functions

In this section, we present an example of the above model with Cobb-Douglas production functions that possesses all the properties discussed above. The example, in addition, demonstrates that for any value of discount factor ρ, it is possible to find parameter values with which period-2 cycles exist and that if period-2 cycles exist for a certain value of discount factor ρ, they persist for any smaller value.

Consider the following production functions.

$$F_K(K_K, L_K) = (K_K)^{1/2}(L_K)^{1/2}; \tag{15}$$

$$F_C(K_C, L_C) = (K_C)^{\alpha}(L_C)^{1-\alpha}. \tag{16}$$

Assume $1/2 < \alpha < 1$, which implies the consumption good sector is capital intensive. In addition, consider

$$u(c) = c. \tag{17}$$

With these specific forms of utility and production functions, by Proposition 2, optimal transition function h is unimodal.

Given the specific form of production and utility functions, function γ can be expressed as

$$\gamma(y;k) = \left(\frac{1-\alpha}{(1-\alpha)k + (2\alpha-1)g(k,y)}\right)^{1-\alpha}[k - g(k,y)], \tag{18}$$

where

$$g(k,y) = \frac{(2\alpha-1)y^2 + \sqrt{(2\alpha-1)^2y^4 + 4\alpha(1-\alpha)y^2k}}{2\alpha}. \tag{19}$$

Note, by (17), $v(k,y) = \gamma(y;k)$. Thus, given (18), system (12) has a unique stationary solution

$$k = \frac{\alpha\rho^2}{2[2(1-\alpha) + \rho(2\alpha-1)]}. \tag{20}$$

The characteristic roots are

$$z' = -\frac{2(1-\alpha)}{\rho(2\alpha-1)} \tag{21}$$

and

$$z" = -\frac{2\alpha-1}{2(1-\alpha)}. \tag{22}$$

Suppose $3/4 < \alpha < 1$. Then, $z" < -1$. Moreover, $z' < -1$ if $\rho < \frac{2(1-\alpha)}{2\alpha-1}$. In short, if this condition is satisfied, the stationary state is totally unstable in the sense that the absolute values of both characteristic roots are larger than 1. Thus, by Proposition 3, the following holds:

Proposition 4 *Suppose* $3/4 < \alpha < 1$. *Then, the two-sector example with the above Cobb-Douglas production functions has a cyclical optimal path of period 2 for any* $\rho < \rho_\alpha \equiv 2(1-\alpha)/(2\alpha-1)$.

If $3/4 < \alpha < 1$, the upper bound of ρ with which a period-2 path appears is smaller than 1 ($\rho_\alpha < 1$). However, as $\alpha \to 1$, $\rho_\alpha \to 1$. This, therefore, establishes the existence of a cyclical optimal path of period 2 for any value of discount factor ρ (provided that the value of α is chosen suitably). Once, for a value of α, period-2 cycles appear for a certain value of discount factor ρ, they appear for any ρ less than that value.

Note that, given $3/4 < \alpha < 1$, $z" < -1$ *and* $-1 < z' < 0$. If, instead, $1/2 < \alpha < 3/4$, $z' < -1$ *and* $-1 < z" < 0$. In these cases, the stationary state is saddle-point stable, *i.e.*, the slope of optimal transition function $h(k)$ at $k = k^\rho$ is between 0 and -1.

6. The Concavity of a Utility Function

In this section, we will consider the relationship between the shape of utility function $u(c)$ and non-linear optimal dynamics. To this end, consider a general utility function, the graph of which is illustrated in the second quadrant of Figure 6; curve P in the first quadrant of Figure 6 is the production possibility frontier.

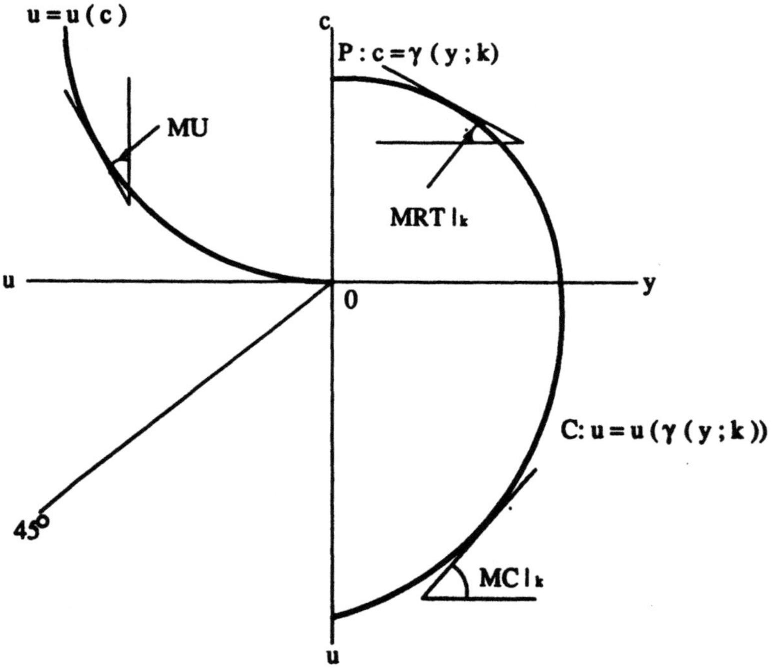

Figure 6

Curve C in the fourth quadrant of Figure 6 may be thought of as a cost curve. That is to say, it illustrates the trade-off between capital output and utility, *i.e.*, the cost of capital output in terms of utility, given a level of capital input, k. The slope of curve C captures the marginal cost of capital output, $MC_{|k} = -v_2(k, \cdot)$. As Figure 1 indicates,

$$MC_{|k} = MU \cdot MRT_{|k}, \tag{23}$$

where MU is the marginal utility, u', and where $MRT_{|k}$ is the marginal rate of transformation along the production possibility frontier, *i.e.*, $-\partial\gamma(y;k)/\partial y$. Given capital output y, therefore, the relative shift in the marginal cost curve that an increase in capital input k causes can be expressed as the sum of the relative changes in MU and $MRT_{|k}$, *i.e.*,

$$(M\hat{C}_{|k}) = M\hat{U} + M\hat{R}T_{|k}, \tag{24}$$

where $\hat{x}$ is the relative change in x when k increases.

Suppose, as in the above sections, that the consumption good sector is capital intensive. In this case, as is noted above, an increase in capital input raises the marginal rate of transformation, *i.e.*, $M\hat{R}T_{|k} > 0$. Moreover, it increases the consumption good output and, as a result, reduces the marginal utility, *i.e.*, $M\hat{U} < 0$. Thus, the actual shift in capital supply curve $S_{|k}$ is determined by the joint effect of these two factors. Therefore, so long as the global concavity of utility function u is sufficiently small, the optimal transition function remains to be unimodal. The introduction of a strict concave utility function does not affect the existence of a cyclical optimal path of period 2. In other words, Proposition 3 remains valid.

7. Concluding Remarks

This study has demonstrated that if the consumption good sector is capital intensive, the optimal transition function exhibits very sharp non-linearity. If, in addition, the stationary state is locally unstable, cyclical optimal paths of period 2 appear from that non-linearity. This is true for any discount value of the discount factor of future utilities. Moreover, once a cyclical optimal path of period 2 appears for a given value of the discount factor, it appears for any value less than that value.

It is important to compare this result with the conventional turnpike theorems that imply that an optimal path converges to a stationary state if the discount factor is chosen sufficiently close to 1 (Scheinkman, 1976). The difference lies in what is fixed and what is allowed to be changed. On the one hand, the existing turnpike theorems establish the convergence in the case in which the technology is fixed. Our result, on the other hand, demonstrates the lack of convergence for a suitably chosen technology in the case in which the discount factor is fixed. Between our result and the conventional turnpike theorems, therefore, there is no contradiction.

The non-linearity in our example is attributable to the upper limit of the capital output from a given capital input and the existence of the capital intensive consumption good sector. Considering a linear utility function like that of this study, Benhabib and Nishimura (1985) demonstrate that if the consumption good sector is capital intensive, the optimal level of end-of-a-period stock is monotone decreasing in beginning-of-a-period stock level in the interior of the feasible set. We have revealed, however, that as the beginning-of-a-period stock level decreases, the optimal level of end-of-a-period stock can increase only until it hits the boundary of the feasible set. As the beginning-of-a-period stock level decreases further, the optimal choice decreases along the boundary of the feasible set.

This implies that, typically, an optimal path alternatingly hits the boundary of the feasible set for relatively low levels of beginning-of-a-period stock and lies in the interior of the feasible set for relatively high stock levels. That an optimal path hits the boundary of the feasible set in a certain period implies that in that period, the maximum level of capital output is produced with no consumption good output. In other words, along an optimal path, the economy is alternatingly completely specialized in the capital good sector and incompletely specialized.

This phenomenon can be compared with international specialization, a central issue in trade theory. In that context, specialization is caused by international trade, which leads to a Pareto optimal allocation. In the present context, in contrast, specialization is caused by "intertemporal trade," which leads to an optimal path of capital accumulation.

Our model can be extended into a large-country dynamic model of international trade. In that context, as the work of Nishimura and Yano (1993a,1993b) indicates, specialization can occur both intertemporally and internationally. The characterization of the interaction between intertemporal and international specialization patterns may reveal the existence of non-linearity more complex than that we consider in this study.

References

[1] J.Benhabib and R. Day (1982), *A Characterization of erratic dynamics in the overlapping generations model*, Journal of Economic Dynamics and Control **4** (1982), 37–55.

[2] J.Benhabib and K.Nishimura (1979), *The Hopf bifurcation and the existence and stability of closed orbits in multisector models of optimal economic growth*, Journal of Economic Theory **21** (1979), 421–444.

[3] ___ and ___ (1985), *Competitive equilibrium cycles*, Journal of Economic Theory **35** (1985), 284–306.

[4] Boldrin.M and R.Deneckere (1990), *Sources of complex dynamics in two-sector growth models*, Journal of Economic Dynamics and Control **14** (1990), 627–653.

[5] M.Boldrin and L. Montrucchio (1986), *On the indeterminacy of capital accumulation paths*, Journal of Economic Theory **40** (1986), 26–39.

[6] R.Deneckere and S.Pelikan (1986), *Competitive chaos*, Journal of Economic Theory **40** (1986), 13–25.

[7] J.Grandmont (1985), *On endogenous competitive business cycles*, Econometrica **53** (1985), 995–1045.

[8] L.Mckenzie (1986), *Optimal economic growth, turnpike theorems and comparative dynamics*, Handbook of Mathematical Economics, Vol III, K. Arrow and M. Intriligator, eds., North-Holland, Amsterdam, 1986.

[9] K.Nishimura and M. Yano (1993a), *Interlinkage in the endogenous real business cycles of international economies*, Economic Theory **3** (1993), 151–168.

[10] ___ and ___ (1993b), *Endogenous real business cycles and international specialization*, Theory, Policy and Dynamics in International Trade: Essays in Honor of Ronald W. Jones, W. Ethier, E. Helpman and P. Neary, eds., Cambridge University Press, Cambridge, 1993.

[11] ___ and ___(1993c), *Optimal chaos, non-linearity and feasibility conditions*, Economic Theory (forthcoming).

[12] ___ and ___(1993d), *Durable capital and chaos in competitive business cycles*, Journal of Economic Behavior and Organaization (forthcoming).

[13] ___ and ___ (1993e), *Optimal chaos when future utilities are discounted arbitrarily weakly*,Research in Non-Linear Analysis and Mathematical Economics (Lecture Note Series), T. Maruyama and W. Takahashi, eds., Kyoto University Institute of Mathematical Analysis, Kyoto (forthcoming).

[14] J.Schinkinkman (1976), *On optimal steady state of n-sector growth models when utility is discounted*,Journal of Economic Theory **12** (1976), 11–20.

[15] H. Uzawa (1961), *On a two-sector model of economic growth*, Review of Economic Studies **28** (1961), 40–47.

Methods of Duals in Nonlinear Analysis

— Lipschitz Duals of Banach Spaces and Some Applications —

Ikuko Sawashima

Department of Mathematics, Ochanomizu University
Otsuka, Bunkyo-ku, Tokyo 112, JAPAN

1. Introduction

In the functional analysis for linear operators, there are many types of useful methods, techniques and tools. On the other hand, for nonlinear operators, it seems that functional analysis does not have so many fundamental methods, though various types of fixed point theorems have been investigated. Such a situation it made us eager to find other applicable methods to the nonlinear analysis.

Here we focus our attention on the method using dual spaces and dual operators. It is one of the most effective methods in functional analysis. The dual space E^* of a Banach space E consists of all continuous linear functionals on E. If T is a continuous linear operator in E, then, for each x^* in $E^*, T^*x^* = x^* \circ T$ ($\circ$ means the composition) is also continuous linear and hence the dual T^* of T maps E^* to E^*. If T is not linear, $x^* \circ T$ is no more linear. Therefore T^*x^* is out of E^*. This is the most simple reason why nobody use the dual of a nonlinear operator. However we look at again the linear case. Both T and x^* have the same continuity and linearity. Hence T^*x^* has also continuity and linearity, that is, T^*x^* belongs to E^*. So, even if the nonlinear case, if an operator T in E has a certain convenient property (P) such as Lipschitz continuity, $T0 = 0$ or homogeneity and if a functional on E has the similar property (P) then the composite of T and f is a functional on E having the property (P). Therefore, if we determine the space of all functionals on E, which have the property (P), as a new dual space $E^\sharp$ and if we define as a new dual operator $T^\sharp$ by $T^\sharp f = f \circ T$, then $T^\sharp$ maps $E^\sharp$ to $E^\sharp$. Further, if the set $E^\sharp$ and the operator $T^\sharp$ hold good properties such as E^* and T^* have, then a method using $E^\sharp$ and $T^\sharp$ as nonlinear duals may be an effective way to solve problems in the nonlinear functional analysis. This is our motivation and a starting point of this study.

In this report, we will deal with nonlinear operators in a Banach space E, which are Lipschitz continuous and map the origin 0 of E to 0, that is, we will pick up these properties of operators as a convenient property (P). Along the above consideration, a Lipschitz dual space and a Lipschitz dual operator will be defined as new duals in the section 2. In the section 3 and 4, the most typical properties of these duals and the second duals will be shown. In the section 5, we will mention some results for the mean ergodicity of Lipschitz dual operators, as applications.

2. Definitions of duals

Let E be a Banach space with the scalar field $\Phi = \boldsymbol{C}$ or $\boldsymbol{R}$ and let E^* be the dual space of E, that is, the space of all continuous linear functionals on E with the dual norm $\| \ \|$:

$$\|x^*\| = \sup\{|x^*(x)|;\ \|x\| \le 1\} \quad \text{for } x^* \in E^*$$

Let S be a mapping from E to a Banach space F. Then S is said to be *Lipschitz continuous* if there exists a positive constant K such that

$$\|Sx - Sy\| \le K\|x - y\| \quad \text{for every } x, y \in E.$$

The infimum of such numbers K is called the *Lipschitz coefficient.*

Definition 1 We define the *Lipschitz dual* $E^\sharp$ of E as the space of all Lipschitz continuous functionals on E with the value 0 at 0. Addition and scalar multiplication in $E^\sharp$ are usual addition and scalar multiplication as functions on E and the norm $\|x^\sharp\|_{\mathrm{L}}$ in $E^\sharp$ is the Lipschitz coefficient of $x^\sharp$. The norm $\| \ \|_{\mathrm{L}}$ is called the *Lipschitz norm.*

It is clear that

$$\|x^\sharp\|_{\mathrm{L}} = \sup\{|x^\sharp(x) - x^\sharp(y)| \,/\, \|x - y\|;\, x \neq y\} \quad \text{for } x^\sharp \in E^\sharp$$

and that $E^* \subset E^\sharp$ and $\|x^*\| = \|x^*\|_{\mathrm{L}}$.

Definition 2 Let T be a Lipschitz continuous operator in E with $T0 = 0$. Then we define the *Lipschitz dual operator* $T^\sharp$ of T in $E^\sharp$
as an operator that maps each element $x^\sharp$ in $E^\sharp$ to the composite T and $x^\sharp$, i.e.

$$T^\sharp x^\sharp(x) = x^\sharp(Tx) \quad \text{for } x \in E.$$

The restriction of $T^\sharp$ to E^* is denoted by ${T^\sharp}_{|E^*}$. ${T^\sharp}_{|E^*}$ maps E^* to $E^\sharp$.

3. Properties of $E^\sharp$ and $T^\sharp$

Proposition 1 *The Lipschitz dual space $E^\sharp$ of E is a Banach space and the dual space E^* of E is a closed subspace of $E^\sharp$.*

Proof. Let $x^\sharp, y^\sharp \in E^\sharp$ and $\alpha \in \Phi$. Then, for $x, y \in E$

$$\begin{aligned} |(x^\sharp + y^\sharp)(x) - (x^\sharp + y^\sharp)(y)| &\le |x^\sharp(x) - x^\sharp(y)| + |y^\sharp(x) - y^\sharp(y)| \\ &\le \|x^\sharp\|_{\mathrm{L}} \|x-y\| + \|y^\sharp\|_{\mathrm{L}} \|x-y\| \\ &= (\|x^\sharp\|_{\mathrm{L}} + \|y^\sharp\|_{\mathrm{L}}) \|x-y\| \end{aligned}$$

and

$$|\alpha x^\sharp(x) - \alpha x^\sharp(y)| = |\alpha|\, |x^\sharp(x) - x^\sharp(y)| \le |\alpha|\, \|x^\sharp\|_{\mathrm{L}} \|x-y\|.$$

Hence $x^\sharp + y^\sharp$ and $\alpha x^\sharp$ are contained in $E^\sharp$ and

$$\|x^\sharp + y^\sharp\|_{\mathrm{L}} \le \|x^\sharp\|_{\mathrm{L}} + \|y^\sharp\|_{\mathrm{L}} \quad \text{and} \quad \|\alpha x^\sharp\| \le |\alpha|\, \|x^\sharp\|_{\mathrm{L}}.$$

Further,

$$\|\alpha x^\sharp\|_{\mathrm{L}} \le |\alpha|\, \|x^\sharp\|_{\mathrm{L}} \le |\alpha|\, |1/\alpha|\, \|\alpha x^\sharp\|_{\mathrm{L}},$$

which implies the equality $\|\alpha x^\sharp\|_{\mathrm{L}} = |\alpha|\, \|x^\sharp\|_{\mathrm{L}}$. If $\|x^\sharp\|_{\mathrm{L}} = 0$ then $x^\sharp(x) = x^\sharp(0) = 0$. The inverse is obvious. Therefore $E^\sharp$ is a normed linear space.

To show the completeness, let $\{x^\sharp_n\}$ be a Cauchy sequence of elements in $E^\sharp$. Then, for arbitrary positive number ε there exists a positive integer N such that $N \le n, m$ implies

$$(*) \qquad |x^\sharp_n(x) - x^\sharp_m(x) - (x^\sharp_n(y) - x^\sharp_m(y))| < \varepsilon \|x-y\|$$

for every $x, y \in E$. Hence it is clear that for every x, $\{x^\sharp_n(x)\}$ is a Cauchy sequence in Φ, so that $\{x^\sharp_n(x)\}$ converges to a number $a(x)$ as $n \to \infty$. Thus

$$|a(x) - a(y)| = \lim |x^\sharp_n(x) - x^\sharp_n(y)| \le \lim \|x^\sharp_n\|_{\mathrm{L}} \|x-y\|,$$

since $\|x^\sharp_n\|_{\mathrm{L}}$ is a convergent sequence and $x^\sharp_n(0) = 0$. Then the functional $a(x)$ on E is Lipschitz continuous and $a(0) = 0$. That is, the functional belongs to $E^\sharp$. In $(*)$, letting $m \to \infty$, $\|x^\sharp_n - a\|_{\mathrm{L}} \to 0$ as $n \to \infty$. Therefore $E^\sharp$ is complete.

Since $E^\sharp$ is a Banach space, it is clear that E^* is closed subspace of $E^\sharp$.

The next theorem corresponds to the Alaoglu Theorem [2].

Theorem 1 *The closed unit ball in the Lipschitz dual space $E^\sharp$ is $\sigma(E^\sharp, E)$-compact and the dual space E^* of E is a $\sigma(E^\sharp, E)$-closed subspace of $E^\sharp$.*

Proof. Let $B^\sharp$ be the closed unit ball, i.e. $B^\sharp = \{x^\sharp; \|x^\sharp\|_L \le 1, x^\sharp \in E^\sharp\}$ and $(B^\sharp, \sigma(E^\sharp, E))$ be the subspace $B^\sharp$ with $\sigma(E^\sharp, E)$-topology. Further let $\Pi_{x\in E}\{\xi; |\xi| \le \|x\|\}$, simply Π, be the Cartesian product space of closed balls in Φ, with the product topology. Then $(B^\sharp, \sigma(E^\sharp, E))$ is topologically embedded into Π, by ϕ; $\phi(x^\sharp) = \{x^\sharp(x)\}_{x\in E}$. By the Tihonov Theorem, Π is compact. Thus, it suffices to show that the image $\phi(B^\sharp)$ is closed in Π. Let a net $\{\phi x^\sharp_\iota\}$ in $\phi(B^\sharp)$ converge to an element $\{\alpha_x\}$ in Π. Then, for every $x \in E, x^\sharp_\iota(x)$ converges to $\alpha_x \in \{\xi; |\xi| \le \|x\|\}$. Thus,

$$|\alpha_x - \alpha_y| = \lim |x^\sharp_\iota(x) - x^\sharp_\iota(y)| \le \lim \|x^\sharp_\iota\|_L \|x - y\| \le \|x - y\|$$

and $\alpha_0 = 0$. Hence $\{\alpha_x\}_{x\in E}$ belongs to $\phi(B^\sharp)$, which shows that $\phi(B^\sharp)$ is closed in Π. Therefore $B^\sharp$ is $\sigma(E^\sharp, E)$-compact.

Since the process of taking the pointwise limit for elements of E^* keeps linearity, it is easily proved that E^* is $\sigma(E^\sharp, E)$-closed.

Let $\mathrm{L}(E)$ be the set of all Lipschitz continuous operators in E, each of which maps 0 to 0 and the norm $\|\ \|_\mathrm{L}$ of T in $\mathrm{L}(E)$ be the Lipschitz coefficient of T. Then,

$$\|T\|_\mathrm{L} = \sup\{\|T_x - T_y\| \,/\, \|x - y\|; x \ne y\}.$$

and $\mathrm{L}(E)$ is a Banach space with usual addition, scalar multiplication of operators in a Banach space E and the norm $\|\ \|_\mathrm{L}$. This can be proved in a similar way to Proposition 1. Further $\mathrm{L}(E)$ is a Banach algebra with respect to multiplication as the composite of operators. Because, the Lipschitz coefficient of the composite of S and $T\,(S, T \in \mathrm{L}(E))$ is smaller than the product of Lipschitz coefficients of S and T, i.e., $\|ST\|_\mathrm{L} \le \|S\|_\mathrm{L}\,\|T\|_\mathrm{L}$.

Proposition 2 *Let $S, T \in \mathrm{L}(E)$ and $\alpha \in \Phi$. Then the following equations hold;*

(a) $(T + S)^\sharp{}_{|E^*} = T^\sharp{}_{|E^*} + S^\sharp{}_{|E^*}$

(b) $(\alpha T)^\sharp{}_{|E^*} = \alpha(T^\sharp)_{|E^*}$

(c) $(ST)^\sharp = T^\sharp S^\sharp$.

Proof. Since $x^* \in E^*$ is linear, (a) and (b) is clear. (c) is also clear by the definition of composite of operators.

The next theorem has an important role in our study, though its proof has no difficulty. By this theorem, we can apply many techniques and tools in the functional analysis for linear operators to Lipschitz dual operators.

Theorem 2 *Let T be a Lipschitz continuous operator in E with $T0 = 0$. Then*

(1) *the Lipschitz dual $T^\sharp$ of T is a bounded linear operator in $E^\sharp$ and*

$$\|T\|_{\mathrm{L}} = \|T^\sharp\| = \|T^\sharp{}_{|E^*}\|,$$

where $\| \ \|$ is the usual norm for bounded linear operators,

(2) $\|T^n\|_{\mathrm{L}}^{1/n}$ *converges to the spectral radius* $\mathrm{r}(T^\sharp)$ *of* $T^\sharp$ *i.e.*

$$\mathrm{r}(T^\sharp) = \sup\{|\lambda|\,;\ \lambda \text{ is in the spectrum of } T^\sharp\} = \lim \|T^n\|_{\mathrm{L}}{}^{1/n}.$$

Proof. To show (1), let, $x^\sharp, y^\sharp \in E^\sharp$ and $\alpha \in \Phi$.
Then, for every $x \in E$,

$$\begin{aligned} T^\sharp(\alpha x^\sharp + y^\sharp)\,(x) &= (\alpha x^\sharp + y^\sharp)\,(Tx) \\ &= (\alpha x^\sharp)\,(Tx) + y^\sharp\,(Tx) \\ &= \alpha(x^\sharp(Tx)) + y^\sharp\,(Tx) \\ &= \alpha(T^\sharp x^\sharp\,(x)) + T^\sharp y^\sharp\,(x) \\ &= (\alpha T^\sharp x^\sharp + T^\sharp y^\sharp)\,(x). \end{aligned}$$

Hence, $T^\sharp(\alpha x^\sharp + y^\sharp) = \alpha T^\sharp x^\sharp + T^\sharp y^\sharp$, namely, $T^\sharp$ is linear.
If $x, y \in E$. then

$$\begin{aligned} |T^\sharp x^\sharp(x) - T^\sharp x^\sharp(y)| &= |x^\sharp(Tx) - x^\sharp(Ty)| \\ &\le \|x^\sharp\|_{\mathrm{L}}\,\|Tx - Ty\| \\ &\le \|x^\sharp\|_{\mathrm{L}}\,\|T\|_{\mathrm{L}}\,\|x - y\|. \end{aligned}$$

Hence, for every $x^\sharp \in E^\sharp$,

$$\|T^\sharp x^\sharp\|_{\mathrm{L}} \le \|x^\sharp\|_{\mathrm{L}}\,\|T\|_{\mathrm{L}},$$

that is, $T^\sharp$ is a bounded linear operator in $E^\sharp$ and

$$\|T^\sharp\| \le \|T\|_{\mathrm{L}}.$$

To have the opposite inequality, let $x, y \in E$. Then,

$$\begin{aligned} \|Tx - Ty\| &= \sup\{|x^*(Tx - Ty)|\,;\ x^* \in E^*, \|x^*\| \le 1\} \\ &= \sup\{|T^\sharp x^*(x) - T^\sharp x^*(y)|\,;\ x^* \in E^*, \|x^*\| \le 1\} \\ &\le \sup\{\|T^\sharp x^*\|_{\mathrm{L}}\,;\ x^* \in E^*, \|x^*\| \le 1\}\,\|x - y\| \\ &= \|T^\sharp{}_{|E^*}\|\,\|x - y\| \\ &\le \|T^\sharp\|\,\|x - y\|, \end{aligned}$$

where the first equality holds by the well known fact that
$\|z\| = \sup\{|x^*(z)|\,;\ \|x^*\| \le 1, x^* \in E^*\}$. Thus,

$$\|T\|_{\mathrm{L}} \le \|T^\sharp{}_{|E^*}\| \le \|T^\sharp\|,$$

which leads to the norm equalities, $\|T\|_L = \|T^\sharp{}_{|E^*}\| = \|T^\sharp\|$.
By Proposition 2 (c) and this norm equalities imply

$$\|T^n\|_L{}^{1/n} = \|(T^n)^\sharp\|^{1/n} = \|(T^\sharp)^n\|^{1/n}.$$

By a fundamental theorem in the spectral theory, $\|(T^\sharp)^n\|^{1/n}$ converges to $r(T^\sharp)$ [2],[3]. Therefore $\|T^n\|_L{}^{1/n}$ converges to $r(T^\sharp)$.

Corollary 1 *Let S and T be in* L*(E). Then*

$$\|T + S\|_L = \|T^\sharp{}_{|E^*} + S^\sharp{}_{|E^*}\| \leq \|T^\sharp + S^\sharp\|.$$

Proof. It is clear by Theorem 2 and Proposition 2 (a).

4. Second duals

In the previous section, we found that the Lipschitz dual $T^\sharp$ of T is a bounded linear operator in the Lipschitz dual $E^\sharp$. Therefore, the dual operator of $T^\sharp$ can be defined in the dual space of $E^\sharp$. Hence it is not necessary to use again the Lipschitz dual of $E^\sharp$. So, the second dual space of a Banach space E in our case is defined by the dual space of the Banach space $E^\sharp$, which is denoted by $E^{\sharp *}$, i.e. $E^{\sharp *}$ is the space of all bounded linear functionals with dual norm $\| \ \|$:

$$\|x^{\sharp *}\| = \sup\{ \ |x^{\sharp *}(x^\sharp)| \ ; \ x^\sharp \in E^{\sharp *}, \ \|x^{\sharp *}\|_L \leq 1\}.$$

In order to discriminate $E^{\sharp *}$ from the usual second dual E^{**}, we call the space $E^{\sharp *}$ *the semi-Lipschitz second dual space* of E. The *semi-Lipschitz second dual operator* of T is defined by the dual operator of the Lipschitz dual operator $T^\sharp$, which is denoted by $T^{\sharp *}$. The well known embedding map from E into E^{**} is denoted by Q, i.e., for each $x \in E$,

$$Q\,x\,(x^*) = x^*(x) \text{ for all } x^* \in E^*.$$

Similarly, the embedding map $\tilde{\ }Q$ from E to $E^{\sharp *}$ is defined as follows ;
for each $x \in E$,

$$\tilde{Q}x\,(x^\sharp) = x^\sharp(x) \quad \text{for all } x^\sharp \in E.$$

By those definitions, it is clear that

$$\tilde{Q}x_{\,|E^*} = Qx, \quad \text{for } x \in E,$$

where $\tilde{Q}x_{\,|E^*}$ means the restriction of $\tilde{Q}x$ to E^*.

Proposition 3

(1) *The following equations hold for $x, y \in E$ and $\alpha \in \Phi$;*

(a) $(\alpha\tilde{Q}x + \tilde{Q}y)_{|E^*} = \tilde{Q}(\alpha x + y)_{|E^*}$,

(b) $\|\tilde{Q}x - \tilde{Q}y\| = \|\tilde{Q}(x-y)_{|E^*}\| = \|x - y\|$,

(c) $\|\tilde{Q}(x-y)\| = \|x-y\|$.

(2) *The linear hull of $\tilde{Q}E$ is $\sigma(E^{\sharp *}, E^{\sharp})$-dense in $E^{\sharp *}$.*

Remark The embedding map $\tilde{Q}$ is not linear though Q is linear. However, as seen in Proposition 3. (1) (b), $\tilde{Q}$ is isometric.

Proof. To show (1), let $x, y \in E$ and $\alpha \in \Phi$. Then, for every $x^* \in E^*$,

$$(\alpha\tilde{Q}x + \tilde{Q}y)(x^*) = \alpha x^*(x) + x^*(y) = x^*(\alpha x + y) = \tilde{Q}(\alpha x + y)x^*,$$

which shows (a). (a) and the well known fact that $\|Q z\| = \|z\|$ for every $z \in E$ imply

$$\|\tilde{Q}x - \tilde{Q}y\| \geq \|(\tilde{Q}x - \tilde{Q}y)_{|E^*}\| = \|\tilde{Q}(x-y)_{|E^*}\| = \|Q(x-y)\| = \|x-y\|.$$

On the other hand,

$$\begin{aligned} \|\tilde{Q}x - \tilde{Q}y\| &= \sup\{|\tilde{Q}x(x^{\sharp}) - \tilde{Q}y(x^{\sharp})| \,;\, \|x^{\sharp}\|_{\mathrm{L}} \leq 1,\ x^{\sharp} \in E^{\sharp}\} \\ &= \sup\{|x^{\sharp}(x) - x^{\sharp}(y)| \,;\, \|x^{\sharp}\|_{\mathrm{L}} \leq 1,\ x^{\sharp} \in E^{\sharp}\} \\ &\leq \|x-y\|. \end{aligned}$$

which leads to (b). As a special case of (b), $\|\tilde{Q}z\| = \|z\|$ holds for every $z \in E$. Hence (c) is clear.

To show (2), let F be the $\sigma(E^{\sharp *}, E^{\sharp})$-closed linear hull of $\tilde{Q}E$. Suppose F is a proper subspace of $E^{\sharp *}$. Then we can take an element $a^{\sharp *}$ in $E^{\sharp *} \cap F^c$. By the Hahn-Banach Theorem, there exists a $\sigma(E^{\sharp *}, E^{\sharp})$-continuous linear functional f such that $f(a^{\sharp *}) \neq 0$ and $f(x^{\sharp *}) = 0$ for all $x^{\sharp *} \in F$. By the duality theorem [2,Chap.V], there exists an element $b^{\sharp} \in E^{\sharp}$ such that $x^{\sharp *}(b^{\sharp}) = f(x^{\sharp *})$ for all $x^{\sharp *} \in E^{\sharp *}$. Thus, $a^{\sharp *}(b^{\sharp}) \neq 0$ and $x^{\sharp *}(b^{\sharp}) = 0$ for all $x^{\sharp *} \in F$. Since F includes $\tilde{Q}E$,

$$b^{\sharp}(x) = \tilde{Q}(x)(b^{\sharp}) = 0 \quad \textit{for all } x \in E.$$

Hence $b^{\sharp} = 0$, which contradicts $a^{\sharp *}(b^{\sharp}) \neq 0$. Therefore $F = E^{\sharp *}$, that is, the linear hull of $\tilde{Q}E$ is $\sigma(E^{\sharp *}, E^{\sharp})$-dense in $E^{\sharp *}$.

5. Applications

By Theorem 2, the Lipschitz dual $T^\sharp$ and the semi-Lipschitz second dual $T^{\sharp *}$ of
$T \in \mathrm{L}(E)$ are bounded linear operators in $E^\sharp$ and $E^{\sharp *}$ respectively. Therefore the all known results for linear operators [3],[4] may be applied to study of $T^\sharp$ and $T^{\sharp *}$. For instance, $T^\sharp$ has following fundamental spectral properties ; if $|\lambda| > r = \mathrm{r}(T^\sharp)$ then

(I) $\lambda - T^\sharp$ has its inverse $(\lambda - T^\sharp)^{-1}$ in $\mathrm{B}(E^\sharp)$,

(II) $(\lambda - T^\sharp)^{-1}$ is represented by the Neumann series, i.e.

$$(\lambda - T^\sharp)^{-1} = \sum_{n=0}^{\infty} (T^\sharp)^n / \lambda^{n+1},$$

where $\mathrm{B}(E^\sharp)$ denotes the Banach algebra of all bounded linear operators in $E^\sharp$. Further, any closed ball of $E^\sharp$ is $\sigma(E^\sharp, E)$-compact by Theorem 1. So it makes us more progress in our study for $T^\sharp$ such as ergodicity, spectral properties, existence of fixed points and so on. In this paper, we will show only several results with respect to ergodicity of $T^\sharp$.

Let S be a Hausdorff space with respect to a topology τ and $s_n, n = 1, 2, \ldots$, be a sequence in a compact subset of S. Further let $\mathcal{F}$ be a *free ultrafilter* of N that is, $\mathcal{F}$ be a maximal filter including all subsets $\{n; n \geq m\}$ $m = 1, 2, \ldots$. Then we define the *filter limit* of $\{s_n\}$ for $\mathcal{F}$ denoted by τ-$\mathcal{F}$-$\lim s_n$, as follows;

$$\tau\text{-}\mathcal{F}\text{-}\lim s_n = \bigcap_{F \in \mathcal{F}} [\text{closure of } \{s_n \; ; \; n \in F\}].$$

Since S is a Hausdorff space and $\{s_n\}$ is in the compact set, τ-$\mathcal{F}$-$\lim s_n$ is determined as an element in the compact set. The *cluster set* of $\{s_n\}$ means the intersection of all closures of sets $\{s_n; m \leq n\}, m = 1, 2, \ldots$. It is easily proved that, for every free ultrafilter $\mathcal{F}$, τ-$\mathcal{F}$-$\lim s_n$ is determined as an element in the cluster point of $\{s_n\}$ and, inversely, every cluster point of $\{s_n\}$ is the filter limit of $\{s_n\}$ for a certain free ultrafilter of N [1].

Hereafter, the topology $\sigma(E^\sharp, E)$ is simply denoted by $\mathrm{w}^\sharp$. Let $\{T_n\}$ be a sequence of operators in $\mathrm{L}(E)$. If $\sup \|T_n\|_{\mathrm{L}} < \infty$, $\{T_n\}$ is said to be norm bounded and if, for every $x \in E$ and $x^* \in E^*, x^*(T_n x)$ is convergent to a scalar as $n \to \infty, \{T_n\}$ is said to be a weak Cauchy sequence. The $\mathrm{w}^\sharp$-cluster set of $\{T_n^\sharp x^\sharp\}$ is the intersection of all $\mathrm{w}^\sharp$-closures of sets $\{T_n^\sharp x^\sharp \; ; \; m \leq n\}, m = 1, 2, \ldots$, which is denoted by $\mathrm{Cl}(T_n^\sharp x^\sharp)$.

Lemma 1 *Let $T_n, n = 1, 2, \ldots$, be operators contained in $\mathrm{L}(E)$. If there exists an element $x^\sharp \in E^\sharp$ such that $\sup \|T_n^\sharp x^\sharp\|_{\mathrm{L}} < \infty$, then*

(1) *for every free ultrafilter $\mathcal{F}, \mathrm{w}^\sharp$-$\mathcal{F}$-$\lim T_n^\sharp x^\sharp$ exists in $\mathrm{Cl}(T_n^\sharp x^\sharp)$,*

(2) $\mathrm{Cl}\,(T_n^\sharp x^\sharp)$ *is represented as the set of the filter limits of* $\{T_n^\sharp x^\sharp\}$ *for all free ultrafilters of* N *with respect to* $\mathrm{w}^\sharp$, *i.e.*

$$\mathrm{Cl}\,(T_n^\sharp x^\sharp) = \{\mathrm{w}^\sharp\text{-}\mathcal{F}\text{-}\lim T_n^\sharp x^\sharp\ ;\ \mathcal{F} \textit{ is a free ultrafilter of } N\}$$

Proof. Notice that any closed ball in $E^\sharp$ is $\mathrm{w}^\sharp$-compact by Theorem 1. Then (1) and (2) are clear by the fact for filter limits, which is mentioned before.

Lemma 2 *Let* $\{T_n\}$ *be a norm bounded sequence of operators contained in* $\mathrm{L}(E)$. *Then*

(1) *for every* $x^\sharp \in E^\sharp$, $\mathrm{Cl}\{T_n^\sharp x^\sharp\}$ *is a non empty* $\mathrm{w}^\sharp$*-compact set,*

(2) $\{T_n\}$ *is a weak Cauchy sequence if and only if, for every* $x^* \in E^*$, $\mathrm{Cl}(T_n^\sharp x^*)$ *consists of one and only one element.*

Proof. Since $\|T_n\|_{\mathrm{L}} = \|T_n^\sharp\|$ by Theorem 2, (1) is clear by Lemma 1. Let $\mathrm{Cl}(T_n^\sharp x^*)$ consists of one and only one element $y^\sharp_{x^*}$ for each $x^* \in E^*$. Then $\mathrm{w}^\sharp\text{-}\mathcal{F}\text{-}\lim T_n^\sharp x^* = y^\sharp_{x^*}$ for every free ultrafilter $\mathcal{F}$ by Lemma 1. Hence $\mathrm{w}^\sharp$-$\lim T_n^\sharp x^* = y^\sharp_{x^*}$ for $x^* \in E^*$, which means that, for every $x \in E$, $x^*(T_n x)$ converges to $y^\sharp_{x^*}(x)$. Namely, $\{T_n\}$ is a weak Cauchy sequence. To show ' only if ' part of (2), let $\{T_n\}$ be a weak Cauchy sequence of operators of $\mathrm{L}(E)$. Then, for every $x^* \in E^*$ and $x \in E, x^*(T_n x)$ converges to a scalar, which is denoted by $Px^*(x)$. Thus

$$\begin{aligned} |Px^*(x) - Px^*(y)| &= \lim |x^*(T_n x) - x^*(T_n y)| \\ &\leq \sup \|T_n\|_{\mathrm{L}}\ \|x - y\|\ \|x^*\|. \end{aligned}$$

Hence, Px^* is Lipschitz continuous and $Px^*(0) = 0$. It follows that $Px^* \in E^\sharp$ and
$\mathrm{w}^\sharp$-$\lim T_n^\sharp x^* = Px^*$. Therefore, by Lemma 1, $\mathrm{Cl}(T_n^\sharp x^*)$ consists of one and only one element Px^*.

Theorem 3 *Let* T *and* $T_n, n = 1, 2, \ldots,$ *be operators contained in* $\mathrm{L}(E)$ *and satisfy the following conditions* (i) *and* (ii) ;

(i) $\{T_n\}$ *be norm bounded,*

(ii) *for every* x *in* $E, T_n x - T_n T x$ *be* $\sigma(E, E^*)$*-convergent to* 0 *as* $n \to \infty$.

Then, for every free ultrafilter $\mathcal{F}$, *an operator* P *defined by*

$$Px^* = \mathrm{w}^\sharp\text{-}\mathcal{F}\text{-}\lim T_n^\sharp x^* \quad \textit{for } x^* \in E^*$$

is a bounded linear operator from E^* *into* $\mathrm{F}\,(T^\sharp)$, *where* $\mathrm{F}\,(T^\sharp)$ *denotes the set of all fixed points of* $T^\sharp$.

Proof. Let $\mathcal{F}$ be any free ultrafilter of N. Then by Lemma 1, for every x^*, $\text{w}^\sharp$-$\mathcal{F}$-$\lim T_n^\sharp x^*$ is an element in $\text{Cl}(T_n^\sharp x^*)$. Hence Px^* is determined. Since the process of taking the filter limit of scalars keeps linearity and modulus, it can be proved easily along the way to the proof of Lemma 2, that $Px^* \in E^\sharp$ and

$$\|Px^*\|_{\text{L}} \leq \sup \|T_n\|_{\text{L}} \|x^*\|.$$

Therefore P is a bounded linear operator from E^* into $E^\sharp$. To prove that Px^* is a fixed point of $T^\sharp$, we need Condition (ii). Let $x^* \in E^*$. Then, for every $x \in E$.

$$\lim\{T_n^\sharp x^*(x) - T_n^\sharp x^*(Tx)\} = \lim\{x^*(T_n x) - x^*(T_n Tx)\} = 0,$$

whence

$$\text{w}^\sharp\text{-}\mathcal{F}\text{-}\lim\{T_n^\sharp x^*(x)\} = \text{w}^\sharp\text{-}\mathcal{F}\text{-}\lim\{T_n^\sharp x^*(Tx)\}.$$

Therefore

$$Px^*(x) = Px^*(Tx) = T^\sharp Px^*(x),$$

which shows that P is an operator from E^* into $\text{F}(T^\sharp)$.

Remark 1 The condition (i) in Theorem 3 is equivalent to the following condition;

(i′) for each $x^* \in E^*$, there exists a positive number K_{x^*} such that

$$\|T_n^\sharp x^*\|_L \leq K_{x^*} \quad \text{for all } n = 1, 2, \ldots$$

Because, if T satisfies (i′) then, by the principle of uniform boundedness [2], $sup\,\|T_n^\sharp{}_{|E^*}\| < \infty$. By Theorem 2 (1), T satisfies Condition (i). That Condition (i) implies (i′) is obvious.

Remark 2 Assume the same with Theorem 3, but substitute the the following condition (ii′) for Condition (ii) and add the condition (iii);

(ii′) for every $x \in E$ and $x^\sharp \in E^\sharp$, $x^\sharp(T_n x) - x^\sharp(T_n Tx)$ converges to 0,

(iii) $\text{F}(T^\sharp) \subset \text{F}(T_n{}^\sharp)$.

Then, for each free ultrafilter $\mathcal{F}$, an operator P defined on $E^\sharp$ as

$$Px^\sharp = \text{w}^\sharp\text{-}\mathcal{F}\text{-}\lim T_n^\sharp x^\sharp$$

is a bounded linear operator from $E^\sharp$ onto $\text{F}(T^\sharp)$.

The remaining part of this section we will apply the results obtained above to the Cesàro-mean of T. Let the Cesàro-mean of T be denoted by $M_n(T)$ or simply by M_n and that of $(T^\sharp)$ by $M_n(T^\sharp)$, i.e.

$$M_n = M_n(T) = (I + T + \cdots + T^{n-1}) / n,$$

$$M_n(T^\sharp) = (I + T^\sharp + \cdots + T^{\sharp n-1}) / n,$$

Then the next Proposition 4 is easily shown by Proposition 2 and Corollary 1.

Proposition 4 *Let $T \in \mathrm{L}(E)$. Then*

$$(M_n(T))^\sharp{}_{|E^*} = M_n(T^\sharp)\ {}_{|E^*},$$

and if $|\lambda| > \mathrm{r}(T^\sharp)$,

$$(\lambda - T^\sharp)^{-1}{}_{|E^*} = (\sum_{n=0}^{\infty}(T^n/\lambda^{n+1}))^\sharp{}_{|E^*}.$$

Theorem 4 *Let T be an operator contained in $\mathrm{L}(E)$ and satisfy the following conditions;*

(i) $\sup \|M_n\|_{\mathrm{L}} < \infty$

(ii) *for every x in E, $(T^n x)/n$ converges weakly to 0 as $n \to \infty$.*

Then, one and only one of the following cases happens

(1) *M_n converges weakly to 0,*

(2) *there exists a nonzero function $p(x, x^*)$ defined on $E \times E^*$, such that*

(a) *for each x^*, $p(x, x^*)$ as a function on E is a fixed point of $T^\sharp$,*

(b) *for each x, $p(x, x^*)$ as a function on E^* is contained in E^{**}.*

Proof. If $\mathrm{Cl}(M_n^\sharp x^*)$ consists of only 0 for each elements $x^* \in E^*$, then the case (1) happens by Lemma 2 (2) with $T_n = M_n$. If the case (1) does not happen then there exists an element $x^*{}_0 \in E^*$ such that $\mathrm{Cl}(M_n^\sharp x^*{}_0)$ contains at least nonzero element $y^\sharp{}_0$. Hence, by Lemma 1 (2) , there exists a free ultrafilter $\mathcal{F}_0$ such that

$$\mathrm{w}^\sharp\text{-}\mathcal{F}_0\text{-}\lim M_n^\sharp x^*{}_0 = y^\sharp{}_0 \neq 0$$

By Proposition 4, $M_n x - M_n T x$ is usually calculated as follows;

$$\begin{aligned} x^*(M_n x - M_n T x) &= x^*\{(x + \cdots + T^{n-1}x) / n \ - \ (Tx + \cdots + T^n x) / n\}. \\ &= \{x^*(x)\} / n \ - \ x^*\{(T^n x) / n\}. \end{aligned}$$

Hence, by Condition (ii), $M_n x - M_n T x$ converges weakly to 0 as $n \to \infty$. Therefore, by Theorem 3, an operator P defined by

$$Px^* = \mathrm{w}^\sharp\text{-}\mathcal{F}_0\text{-}\lim M_n^\sharp x^*$$

is a bounded linear operator from E^* into $\mathrm{F}(T^\sharp)$. Put $p(x, x^*) = Px^*(x)$ for every $x^* \in E^*$ and $x \in E$. Then, the function $p(x, x^*)$ is not identically 0 and has the properties (a) and (b).

Corollary 2 *Under the same assumption of Theorem* 4, *if the case* (1) *does not happen then there exists a nonzero fixed point of* $T^\sharp$.

Theorem 5 *Let* T *be an operator contained in* $\mathrm{L}(E)$ *and satisfy the following conditions;*

(i) $\sup \|M_n(T^\sharp)\| < \infty$

(ii) $\|(T^n x)/n\| \to 0$ *for every* $x \in E$.

Then, there exists a bounded linear operator P *in* $E^\sharp$, *which has the following properties;*

(1) $P = T^\sharp P = PT^\sharp$, $P^2 = P$,

(2) *the range of* P *coincides with* $\mathrm{F}(T^\sharp)$,

(3) $Px^\sharp \in \mathrm{Cl}(M_n(T^\sharp)x^\sharp) \subset$ w$^\sharp$-*closed convex hull of* $\{(T^\sharp)^n x^\sharp; n = 0, 1, \ldots,\}$

Remark If T is non-expansive then T clearly satisfies the conditions (i) and (ii). If

$$\sup\{(\|I\| + \|T\|_{\mathrm{L}} + \cdots + \|T^{n-1}\|_{\mathrm{L}})\,/\,n\} < \infty,$$

then

$$\sup \|M_n(T^\sharp)\| < \infty.$$

Condition (ii) is equivalent to a condition that, for every $x^\sharp \in E^\sharp$, $((T^\sharp)^n x^\sharp)/n$ converges to 0 in w$^\sharp$-topology. Because, the functional $e^\sharp$ defined as $e^\sharp(x) = \|x\|$ is an element of $E^\sharp$.

Proof. Let $\mathcal{F}$ be a free ultrafilter of $\boldsymbol{N}$. Then, for each $x^\sharp \in E^\sharp$, w$^\sharp$-$\mathcal{F}$-$\lim M_n(T^\sharp)x^\sharp$ exists in $\mathrm{Cl}(M_n(T^\sharp)x^\sharp)$ by Condition (i) and Theorem 1. Define an operator P as

$$Px^\sharp = \mathrm{w}^\sharp\text{-}\mathcal{F}\text{-}\lim M_n(T^\sharp)\,x^\sharp \quad \text{for } x^\sharp \in E^\sharp.$$

Then P is an operator in $E^\sharp$ and P clearly has property (3). Since

$$\begin{aligned}\|\{M_n(T^\sharp)x^\sharp - M_n(T^\sharp)T^\sharp x^\sharp\}(x)\| &= |\{x^\sharp/n - (T^{\sharp n}x^\sharp)\,/\,n\}(x)| \\ &= |x^\sharp(x)\,/\,n - x^\sharp(T^n x)\,/\,n| \\ &\le |x^\sharp(x)\,/\,n| + |x^\sharp(T^n x)\,/\,n| \\ &\le \|x^\sharp\|_{\mathrm{L}}\,(\|x\,/\,n\| + \|(T^n x)\,/\,n\|),\end{aligned}$$

$\{M_n(T^\sharp)x^\sharp - M_n(T^\sharp)T^\sharp x^\sharp\}$ w$^\sharp$-converges to 0 as $n \to \infty$ by Condition (ii). And $\{M_n(T^\sharp)x^\sharp - T^\sharp M_n(T^\sharp)x^\sharp\}$ also w$^\sharp$-converges to 0 as $n \to \infty$. Because,

$T^\sharp$ and $M_n(T^\sharp)$ are commutative by the linearity of $T^\sharp$. Hence, for each $x^\sharp \in E^\sharp$,

$$\begin{aligned} \text{w}^\sharp\text{-}\mathcal{F}\text{-}\lim M_n(T^\sharp)\, x^\sharp &= \text{w}^\sharp\text{-}\mathcal{F}\text{-}\lim T^\sharp M_n(T^\sharp)\, x^\sharp \\ &= \text{w}^\sharp\text{-}\mathcal{F}\text{-}\lim M_n(T^\sharp)\, T^\sharp x^\sharp \end{aligned}$$

which implies the first equality of (1) and $Px^\sharp \in \mathrm{F}(T^\sharp)$ for all $x^\sharp \in E^\sharp$. For every
$y^\sharp \in \mathrm{F}(T^\sharp)$, clearly holds $M_n(T^\sharp)\, y^\sharp = y^\sharp$. Therefore, the range of P coincides with $\mathrm{F}(T^\sharp)$ and $P^2 = P$.

References

[1] N.Bourbaki, *Topologie générale*, Chap. I, Hermann, Paris, 1951.

[2] John B.Conway, *A Course in Functional Analysis*, Springer, New York, 1985.

[3] N.Dunford and J.Schwartz, *Linear Operators*, I, Interscience, New York, 1958.

[4] U.Krengel, *Ergodic Theorems*, Walter de Gruyter, Berlin New York, 1985.

Swimming below Icebergs

S. Simons

Department of Mathematics, University of California
Santa Barbara, CA 93106-3080, U.S.A.

Introduction

You are swimming close to an iceberg in the ocean. You calculate at what slope you have to swim down so that, whatever the direction in which you swim, you can be sure that you will not collide with the iceberg. We shall see that, provided that the lower surface of the iceberg is convex, this limiting slope is intimately related to the existence of subtangents to the iceberg that satisfy various conditions. These considerations lead to generalizations of Rockafellar's Maximal Monotonicity Theorem, and also of recent results on the existence of subtangents separating the epigraphs of proper convex lower semicontinuous functions from nonempty bounded closed convex sets.

Notation

Let E be a real Banach space with adjoint E^* and $\phi, \psi : E \to \mathbb{R} \cup \{\infty\}$ be convex, lower semicontinuous functions which are each finite at least two points.

If $x \in E$, the ***subdifferential*** of ψ at x is defined by

$$\partial\psi(x) := \{x^* : x^* \in E^*, \text{ for all } y \in E, \psi(x) + \langle y - x, x^* \rangle \le \psi(y)\}.$$

We write $\partial\psi := \{(x, x^*) : x \in E, x^* \in \partial\psi(x)\} \subset E \times E^*$. If $(x, x^*) \in \partial\psi$, we write $\sigma\psi(x, x^*)$ for the ***subtangent to*** ψ ***at*** $(x, \psi(x))$ ***with slope*** x^*, that is to say

$$\sigma\psi(x, x^*)(y) := \psi(x) + \langle y - x, x^* \rangle.$$

Iceberg Theorem Let $q \in E$ (possibly with $\phi(q) = \infty$) and $\inf_E \phi < h < \phi(q)$. Define

$$K := \sup_{y \in E, \phi(y) < h} \frac{h - \phi(y)}{\|q - y\|}.$$

Then $0 < K < \infty$ and, for all $\varepsilon \in (0, 1)$, there exists $(z, y^*) \in \partial\phi$ such that

$$\langle q - z, y^* \rangle \ge (1 - \varepsilon)K\|q - z\| > 0 \text{ and } \|y^*\| \le (1 + \varepsilon)K,$$

from which

$$(1 - \varepsilon)K \le \|y^*\| \le (1 + \varepsilon)K.$$

Proof The proof of this result relies on an application of Ekeland's variational principle ([4], Theorem 1, p. 444), the definition of directional derivative, various elementary manipulations with sublinear functionals and the Hahn-Banach theorem. See [11], Théorème 2 and [12], Main Theorem for more details.

Interpretation in terms of the "iceberg" metaphor Suppose that the function ϕ above represents the surface of an iceberg below the water. Then the K is the slope at which a swimmer at (q, h) has to swim (down) in any direction in order to be sure not to collide with the iceberg. The Iceberg Theorem then says that there is a subtangent f at a point $(z, \phi(z))$ *on the iceberg* such that the line going from $(z, f(z))$ to $(q, f(q))$ has slope *arbitrarily close* to K and also *arbitrarily close* to the slope (that is to say the Lipschitz constant) of f. Specifically, $f = \sigma\phi(z, y^*)$.

Maximal Monotonicity Theorem $\partial\psi$ is *maximal monotone*, that is to say: if $(q, q^*) \in E \times E^*$ and, for all $(z, z^*) \in \partial\psi$, $\langle z - q, z^* - q^* \rangle \geq 0$ then $(q, q^*) \in \partial\psi$.

Proof Let $(q, q^*) \in E \times E^* \backslash \partial\psi$. Then $q^* \notin \partial\psi(q)$, from which $0 \notin \partial(\psi - q^*)(q)$, hence $\inf_E(\psi - q^*) < (\psi - q^*)(q)$. From the Iceberg Theorem, there exists $(z, y^*) \in \partial(\psi - q^*)$ such that $\langle z - q, y^* \rangle < 0$. Now there exists $z^* \in \partial\psi(z)$ such that $y^* = z^* - q^*$, hence $\langle z - q, z^* - q^* \rangle < 0$.

Generalizations of the Maximal Monotonicity Theorem The techniques of the Iceberg Theorem can actually be used to obtain generalizations of the Maximal Monotonicity Theorem. In [9], Theorems 5.1 and 5.2, it was shown that q or q^* above can be replaced by appropriate convex sets. Specifically:

$Q - q^*$ Theorem If Q is a nonempty weakly compact convex subset of E, $q^* \in E^*$ and,

for all $(z, z^*) \in \partial\psi$, there exists $q \in Q$ such that $\langle z - q, z^* - q^* \rangle \geq 0$

then

$$(Q \times \{q^*\}) \cap \partial\psi \neq \emptyset.$$

$q - Q^*$ Theorem If Q^* is a nonempty weak* compact convex subset of E^*, $q \in E$ and,

for all $(z, z^*) \in \partial\psi$, there exists $q^* \in Q^*$ such that $\langle z - q, z^* - q^* \rangle \geq 0$

then

$$(\{q\} \times Q^*) \cap \partial\psi \neq \emptyset.$$

Taken together, these two results lead to the following question:

$Q - Q^*$ Question If Q is a nonempty weakly compact convex subset of E, Q^* is a nonempty weak* compact convex subset of E^* and,

for all $(z, z^*) \in \partial\psi$, there exists $(q, q^*) \in Q \times Q^*$ such that $\langle z - q, z^* - q^* \rangle \geq 0$.

Does it follow that

$$(Q \times Q^*) \cap \partial\psi \neq \emptyset\,?$$

If Q^* is the unit ball of E^*, then the question above becomes:

Special case of the $Q - Q^*$ Question Let Q be a nonempty weakly compact convex subset of E and, for all $(z, z^*) \in \partial\psi$, there exists $q \in Q$ such that $\langle q - z, z^* \rangle \leq \|q - z\|$. Does there necessarily exist $q^* \in \partial\psi(Q)$ such that $\|q^*\| \leq 1$?

Are the above results special properties of subdifferentials, or are they, in fact, cases of more general properties of maximal monotone operators? Specifically:

Question Let M be a maximal monotone subset of $E \times E^*$, Q be a nonempty weakly compact convex subset of E, $q^* \in E^*$ and,

for all $(z, z^*) \in M$, there exists $q \in Q$ such that $\langle q - z, q^* - z^* \rangle \geq 0$.

Does it follow that

$$(Q \times q^*) \cap M \neq \emptyset\,?$$

Finally, in a recent paper, Fitzpatrick and Phelps [5] introduced a new class of operators on a Banach space, the *locally maximal monotone* operators, and showed that this kind of operator has important approximation properties. In [10], the techniques of the Iceberg Theorem were applied to give a positive answer to a question posed in [5]:

Local Maximal Monotonicity Theorem If U is a convex open subset of E^* with $\partial\psi \cap (E \times U) \neq \emptyset$, $(q, q^*) \in E \times U$ and,

$$\text{for all } (z, z^*) \in \partial\psi \cap (E \times U), \langle q - z, q^* - z^* \rangle \geq 0$$

then $(q, q^*) \in \partial\psi$.

On the various proofs of the Maximal Monotonicity Theorem Rockafellar's original proof appeared in [7]. Simpler proofs of Rockafellar's result were given by Taylor [13], Borwein [2], Simons [8] and Zagrodny. See [6], p. 50-59 for some history. It is worth pointing out that the Brøndsted-Rockafellar lemma [3], which was used in earlier proofs of this result was, in fact, an antecedent of Ekeland's variational principle. The proof outlined here using the Iceberg Theorem is substantially shorter and simpler then any previously given.

Separating subtangents

We now extend the definition of K to a situation more general than that considered above. Specifically, let B be a nonempty bounded closed convex subset of $E \times \mathbb{R}$, $\operatorname{gap}(B, \operatorname{epi}\phi) > 0$ and $\sup_B pr_2 > \inf_E \phi$. ($\operatorname{gap}(B, \operatorname{epi}\phi)$ stands for the distance between B and the epigraph of ϕ, pr_2 stands for the projection map from $E \times \mathbb{R}$ to $\mathbb{R}$, and the distances in $E \times \mathbb{R}$ are measured by any norm that gives the product topology.) Write

$$K := \sup_{(x,\lambda)\in B, y\in E, \lambda>\phi(y)} \frac{\lambda - \phi(y)}{\|x - y\|}.$$

As in the Iceberg Theorem, one can prove that $K \in (0, \infty)$. Here is the interpretation of this more general situation in terms of our "iceberg" metaphor: let B be a rock in the ocean and ϕ be the surface of an iceberg below the water. Then K is the smallest number such that if one swims down from an arbitrary point of B with slope greater than K then one can be sure that one is not going to hit the iceberg. If $(z, z^*) \in \partial\phi$, we shall say that $\sigma\phi(z, z^*)$ *lies above* B if, for all $(x, \lambda) \in B$, $\sigma\phi(z, z^*)(x) > \lambda$.

Lemma If $(z, z^*) \in \partial\phi$ and $\sigma\phi(z, z^*)$ lies above B then $\|z^*\| \geq K$.

Proof Let $(z, z^*) \in \partial\phi$ and $\sigma\phi(z, z^*)$ lie above B. If $(x, \lambda) \in B$ and $y \in E$ then

$$\phi(y) + \langle x - y, z^* \rangle \geq \sigma\phi(z, z^*)(y) + \langle x - y, z^* \rangle = \sigma\phi(z, z^*)(x) \geq \lambda.$$

Thus

$$\|x - y\| \|z^*\| \geq \langle x - y, z^* \rangle \geq \lambda - \phi(y).$$

The desired inequality follows by dividing by $\|x - y\|$ and taking the supremum over the appropriate values of (x, λ) and y. The following result was proved in [9], Theorem 4.4:

Sharp Separation Theorem For all $\varepsilon > 0$, there exists $(z, z^*) \in \partial\phi$ such that

$$\|z^*\| \leq (1 + \varepsilon)K \text{ and } \sigma\phi(z, z^*) \text{ lies above } B.$$

Beer ([1], Lemma 4.10) proved the existence of a subtangent to ϕ lying above B. What is new about the Sharp Separation Theorem is the sharp bound on the possible slopes of such subtangents. Another way of putting the Sharp Separation Theorem is:

Alternative Formulation of the Sharp Separation Theorem

$$\sup_{(x,\lambda)\in B, y\in E, \lambda>\phi(y)} \frac{\lambda - \phi(y)}{\|x - y\|} = \inf_{(z,z^*)\in\partial\phi, \sigma\phi(z,z^*) \text{ lies above } B} \|z^*\|,$$

and the common value $\in (0, \infty)$. We do not know if this result can be considered as a *minimax relation.*

There are interesting open questions on the existence of separating subtangents. Here is one example: let us suppose for simplicity that $B = \{(0, 0)\}$, so that $\phi(0) > 0 > \inf_E \phi$ and

$$K = \sup_{\phi(y)<0} \frac{-\phi(y)}{\|y\|}.$$

Then, from the Sharp Separation Theorem,

for all $\varepsilon > 0$, there exists $(z, z^*) \in \partial\phi$ such that $\|z^*\| \leq (1 + \varepsilon)K$ and $\langle z, z^* \rangle < \phi(z)$.

It also follows from the Iceberg Theorem that

for all $\varepsilon > 0$, there exists $(z, z^*) \in \partial\phi$ such that $\|z^*\| \leq (1 + \varepsilon)K$ and $\langle z, z^* \rangle < 0$.

We do not know if it is possible to prove a "combination" result such as:

Conjecture For all $\varepsilon > 0$, there exists $(z, z^*) \in \partial\phi$ such that

$$\|z^*\| \leq (1 + \varepsilon)K, \langle z, z^* \rangle < \phi(z) \text{ and } \langle z, z^* \rangle < 0.$$

References

[1] G. Beer, *The slice topology: A viable alternative to Mosco convergence in nonreflexive spaces*, Nonlinear Analysis **19**(1992), 271-290.

[2] J. M. Borwein, *A Note on ε-Subgradients and Maximal Monotonicity*, Pacific Journal of Mathematics **103**(1982), 307- 314.

[3] A. Brøndsted and R.T. Rockafellar, *On the Subdifferentiability of Convex Functions*, Proc. Amer. Math. Soc. **16**(1965), 605-611.

[4] I. Ekeland, *Nonconvex minimization problems*, Bull. Amer. Math. Soc. **1**(1979), 443-474.

[5] S. Fitzpatrick and R. R. Phelps, *Bounded approximants to monotone operators on Banach Spaces*, Ann. Inst. Henri Poincaré, Analyse non linéaire **9**(1992), 573-595.

[6] R. R. Phelps, *Convex Functions, Monotone Operators and Differentiability*, Springer-Verlag, Berlin, Germany, 1989.

[7] R. T. Rockafellar, *On the Maximal Monotonicity of Subdifferential Mappings*, Pacific Journal of Mathematics **33**(1970), 209-216.

[8] S. Simons, *The least slope of a convex function and the maximal monotonicity of its subdifferential*, Journal of Optimization Theory **71**(1991), 127-136.

[9] S. Simons, *Subtangents with controlled slope*, Nonlinear Analysis, to appear.

[10] S. Simons, *Subdifferentials are locally maximal monotone*, Bull. Australian Math. Soc. **47**(1993), 465-471.

[11] S. Simons, *Les dérivées directionnelles et la monotonicité maximale des sous-différentiels*, Séminaire d'Initiation à l'Analyse (Séminaire Choquet) 1991/92, Publications Mathématiques de l'Université Paris 6(1993).

[12] S. Simons, *Swimming below icebergs*, Set-Valued Analysis, to appear.

[13] P. D. Taylor, *Subgradients of a Convex Function Obtained from a Directional derivative*, Pacific Journal of Mathematics **44**(1973), 739-747.

Comparative Statics and Algorithms for Finding Economic Equilibria**

Steve Smale

Department of Economics, University of California, Berkeley
California 94720, U.S.A.

Part I. Statics.

Consider an excess demand function $Z : \mathbb{R}^l_+ \to \mathbb{R}^l$, $p \to Z(p)$, where $\mathbb{R}^l_+$ is the set of price vectors $p = (p_1, \ldots, p_l)$, $p_i \geq 0$ and the value of Z are taken in commodity space $\mathbb{R}^l$. For example, $Z = D - S$, demand less supply, and D, S are derived from a microeconomical setting. This is the approach in [Smale], where some background for this note may be found.

It is natural to suppose that Z satisfies some diffentiability condition and the following axioms:

a) $Z(\lambda p) = Z(p)$, all $\lambda > 0$ and $p \in \mathbb{R}^l_+$ (homogeneity)

b) $p \cdot Z(p) = 0$ (or $\sum p_i Z_i(p) = 0$) (Walras Law).

With an appropriate boundary condition one can show the existence of an equilibrium p^*, so that $Z(p^*) = 0$. See the previously mentioned reference for example (that proof needs only the continuity of Z).

It is convenient to remove the conditions a) and b). For that let $\Delta_1 = \{p \in \mathbb{R}^l_+ \mid \sum p_i = 1\}$ and $\Delta_0 = \{z \in \mathbb{R}^l \mid \sum z_i = 0\}$. It follow that $\Phi : \Delta_1 \to \Delta_0$ defined by

$$\Phi(p) = Z(p) - (\sum Z_i(p))p$$

is well-defined. Moreover it is not difficult to see, using a) and b),

Proposition. *For* $p \in \mathbb{R}^l_+$, $Z(p) = 0$ *if and only if* $\Phi(p) = 0$.

*Written for the July 1993 T.I.Tec/K.E.S. Conference in Tokyo on Economics and Mathematics.

**Partially supported by NSF funds.

The introduction of Φ has the effect of putting the problem of economic equilibrium into the context of studying the zeros of a map of a domain in Euclidean space into a second Euclidean space of the same dimension.

We now allow parameters in our problem as endowments, say $e \in E$. This leads to the problem of analyzing the solutions of the equation (system)

$$F(e,p) = 0, \; F : E \times \Delta_1 \to \Delta_0.$$

If the parameters are sufficiently rich then $\Sigma = \{(e,p) \mid F(e,p) = 0\}$ is a smooth submanifold of the product space $E \times \Delta_1$, and is called the *equilibrium manifold.*

The derivative with respect to p, $D_pF(e,p) : \overline{\Delta}_1 \to \Delta_0$, is a linear map where $\overline{\Delta}_1$ is the linear space containing Δ_1. This could be expressed in terms of a matrix of partial derivatives for a choice of coordinates. If $DF_p(e^*,p^*)$ is non-singular at $(e^*,p^*) \in \Sigma$, then the implicit function theorem, in a neighborhood of e^*, yields a map $G(e)$ with values in Δ_1, such that $G(e^*) = p^*$ and $F(e, G(e)) = 0$.

Thus $G(e)$ is an economic equilibrium corresponding to parameters e; in fact, it is "the natural" equilibrium if p^* is.

Inspired by Paul Samuelson [Samuelson] we call the derivative

$$DG(e^*) : E \to \overline{\Delta}_1$$

the *matrix* of *comparative statics.* Thus $DG(e^*)$ measures the infinitesimal change in prices corresponding to an infinitesimal change in endowments (for example). It is important to note that while $G(e)$ is only given implicitly, $DG(e^*)$ can be calculated from the derivative of F at (e^*,p^*).

Here is how that is done.

Let $(\overline{e},\overline{p})$ be tangent to Σ at (e^*,p^*). Then

$$D_pF(e^*,p^*)(\overline{p}) + D_eF(e^*,p^*)(\overline{e}) = 0.$$

So

$$\overline{p} = -D_pF(e^*,p^*)^{-1}D_eF(e^*,p^*)\overline{e} = DG(e^*)\overline{e}.$$

After [Wilkinson], one may well call $DG(e^*)$, the *condition matrix* in the setting of numerical analysis. Then the *condition number* is $\mu(e^*,p^*) = \|DG(e^*)\|$, operator norm. In the papers of Shub–Smale [Shub–Smale], I–V, this point of view is intensively developed. Part of these papers are in the context of the implicit function theorem as just described.

Part II. Algorithms.

Along with Herb Scarf and Curtis Eaves, in the mid-1970's, I was much concerned with algorithms for computing economic equilibria. My point of view has now shifted a bit in that I would see the algorithms more in the context of the historic process. This is a kind of mix of comparative statics and dynamics. Roughly speaking, economic history gives a starting point of a continuation algorithm.

Consider now a path $e(t)$, $0 \le t \le 1$, in E where $e(0) = e^*$, and (e^*, p^*) is known point on the equilibrium manifold. Supposing for the moment that the hypothesis of the implicit function theorem remains valid, there is an implicitly given path $p(t)$, unique, such that $p(0) = p^*$, and $F(e(t), p(t)) \equiv 0$. An algorithm which yields $p(t)$ approximately goes as follows.

Suppose $t_0, \ldots, t_k$ is a partition of $[0,1]$ so that $t_0 = 0$, $t_k = 1$ and $t_i < t_{i+1}$. Then in this format, we write Newton's method as:

$$
\begin{aligned}
p_{i+1} &= p_i - D_pF(e_{i+1}, p_i)^{-1}F(e_{i+1}, p_i) \\
p_0 &= p^* \text{ and } e_i = e(t_i).
\end{aligned}
$$

It is easy to see that if $\max_i(t_{i+1} - t_i) = \Delta t$ is small enough, then the p_i are uniformly close to $p(t_i)$. On the other hand the speed of the algorithm is proportional to $\frac{1}{\Delta t}$. One may say that the best k is the complexity of the problem. Finding the best choice of Δt in various contexts is a fundamental problem of complexity theory.

Much of my effort in the last 15 years has been on that problem and a good part of that is my joint work with Mike Shub.

It turns out that in the complexity analysis of Shub–Smale I–V, the condition numbers $\mu(e(t), p(t))$ play an important role. This becomes even more important in the ideal situation of solving a complex polynomial system of equations. Here we are able to prove a *condition number theorem* which gives the condition number as the reciprocal of the distance to the singular set of (e, p) in Σ. The singular set is just the set of (e, p) such that the linear map $D_pF(e, p)$ is singular.

I have just given a few words of introduction to the papers [Smale(Berkeley)], [Smale–Shub I–V].

References

[1] P. Samuelson, *Foundations of Economic Analysis*, Atheneum, NY, 1971.

[2] M. Shub and S. Smale, *Complexity of Bezout's Theorem I: Geometric aspects*, J. of the Amer. Math. Soc. **6**, I–V (1993), 459–501.

[3] M. Shub and S. Smale, *Complexity of Bezout's Theorem II: Volumes and probabilities*, Computational Algebraic Geometry (F. Eyssette and A. Galligo, Eds.) Progress in Mathematics, **109**, Birkhäuser (1993) 267–285.

[4] M. Shub and S. Smale, *Complexity of Bezout's Theorem III: Condition number and packing*, J.of Complexity, **9** (1993), 4–14.

[5] M. Shub and S. Smale, *Complexity of Bezout's Theorem IV: Probability of Success, Extensions*, to appear in SIAM J. on Numerical Analysis.

[6] M. Shub and S. Smale, *Complexity of Bezout's Theorem V: Polynomial Time*, to appear in "Theoretical Computer Science" **133** (1994).

[7] S. Smale, *Global analysis and economics*, in Handbook of Math. Economics, Vol. 1, K. Arrow, M. Intrilligator Eds., North Holland, NY, 1981.

[8] S. Smale, *Algorithms for solving equations*, Proceedings of the International Congress of Mathematicians, Berkeley, CA, American Mathematical Society, Providence, RI, pp.172–195 (referred to as Berkeley).

[9] J. Wilkinson, *Rounding Errors in Algebraic Processes*, Prentice Hall, Englewood Hills, NJ, 1963.

Classification of chaos and a large deviation theory for compact dynamical systems

Yoichiro Takahashi
Department of Mathematical Sciences, University of Tokyo
Komaba, Tokyo, 153 JAPAN

A survey will be given on the large deviation theory aspect of the classification of chaos with a slight generalization of previous results.

1. Introduction

The phase portraits changes drastically with parameter of a family of interval maps, $f(x) = ax(1-x)$, say, The first example of a family of map that was completely investigated with mathematical rigor was the unimodal linear maps [1] (except for the β-transformation [2] ; cf. [5]) and for such families one can observe several kinds of statistical limiting behaviors:

(a) (Observable chaos) The asymptotic measure μ for f is absolutely continuous and its density function is supported by

(a_1) a single interval, (a_n) 2^m intervals $(m > 0)$.

(b) The asymptotic measure μ is singular and is supported by

(b_p) a periodic orbit of period p, (b_∞) a continuum.

It should be emphasized that those ergodic properties (a) - (b) does not directly correspond to the topological properties:

(c) The map f exhibits topological chaos in the sense that every trial of the coin tossing can be embedded into the orbits of some interate f^m.

(d) The map f is not chaotic and admits an attractor which consists of

(d_p) a periodic orbit of period p, (d_∞) a continuum.

Note that (d) implies (b) but (c) does not imply (a). There does exists the case (c) & (b). In other words, stable periodic motions do appear in chaotic parameter region, as is observed in forced Duffing equation etc.

2. Classification of chaos

In 1980 [3] gave a classification of chaos in terms of the metrical entropy $h(\mu, f)$ and topological entropy $h_{top}(f)$:

$$(a) \longleftrightarrow h(\mu, f) > 0, \qquad (b) \longleftrightarrow h(\mu, f) = 0$$
$$(c) \longleftrightarrow h_{top}(f) > 0, \qquad (d) \longleftrightarrow h_{top}(f) = 0.$$

Moreover, it is conjectured (from mathematicians' point of view) that the asymptotic measure μ can be characterized by a variational principle:

$$\mu \text{ maximizes the functional } h(\nu, f) - \int log|f'|d\nu$$

under the assumption that f is piecewise smooth with ess inf $|f'| > 0$.

3. Large deviation theory for compact dynamical systems

Such a variational principle was known for expanding maps and Axiom A diffeomorphisms. But it was rather a surprise that it also holds for such a wider class of maps. The reason is then clarified by introducing a large deviation theory to the study of dynamical systems.

The following is a slight generalization of the results in [4].

Let X be a compact Hausdorff space and $f : X \to X$ be a continuous map and fix a capacity m on the Borel σ-algebra $B(X)$ of X (for instance, a probability Borel measure on $B(X)$). Set

$$Q(E) = \lim \sup_{n \to \infty} (1/n) \ \log m\{x \in X; \xi_n(x) \in E\} \tag{1}$$

where $\xi_n(x)$ stands for the empirical distribution on the finite orbit $x, fx, \ldots, f^{n-1}x$, i.e., $\xi_n(x)$ is the probability measure which assigns mass $1/n$ to each of the points $x, fx, \ldots, f^{n-1}x$. Then define a functional $q(\mu)$ for probability measure μ on $B(X)$ by

$$q(\mu) = \inf\{Q(G); G \text{ is a neighborhood of } \mu\} \tag{2}$$

and for each weakly continuous function $F(\mu)$, set

$$P(F) = \lim \sup_{n \to \infty} (1/n) \log \int m(dx) \exp\{nF(\xi_n(x))\}. \tag{3}$$

Theorem 1. There hold the large deviation principle and the variational principle:

$$P(F) = \max\{q(u) + F(\mu); \mu \text{is a probability measure on } B(X)\}. \tag{4}$$

$$q(\mu) = \inf\{P(F) - F(u); F(\mu) \text{ is a weakly continuous function of } \mu\}. \quad (5)$$

In particular,

$$\max\{q(\mu); \mu \text{ is a probability measure on } B(X)\} = 0. \quad (6)$$

Now the problem is reduced to compute the functional $q(\mu)$.

Theorem 2. Let (X, f) be the shift over alphabet A and m be a Gibbs measure (i.e., the jacobian j_m of m be positive and continuous), Then

$$q(\mu) = h(\mu, f) + \int m(dx)(log j_m) \quad (8)$$

In particular, if m is the uniform Bernoulli measure,

$$q(\mu) = h(\mu, f) - \log |A|. \quad (9)$$

Corollary. Let (X, f) be an Anosov diffeomorphism and m be the Lebesque measures. Then,

$$q(\mu) = h(\mu, f) - \chi^+(\mu, f) \quad (10)$$

where $\chi^+(\mu, f)$ is the sum of positive Liapunov exponents.

Remark. For a general diffeomorphism (X, f) and the Lebesque measure m, it is also proved in [4]:

$$q(\mu) \geq h(\mu, f) - \chi^+(\mu, f). \quad (11)$$

From this together with (6) follows the Pesin-Ruelle inequality:

$$\chi^+(\mu, f) \geq h(\mu, f). \quad (12)$$

References

[1] S. Ito, H. Nakada and S. Tanaka, *Unimodal linear transformations I, II*, Tokyo J. Math., **2** (1979), 221–259.

[2] S. Ito and Y. Takahashi, *Markov subshifts and realization of β-transformations*, J. Math. Soc. Japan, **26**, (1974), 33–55.

[3] Y. Oono and Y. Takahashi, *Chaos, noise and Fredholm theory*, Prog. Theor. Physics, **63** (1980), 1804–1807.

[4] Y. Takahashi, *Entropy functional (Free energy) for dynamical systems and their random perturbations*, Proc. Taniguchi Intern. Symp. on Stochastic Analysis at Katata and Kyoto 1982, Kinokuniya/North Holland, 1993.

[5] Y. Takahashi, *An ergodic-theoretical approach to the chaotic behaviour of dynamical systems*, Publ. RIMS Kyoto Univ., **19** (1983), 1265–1282.

On a Perturbation of Dynamic Programming

Kensuke Tanaka
Department of Mathematics, Faculty of Science,
Niigata University
Ikarashi, 2-chō, Niigata 950-21, JAPAN

1. Introduction

Recently, in view of practical problems, dynamic control problems have been investigated by many authors. In general, a mathematical formulation for these problems is given and, in this formulation, an optimal value and an optimal policy are mainly studied. Then, under some conditions, it will be necessary to find out an optimal value and an optimal policy. But, in order to show the existence of an optimal policy, we need to assume the stronger condition such that the control space is compact. So, when the compactness condition is excluded, the ε−optimal policies will be mainly studied.

In this paper, excluding the compactness condition of the action space, we study dynamic control systems. In order to characterize the ε−optimal policies, we make use of the properties in the convex analysis to the control systems. Especially, using properties of the subdifferential and the ε−sub -differential of an operator defined on the control space, we propose to make a given policy to be an ε− optimal policy for the system with one-step cost functions perturbed by some duality pairing. Then, under particular conditions, we show that the given policy is an optimal one for the original system. Further, in view of the convex analysis, the properties of ε−optimal policies are discussed.

This paper is organized in the following way. In Section 2, we give a formulation of the dynamic control systems. In Section 3, the necessary preliminaries and notations in the convex analysis are given. Section 4 is the main parts of this paper. Especially, using the properties of the subdifferential and the infimal convolution of the operator given on the action space, the properties of optimal policies and ε−optimal policies are discussed.

2. Formulation of a dynamic programming problem

A dynamic programming model is specified by a set of six elements

$$(S, C, U, q, r, \beta), \tag{2.1}$$

where

(i) S is a nonempty Borel subset of a Polish (i.e., complete, separable, metric) space with Borel σ-algebra β_S, the *state space* of the control system.

(ii) C is a nonempty Borel subset of a Polish space with Borel σ-algebra β_C, namely, the *control space*.

(iii) U is a multifunction which assigns to each state $x \in S$ a non-empty, closed and convex permissible set of controls $U(x) \subset C$. We assume that $GrU = \{(x,u)|x \in S, u \in U(x)\}$ is analytic in SC, where SC denotes the Cartesian product of sets S, C.

(iv) q is a Borel measurable stochastic kernel on S given SC, that is, $q(B|x,u)$ is a probability of a Borel subset $B \subset S$ for each $(x,u) \in GrU$ and a Borel measurable function of $(x,u) \in GrU$ for each Borel subset B. The law of motion of the control system is given by q.

(v) r is a proper extended real valued function, $GrU \to R^*$, where $R^* = R \cup \{\infty\}$, which is lower semianalytic, i.e., for any $c \in R^*$, $\{(x,u) \in GrU | r(x,u) < c\}$ is analytic, the *one-stage cost function*.

(vi) β is a *discount factor*.

In the specification, we should note that the permissible set of controls $U(x)$ depends on the state $x \in S$ and $q(\cdot|x,u)$ is independent of the time.

Then, a policy π for the model is defined as an infinite sequence $\pi = (\mu_0, \mu_1, \cdots, \mu_k, \cdots)$ such that, for each k, $\mu_k(du_k|x_0, u_0, x_1, \cdots, u_{k-1}, x_k)$ is a universally measurable stochastic kernel on C given $SCS \cdots CS$ satisfying

$$\mu_k(U(x_k)|x_0, u_0, x_1, \cdots, u_{k-1}, x_k) = 1$$

for every $(x_0, u_0, x_1, \cdots, u_{k-1}, x_k)$, where x_k and u_k denote k-th state and k-th control, respectively. If, for each k, μ_k is parametrized only by x_k, π is said to be a *Markov policy*. If, for each k and $(x_0, u_0, x_1, \cdots, u_{k-1}, x_k)$, $\mu_k(du_k|x_0, u_0, x_1, \cdots, u_{k-1}, x_k)$ assigns mass one to some point in C, π is said to be *nonrandomized.* In this case, π can be considered to be an infinite sequence $\pi = (\mu_0, \mu_1, \cdots)$, each component μ_k of which is an universally measurable mapping from $SCS \cdots CS \to C$ such that, for every $(x_0, u_0, x_1, \cdots, u_{k-1}, x_k)$,

$$\mu_k(x_0, u_0, x_1, \cdots, u_{k-1}, x_k) \in U(x_k).$$

If $\mathcal{F}$ is a type of σ-algebra on Polish spaces and each stochastic kernel component of a policy is $\mathcal{F}$-measurable, we say the policy is $\mathcal{F}$-measurable. We denote by Π the set of all policies and by Π_M the set of all Markov policies. If π is a Markov policy of the form $\pi = (\mu, \mu, \cdots)$, it is said to be *stationary.*

Thus, the control system is interpreted as following. If a policy $\pi = (\mu_0, \mu_1, \cdots, \mu_k, \cdots)$ is employed, at the successive k-th stage, $k = 0, 1, 2, \cdots$, we observe the k-th state of the control system and classify it to a possible state $x_k \in S$. So, we choose a control $u_k \in U(x_k)$ by k-th universally measurable stochastic kernel μ_k depending on a *history* $(x_0, u_0, x_1, \cdots, u_{k-1}, x_k)$ up to k-th stage. As a result of state x_k and control u_k at the k-th stage, we will incur a cost $r(x_k, u_k)$. Then, the control system moves to a new state $x_{k+1} \in S$ according to the stochastic kernel $q(\cdot|x_k, u_k)$. After that, the process of the system is analogously developed from x_{k+1}. Since GrU is analytic, from the Jankov von-Neumann theorem, it follows that there exists at least one nonrandomized Markov policy.

In subsequent discussion, we will often use S_k as a copy of S and C_k as a copy of C respectively. So, given an initial distribution p on S and any policy $\pi = (\mu_0, \mu_1, \mu_2, \cdots)$ together with the stochastic kernel q there is a sequence of unique probability measure $P_t^{\pi,p}$ on $S_0C_0S_1C_1 \cdots S_{t-1}C_{t-1}$, $t = 1, 2, \cdots$, of future up to the t-th stage in the control system, i.e.,

$$P_t^{\pi,p} = p\mu_0 q\mu_1 q \cdots \mu_{t-2} q\mu_{t-1}.$$

Further, there exists a unique probability measure $P^{\pi,p}$ on $S_0C_0S_1C_1 \cdots$ such that for each t, marginal measure of $P^{\pi,p}$ on $S_0C_0S_1C_1 \cdots S_{t-1}C_{t-1}$, is $P_t^{\pi,p}$, (see K.Hinderer [8, page 80]).

Using $P_t^{\pi,p}$ and $P^{\pi,p}$ determined in this manner, we will give the expected cost corresponding to a policy. Let $h_t = (x_0, u_0, x_1, u_1, \cdots, x_t, u_t)$ denotes a history of the control system up to the $(t+1)$-th stage. The expected cost at the $(t+1)$-th stage is given by

$$E_\pi [r(x_t, u_t)] = \int_{S_0C_0 \cdots S_tC_t} r^+(x_t, u_t) P_{t+1}^{\pi,p}(dh_t) - \int_{S_0C_0 \cdots S_tC_t} r^-(x_t, u_t) P_{t+1}^{\pi,p}(dh_t),$$

where $r^+ = \max\{r, 0\}$ and $r^- = \max\{-r, 0\}$.

So, if we assume that a policy $\pi = (\mu_0, \mu_1, \mu_2, \cdots)$ is used under the discount factor β, the *total expected discounted cost* is given by

$$I(\pi) = \sum_{t=0}^{\infty} \beta^t E_\pi [r(x_t, u_t)]. \tag{2.2}$$

Then, assuming that the expectation and the integral in (2.2) are interchanged, we have

$$\begin{aligned} I(\pi) &= E_\pi\left[\sum_{t=0}^{\infty}\beta^t r(x_t,u_t)\right] \\ &= \int_{S_0C_0S_1C_1\cdots}\left(\sum_{t=0}^{\infty}\beta^t r(x_t,u_t)\right)P^{\pi,p}(dh), \end{aligned}$$

where $h=(x_0,u_0,x_1,u_1,\cdots)$ denotes an infinite sequence in $S_0C_0S_1C_1\cdots$. If $\pi=(\mu,\mu,\cdots)$ is stationary, we often write $I(\mu)$ in place of $I(\pi)$. Further, if an initial distribution p assigns mass one to a point $x\in S$, p is written as p_x and if p_x is used, we write $I(\pi)(x)$ in place of $I(\pi)$.

Then, we consider a basic minimization problem (MP) for the dynamic control system :

$$\text{(MP)} \qquad \text{minimize} \quad I(\pi)(x) \quad \text{subject to} \quad \pi\in\Pi.$$

Definition 2.1 The *optimal discounted cost* $I^*(x)$ at an initial state x is defined by

$$I^*(x)=\inf_{\pi\in\Pi}I(\pi)(x). \tag{2.3}$$

Definition 2.2 If $I^*(x)=I(\pi^*)(x)$, the policy $\pi^*\in\Pi$ is said to be *optimal* at an initial state x.

In this paper we will study the problem (MP) in only the case such that the optimal cost is finite for all initial states, that is, $I^*(x)\in R$ for all $x\in S$. So, we define an ε-optimal policy as follows.

Definition 2.3 For a given $\varepsilon>0$, a policy $\pi_\varepsilon\in\Pi$ is said to be *ε-optimal* at an initial state x if

$$I(\pi_\varepsilon)(x)\le I^*(x)+\varepsilon.$$

3. The necessary preliminaries and the notations in the convex analysis

In order to give the main theorems in Section 4, we need to mention the necessary contents and notations in the convex analysis. Let B be a Banach space with norm $\|\cdot\|$ and B^* its continuous dual space. The duality pairing between $b\in B$ and $b^*\in B^*$ is denoted by $<b,b^*>$. A function g defined on B is said to be an extended proper function if g is not identically equal to

$+\infty$ and $g(b) > -\infty$ for all $b \in B$. The subdifferential of an extended proper function g at $b_0 \in dom(g)(dom(g)$ is the set where g is finite), $\partial g(b_0)$ is the set of $b^* \in B^*$ satisfying

$$g(b) \geq g(b_0) + < b^*, b - b_0 > \text{ for all } b \in B, \tag{3.1}$$

and the ε-subdifferential of g at $b_0 \in dom(g)$, $\partial_\varepsilon g(b_0)$ is the set of $b^* \in B^*$ satisfying

$$g(b) \geq g(b_0) + < b^*, b - b_0 > -\varepsilon \text{ for all } b \in B. \tag{3.2}$$

From the above constructions, it follows that for all $\varepsilon > 0$,

$$\partial g(b_0) \subset \partial_\varepsilon g(b_0)$$

and, if for some $\varepsilon \geq 0$, $\theta^* \in \partial_\varepsilon g(b_0)$, in which θ^* is a zero vector in B^*,

$$g(b) \geq g(b_0) - \varepsilon \text{ for all } b \in B \tag{3.3}$$

Further, let $Cv(B)$ be the set of all convex extended proper functions on B and $\Gamma_0(B)$ be the set of all functions in $Cv(B)$ which are lower-semicontinuous (l.s.c.). The ε-subdifferential $\partial_\varepsilon g(b_0)$ is a convex, weak*– closed subset in B^*, which reduces to the subdifferential for $\varepsilon = 0$.

Lemma 3.1 *If $g \in \Gamma_0(B)$, $\partial_\varepsilon g(b_0)$ is nonempty whenever $b_0 \in dom(g)$ and $\varepsilon > 0$.*

Lemma 3.2 *If $g \in \Gamma_0(B)$ and $int(dom(g)) \neq \phi$, which denotes interior of $dom(g)$, g is continuous on $int(dom(g))$.*

Thus, under the same conditions as Lemma 3.2, g is locally Lipschitz on $int(dom(g))$.

Lemma 3.3 *If $g \in Cv(B)$ and g is continuous at $b_0 \in dom(g)$, $\partial g(b_0)$ is nonempty.*

If $g \in \Gamma_0(B)$, it follows from Lemma 3.2 and Lemma 3.3 that for all $b_0 \in int(dom(g))$, $\partial g(b_0)$ is nonempty. So, for all $\varepsilon \geq 0$, $\partial_\varepsilon g(b_0)$ is nonempty because $\partial g(b_0) \subset \partial_\varepsilon g(b_0)$.

In [11], the subdifferential $\partial g(b_0)$ of g for $\varepsilon = 0$ is discussed in more details. Then, we are interested in the function $g^{[\xi]} = g \bigtriangledown \xi \parallel \cdot \parallel$ such that

$$g \bigtriangledown \xi \parallel \cdot \parallel (b) = inf_{c \in B}\{g(c) + \xi \parallel b - c \parallel\} \tag{3.4}$$

and $g^{[\xi]}$ is said to be the infimal convolution of g and $\xi \parallel \cdot \parallel$. $g^{[\xi]}(b) < \infty$ for all $b \in B$. If $g \in Cv(B), g^{[\xi]}$ is convex as the infimal convolution of two convex functions. So, there are two posibilities : either $g^{[\xi]}$ is finite everywhere or $g^{[\xi]}$ identically equals $-\infty$.

Lemma 3.4 *If there exists $\bar{b} \in B$ at which $g^{[\xi]}(\bar{b}) > -\infty$, $g^{[\xi]}$ is Lipschitz on B with Lipschitz constant ξ.*

This proof is given in [10, Corollary 1].

Now, we define $C^{\xi}(g)$ as the coincidence set of g and $g^{[\xi]}$, that is,

$$C^{\xi}(g) = \{b \in B | g(b) = g^{[\xi]}(b)\}. \tag{3.5}$$

Lemma 3.5 *If there exists some positive number $\xi > 0$ such that*

$$\{b \in B | \partial g(b) \cap \xi D^{*} \neq \phi\} \neq \phi,$$

we have

$$C^{\xi}(g) = \{b \in B | \partial g(b) \cap \xi D^{*} \neq \phi\}. \tag{3.6}$$

Moreover, for all $b_0 \in C^{\xi}(g)$ and all $\varepsilon \geq 0$,

$$\partial_{\varepsilon} g^{[\xi]}(b_0) = \partial_{\varepsilon} g(b_0) \cap \xi D^{*}, \tag{3.7}$$

where D^{*} denotes the unit closed ball in B^{*}.

This proof is given in [10, Prop. 2.3] and [2, Prop. 2.13].

Lemma 3.6 *Let $g : B \rightarrow R$ be a convex Lipschitz function. Then, there exists k such that, for all $b, b' \in B$ and all $\varepsilon, \varepsilon' > 0$,*

$$H(\partial_{\varepsilon} g(b), \partial_{\varepsilon'} g(b')) \leq \frac{k}{\min(\varepsilon, \varepsilon')}\{\| b - b' \| + |\varepsilon - \varepsilon'|\}, \tag{3.8}$$

where H is the Hausdorff-distance on the set of all nonempty, convex, weak$^{*}-$closed and bounded subsets in B^{*}.

Lemma 3.7 *Let $g \in Cv(B)$, let S be a subset of dom(g). Suppose that, for some ξ_0, $\partial g(b) \cap \xi_0 D^{*}$ is nonempty for all $b \in S$. Then, for all $\xi \geq \xi_0$, there exists k_{ξ} such that, for all $b, b' \in S$ and all $\varepsilon, \varepsilon' > 0$,*

$$H(\partial_{\varepsilon} g(b) \cap \xi D^{*}, \partial_{\varepsilon'} g(b') \cap \xi D^{*}) \leq \frac{k_{\xi}}{\min(\varepsilon, \varepsilon')}\{\| b - b' \| + |\varepsilon - \varepsilon'|\}, \tag{3.9}$$

See J.B.Hiriart-Urruty [9] for the proofs of Theorem 3.6 and Theorem 3.7 in detail.

4. A perturbation of the decision model and an ε-optimal policy

When the dynamic decision models are generally discussed, we will treat the following three cases for the one-stage cost function on GrU(see (iii) in Section 2):

(D) $0 \le \beta < 1$ and for some positive number $M > 0$,
$|r(x,u)| \le M$ for every $(x,u) \in GrU$

(P) $\beta = 1$ and $0 \le r(x,u)$ for every $(x,u) \in GrU$

(N) $\beta = 1$ and $r(x,u) \le 0$ for every $(x,u) \in GrU$.

In the these cases, we can get the following lemma.

Lemma 4.1 *For any initial state $x \in S$ and any policy $\pi \in \Pi$, there exists a Markov policy $\pi_M \in \Pi_M$ such that*

$$I(\pi_M)(x) = I(\pi)(x).$$

The proof of the lemma is given in D.P.Bertsekas and S.E.Shreve [3, Prop.8.1 and Prop.9.1].

Further, in the cases of (D) and (P), D.P.Bertsekas and S.E.Shreve [3, Prop. 9.19] give the following lemma with respect to an ε-optimal policy.

Lemma 4.2 *For each $\varepsilon > 0$, there exists an ε-optimal nonrandomized Markov policy and, if $0 \le \beta < 1$, it can be taken to be stationary.*

In view of the above lemmas, throughout this paper we will treat only the nonrandomized Markov policies, the set of which is denoted by $\Pi_N \subset \Pi_M$(see Section 2), in the cases of (D) and (P).

Thus, we will treat the following minimization problem (M$\overline{\text{P}}$).

(M$\overline{\text{P}}$) minimize $I(\pi)(x)$ subject to $\pi \in \Pi_N$.

So, let $M(S)$ be the set of all proper extended real valued functions defined on S, which are lower semianalytic and bounded from below. Further, let $N(C|S)$ be the set of all universally measurable controls on C given S. Then, the set of nonrandomized Markov policies is considered as $\Pi_N = N(C|S)N(C|S)\cdots$. So, for each $f \in N(C|S)$, we define an operator T_f on $M(S)$ as follows : for each $u \in M(S)$ and $x \in S$,

$$T_f u(x) = r(x, f(x)) + \beta \int_S u(x')q(dx'|x, f(x)). \tag{4.1}$$

The operator T_f is a mapping from $M(S)$ into $M(S)$, because we treat only of the decision models on (D) and (P). Let $v_0 \in M(S)$ be identically zero. Then, using the composition of the operators $T_{f_0}, T_{f_1}, \cdots, T_{f_{t-1}}, \cdots$ for a Markov policy $\pi = (f_0, \cdots, f_{t-1}, \cdots)$, the *t-stage cost function* $I_t(\pi)(x)$ and the cost function $I(\pi)(x)$ corresponding to π are given by

$$I_t(\pi)(x) = (T_{f_0}T_{f_1} \cdots T_{f_{t-1}})v_0(x) \tag{4.2}$$

and

$$I(\pi)(x) = \lim_{t\to\infty}(T_{f_0}T_{f_1} \cdots T_{f_{t-1}})v_0(x). \tag{4.3}$$

In order to use the properties of the convex analysis, we define the extended function $G(\cdot)u(s)$ of $T_{(\cdot)}u(s)$ on C as follows:

$$G(a)u(s) = \begin{cases} T_a u(s) & \text{if } a \in U(s) \subset C \\ \infty & \text{if } a \notin U(s) \text{ and } a \in C. \end{cases}$$

Then, in view of the convex analysis, we have need to give some assumptions in order to characterize the properties of the ε-optimal Markov policies in the decision model. So, we give the following assumptions for the models on (D) and (P):

(A1) For each $x \in S$, $r(x, \cdot) \in \Gamma_0(U(x))$, that is, $r(x, \cdot)$ is convex and l.s.c. on $U(x)$.

(A2) For any $u \in M(S)$ and $x \in S$, the integral operator

$$\int_S u(y)q(dy|x, a)$$

is a convex and l.s.c. function with respect to $a \in U(x)$.

Lemma 4.3 *Suppose that $G(\cdot)u(s)$ is finite at a given $f_* \in N(C|S)$. If C^*, which denotes the dual space of C, is separable and $\partial_\varepsilon G(f_*)u(\cdot)$ is analytic in SC^*, then for any $u \in M(S)$ and $s \in S$, there exists an analytic function $f^* : S \to C^*$ such that for all $f \in N(C|S)$*

$$T_f u(s) \geq T_{f_*}u(s) + \langle f^*(s), f(s) - f_*(s) \rangle - \varepsilon, \tag{4.4}$$

that is,

$$T_f u(s) - \langle f^*(s), f(s) \rangle \geq T_{f_*}u(s) - \langle f^*(s), f_*(s) \rangle - \varepsilon, \tag{4.5}$$

Proof. From (A1), (A2), for each $s \in S$, the extended function $G(\cdot)u(s) : C \to R^*$, is convex, and l.s.c. at $f_*(s) \in U(s) \subset C$. From $f_* \in dom(G(\cdot)u(s))$, it follows that $\partial_\varepsilon G(f_*)u(s)$ is a nonempty, convex and weak*-closed subset of C^* for each $s \in S$ and each $\varepsilon > 0$. So, we get for all $a^* \in \partial_\varepsilon G(f_*)u(s)$

$$T_f u(s) \geq T_{f_*}u(s) + < a^*, f(s) - f_*(s) > -\varepsilon \text{ for all } f \in N(C|S) \qquad (4.6)$$

Then, C^* is a Polish space because it is assumed that C^* is separable. Since $\partial_\varepsilon G(f_*)u(s)$ is an analytic subset on SC^*, from Jankov-von Neumann theorem, it follows that there is an analytic function $f^*(s) \in \partial_\varepsilon G(f_*)u(s)$. Thus, there is an analyticaly measurable function $f^* : S \to C^*$ satisfying (4.4) and (4.5) for the given policy $f_* \in N(C|S)$. Thus, the proof of the lemma is completed. □

Then, in order to show that a given stationary policy, f_* is an ε-optimal one for a dynamic model with the one-stage cost functions perturbed by using the functions $f_\cdot^* : S \to C^*$ as follows:

$$(S, C, U, q, r - < f_\cdot^*, \cdot >, \beta), \qquad (4.7)$$

we introduce, for each control function $f \in N(C|S)$, a perturbed operator $T(f^*, f)$ of T_f on $M(S)$ as follows: for each $u \in M(S)$ and $s \in S$

$$T(f^*, f)u(s) = T_f u(s) - < f^*(s), f(s) >, \qquad (4.8)$$

that is,

$$T(f^*, f)u(s) = r(s, f(s)) - < f^*(s), f(s) > + \beta \int_S u(y) q(dy|s, f(s)). \qquad (4.9)$$

Since $G(\cdot)u(s) \in \Gamma_0(U(s))$, if $f_*(s) \in int(domG(\cdot)u(s))$, it follows from Lemma 3.2 that $G(\cdot)u(s)$ is locally Lipschitzian at $f_*(s) \in U(s)$. So, there exist $M(s) > 0$ and a neighborhood U of $f_*(s)$ such that

$$|G(f)u(s) - G(f_*)u(s)| \leq M(s) \parallel f(s) - f_*(s) \parallel, \qquad (4.10)$$

whenever $f(s) \in U$, where $\parallel \cdot \parallel$ denotes the norm on C. Since $f_*(s) \in U$ and $f^*(s) \in \partial G(f_*)u(s)$, we have, for all $f(s) \in C$

$$< f^*(s), f(s) - f_*(s) > \leq M(s) \parallel f(s) - f_*(s) \parallel, \qquad (4.11)$$

which shows that $\parallel f^*(s) \parallel_* \leq M(s)$, where $\parallel \cdot \parallel_*$ denotes the norm on C^*.

Then, using Lemma 4.3, we can prove the following theorem.

Theorem 4.1 *Let everything be as in Lemma 4.3. For a specified stationary policy $f_* \in N(C|S)$ with $\sup_{s\in S} \| f_*(s) \| = \| f_* \| < \infty$, there exists a sequence of infinite analyticaly measurable functions $\pi^* = (f_0^*, f_1^*, \cdots, f_t^*, \cdots)$, $f_t^* : S \to C^*$. Assume moreover that $\sup_{s\in S} \| f_t^*(s) \|_* \leq M$, that is, bounded on S.*

Then, we have for any policy $\pi \in \Pi_N$,

$$I^*(\pi^*, \pi)(x) \geq I^*(\pi^*, f_*)(x) - \varepsilon, \tag{4.12}$$

where, for a policy $\pi = (f_0, f_1, \cdots, f_t, \cdots) \in \Pi_N$,

$$\begin{aligned} I^*(\pi^*, \pi)(x) &= \sum_{t=0}^{\infty} \beta^{t-1} E_\pi[r(x_t, f_t(x_t)) - < f_t^*(x_t), f_t(x_t) > |x_0 = x] \\ &= \sum_{t=0}^{\infty} \beta^{t-1} \int_S [r(s, f_t(s)) - < f_t^*(s), f_t(s) >] p_t^\pi(ds|x_0 = x). \end{aligned}$$

Proof. For any policy $\pi = (f_0, f_1, \cdots, f_t, \cdots)$ and the initial state $x \in S$, we define a notation as follows

$$V^{t+1}(\pi^*, \pi)(x) = T(f_0^*, f_0)T(f_1^*, f_1) \cdots T(f_t^*, f_t)v_0(x), \tag{4.13}$$

where $v_0 \in M(S)$ is identically zero.

Then, from the condition of the theorem, we have for each $s \in S$ and $t = 0, 1, 2, \cdots$,

$$\sup_{s\in S} | < f_t^*(s), f_*(s) > | \leq \sup_{s\in S} \| f_t^*(s) \|_* \| f_*(s) \| \leq M \| f_* \| . \tag{4.14}$$

So, from (4.3), it follows that for all $x \in S$ and $\pi \in \Pi_N$,

$$\lim_{t\to\infty} V^{t+1}(\pi^*, \pi)(x) = I(\pi^*, \pi)(x). \tag{4.15}$$

Thus, to prove the theorem, for sufficiently large t, it is sufficient to show the result of the theorem for $V^{t+1}(\pi^*, \pi)(\cdot) \in M(S)$. From (4.2), $V^t(\pi^*, \pi)(x)$ is successively constructed by the perturbed operators $T(f^*, f)$. So, from Lemma 4.3 and the conditions of the theorem, it follows that, for the given f_*, there exists a function $f_t^* : S \to C^*$, such that, for each $s \in S$ and $f \in N(C|S)$,

$$T(f_t^*, f)v_0(x) \geq T(f_t^*, f_*)v_0(x) - \varepsilon. \tag{4.16}$$

Thus, applying Lemma 4.3 to $T(f_t^*, f_t)u^*(\cdot) \in M(S)$, with the t-th control f_t in π instead of f in (4.16), we obtain a function $f_{t-1}^* \in \partial_\varepsilon G(f_*)T(f_t^*, f_t)v_0(x)$ such that, for each $x \in S$ and $f \in N(C|S)$

$$T(f_{t-1}^*, f)T(f_t^*, f_t)v_0(x) \geq T(f_{t-1}^*, f_*)T(f_t^*, f_t)v_0(x) - \varepsilon. \tag{4.17}$$

Then, since $T(\cdot,\cdot)$ is a monotone operator on $M(S)$, using the $(t-1)$–th control f_{t-1} in π instead of f in (4.17) and combining (4.16) with (4.17), we obtain for the policy $\pi = (f_0, f_1, \cdots, f_t, \cdots)$,

$$T(f_{t-1}^*, f_{t-1})T(f_t^*, f_t)v_0(x) \geq T(f_{t-1}^*, f_*)T(f_t^*, f_t)v_0(x) - \varepsilon$$
$$\geq T(f_{t-1}^*, f_*)T(f_t^*, f_*)v_0(x) - \beta\varepsilon - \varepsilon. \tag{4.18}$$

Further, applying Lemma 4.3 to (4.18) repeatedly, we arrive at

$$T(f_0^*, f_0)T(f_1^*, f_1)\cdots T(f_{t-1}^*, f_{t-1})T(f_t^*, f_t)v_0(x) - \sum_{i=0}^{t-1}\beta^i\varepsilon$$

$$\geq T(f_0^*, f_*)T(f_1^*, f_*)\cdots T(f_{t-1}^*, f_*)T(f_t^*, f_*)v_0(x) - \sum_{i=0}^{t}\beta^i\varepsilon. \tag{4.19}$$

Thus, from (4.19), we get for any sufficiently large t

$$V^t(\pi^*, \pi)(x) \geq V^t(\pi^*, f_*)(x) - \frac{\varepsilon}{1-\beta}. \tag{4.20}$$

So, in (4.20), taking $\varepsilon(1-\beta)^{-1}$ as ε and, then, letting t be ∞, the proof of the theorem is completed. □

Remark 4.1 If the given $f_* \in N(C|S)$ belongs always to $int(domG(\cdot)u(s))$ for all $u \in M(S)$ and $s \in S$, we can get the result of this theorem as $\varepsilon = 0$.

Theorem 4.2 *Let everything be as in Lemma 4.3 and assume that, for each $u \in M(S)$, the zero vector, θ^* belongs to $\partial_\varepsilon G(f_*)u(s)$ for all $s \in S$, that is, $\theta^* \in \partial_\varepsilon G(f_*)u(s)$ for all $u \in M(S)$ and $s \in S$.*

Then, for any policy π, we have

$$I(\pi)(x) \geq I(f_*)(x) - \varepsilon.$$

Proof. Since $\theta^* \in \partial_\varepsilon G(f_*)u(s)$ for all $u \in M(S)$ and $s \in S$, we can choose $f_t^* = \theta^*, t = 0, 1, 2, \cdots$, as each function f_t^* of π^* in Theorem 4.1. So, for any policy π, $I^*(\pi^*, \pi)(x)$ is equal to $I(\pi)(x)$. Thus, the proof is completed. □

Remark 4.2 As Remark 4.1, if $\theta^* \in int(domG(\cdot)u(s))$ for all $u \in M(S)$ and $s \in S$, we can get thr result of this theorem as $\varepsilon = 0$.

Theorem 4.3 *For any $\bar{\varepsilon} > 0$ and $u \in M(S)$, there exist $\delta > 0$ and k such that, for any $f_t, f_t' \in N(C|S)$ with $f_t(x_t), f_t'(x_t) \in f_*(x_t) + \delta D \subset int(domG(f_*(x_t))u(x_t)$ and any $\varepsilon, \varepsilon' \in (0, \bar{\varepsilon}]$,*

$$H(\partial_\varepsilon G(f_t)u(x_t), \partial_{\varepsilon'} G(f_t')u(x_t)) \leq \frac{k(x_t)}{\min(\varepsilon, \varepsilon')}\{\| f_t(x_t) - f_t'(x_t) \| + |\varepsilon - \varepsilon'|\},$$

where H is the Hausdorff-distance on C^* and D is the closed unit ball on C.

Proof. Since $G(\cdot)u(x_t) \in \Gamma_0(U(x_t))$, $G(\cdot)u(x_t)$ is Lipschitz in some neighborhood $U \subset int(domG(f_*(x_t))u(x_t))$. So, using Lemma 3.6 for $f_t(x_t), f_t'(x_t) \in U$, we complete the proof of the theorem. □

Also, we can obtain a result similar to Lemma 3.7.

Remark 4.3 Using $G^{[\xi]}(\cdot)u(s)$ perturbed by the infimal convolution $G(\cdot)u(s) \bigtriangledown \xi\|\cdot\|$ as $G(\cdot)u(s)$, from (A1) and (A2), it follows that $G^{[\xi]}(\cdot)u(s)$ is bounded from below. From Lemma 3.4, $G^{[\xi]}(\cdot)u(s)$ is Lipschitz on C with Lipschitz constant ξ. So, for the policies in $C^{\xi}(G(\cdot)u(s))$(see (3.4) in Section 3), the results similar to Lemma 3.6 and Lemma 3.7 as Theorem 4.3 are discussed.

References

[1] J.P. Aubin, *Mathematical Methods of Game and Economic Theory,* Revised edition, North-Holland Amsterdam (1982).

[2] H. Attouch and H. Riahi, *Stability results for Ekeland's ε−varitional principle and cone extremal solutions,* Math. Oper. Res., **18** (1993),173–201.

[3] D.P. Bertsekas and S.E. Shreve, *Stochastic Optimal Control,* The Discrete Time Case, Academic Press (1979).

[4] D. Blackwell, *Discounted dynamic programming,* Ann. Math. Statist.,**36** (1965), 226–235.

[5] R.M. Dudley, *Real Analysis and Probability,* Wadsworth & Brooks (1989).

[6] E.B. Dynkin, *Markov processes-I,* Springer-Verlag, Berlin (1965).

[7] E.B. Dynkin and A.A. Yushkevich, *Controlled Markov Processes,* Springer-Verlag, Berlin (1979).

[8] K. Hinderer, *Foundations of Non-stationary Dynamic Programming with Discrete Time Parameter,* Springer-Verlag, Berlin (1970).

[9] J.B. Hiriart-Urruty, *Lipschitz r-continuty of the approximate subdifferential of a convex function,* Math, Scand., **47** (1980),123–134.

[10] J.B. Hiriart-Urruty, *Extension of Lipschitz functions,* J. Math. Anal. Appl., **77** (1980),539–554.

[11] R.R. Phelps, *Convex Functions, Monotone Operators and Differentiability,* Lecture notes in Mathematics 1364, Springer-Verlag, Berlin (1989).

[12] K. Tanaka, *On discounted dynamic programming with constraints,* J. Math. Anal. Appl., **155** (1991), 264–277.

Bargaining Sets in Continuum Economies

Akira Yamazaki
Department of Economics, Hitotsubashi University
Kunitachi, Tokyo 186, JAPAN

1. Introduction

Aumann and Maschler [3] originally introduced the concept of the bargaining set. The bargaining set differs from the core in that it takes account of the reaction of others when a coalition engages on an improving move as an objection to a proposed allocation. Mas-Colell [9] proposed a redefinition of the original concept so that it does not depend upon distinguished individual agents and becomes well defined in continuum economies.

An allocation is an *objection* to another allocation if some nonnegligible set of agents are all made no worse off using only the initial resources they are endowed with, and a nonnegligible subset is better off. A *counterobjection* to an objection is another objection which is no worse than the first objection for almost all its members and is strictly better for a nonnegligible subset of the intersection of the two objecting coalitions. An objection is *justified* if there is no counterobjection to it. The *bargaining set* is the set of all feasible allocations that do not admit justified objections.

A feasible allocation is *Walrasian* if there is a price vector satisfying the following condition: for almost every agent, any bundle which is preferred to the allocated bundle, costs more than the agent's wealth. The core is the set of all feasible allocations that do not admit objections, justified or not. The well-known core equivalence theorem states that the core is identical to the set of Walrasian allocations. (See Aumann [2] and Hildenbrand [7, Theorem 1, p.133].) The bargaining set contains the core since objections by coalitions in the case of the bargaining set are made more difficult than in the case of the core. However, Mas-Colell showed that in continuum exchange economies the bargaining set is equivalent to the set of Walrasian allocations when all the commodities are perfectly divisible and the preferences of agents are completely preordered and strictly monotone.

Shitovitz [12] investigated the Mas-Colell's equivalence theorem when there is a lumpiness (i.e., an atom) in the population of agents, and found that the equality of the bargaining set and Walrasian allocations need not hold whereas the core can equal the set of Walrasian allocations. It was pointed out in [16] that the exact equivalence result need not hold in general when indivisible commodities exist, in other words, when there is a lumpiness in consumption. Anderson and Trockel [1] recently investigated the problem of convergence of the bargaining set to Walrasian allocations in large but finite economies. They found that the Mas-Colell's bargaining set is too large to be able to establish a convergence result and utilized the notion of the Geanakoplos' bargaining set [5]. This notion restricts counterobjections by requiring the group of agents, called a leader, who raises objection, to be excluded from counterobjections. This makes bargaining set smaller by making couterobjections more difficult. In [16] also, a smaller bargaining set had to be introduced due to the possible indivisibility of commodities.

In this paper we would like to discuss several notion of bargaining sets. Bargaining sets are distinguished according as whether objections and/or counterobjections are weakly or strictly preferred, and whether feasibility requirement is weak or strict. Three equivalence results including the Mas-Colell's original one are stated in the last section.

We would like to mention some of the related works of others that have not been referred to in this introduction. Earlier works on the equivalence principle for a bargaining set in the case of differentiable economies with transferable utilities are Shapley and Shubik [11] and Geanakoplos [5]. Grodal [6] investigated the extent to which the Mas-Colell's equivalence theorem remains valid when preferences are unordered. Vind [13] discussed related concepts of bargaining sets for economies. Nomura [10] gives asymptotic equivalence theorems for bargaining sets with infinitely many commodities. A refinement of the bargaining set is given in Dutta, Ray, Sengupta and Vohra [4]. Vohra [14] provides sufficient conditions for the existence of the bargaining set in non-transferable utility games with a finite number of players.

2. Bargaining Sets

We assume that there is a finite number $\ell > 0$ of commodities. A Euclidean space R^ℓ is taken to be *the commodity space* without any presumptions made on the physical characteristics of commodities. They will be expressed through *consumption sets* $X \subset R^\ell$ of individual agents. The usual convention states that quantities made available to agents are positive while quantities made available by agents are negative. Since an indefinite quantity of any commodity cannot be made available by economic agents, let $\mathcal{X}$ denote the set of all nonempty closed subsets X of R^ℓ each having an identical lower bound $\beta \in R^\ell$. $\mathcal{X}$ represents the universal set for consumption sets.

A preference $(X, \succeq)$ is a binary relation on X. It is called: *continuous* if $\succeq$ is closed in $X \times X$; *reflexive* if $x \succeq x$ for any $x \in X$; *transitive* if $x \succeq y$ and $y \succeq z$ imply $x \succeq z$; *complete* if $x \succeq y$ or $y \succeq x$ for any $x, y \in X$;

strictly monotone if $x \geq y$ and $x \neq y$ imply $x \succ y$ where $\succ$ is defined by $x \succ y \iff x \succeq y$ and not $y \succeq x$; *locally nonsatiated* if, given any $x \in X$ and any neighborhood U of x, there exists $y \in U \cap X$ such that $y \succ x$.

Let $\mathcal{P}$ denote the space of continuous preferences $(X, \succeq), X \in \mathcal{X}$, which are reflexive, complete, and transitive (see, e.g. Grodal [6]). The subset $\mathcal{P}_{\ell ns}$ of $\mathcal{P}$ will denote the set of all preferences in $\mathcal{P}$ that are locally nonsatiated.

The *population of economic agents* is described by an atomless measure space $(A, \mathcal{A}, \lambda)$ with $\lambda(A) = 1$ as in Aumann [2] and Hildenbrand [7]. Let $\mathcal{E} : A \to \mathcal{P}_{lns} \times R^\ell$ be a measurable mapping $a \mapsto (X_a, \succeq_a, e(a))$. $\mathcal{E}$ is called an *economy* if $\int e d\lambda < \infty$. Since the measure space is fixed, we will omit the reference to the measure λ in writing integrals.

An economy $\mathcal{E}$ is fixed below. When we say "*if all the commodities are perfectly divisible and if preferences of individual agents are strictly monotone with convex consumption sets,*" it will mean that, for a.e. $a \in A$, $X_a = R^\ell_+$ and $(X_a, \succeq_a)$ is strictly monotone.

Allocation f is an integrable mapping from A to R^ℓ such that, *a.e.* $a \in A$, $f(a) \in X_a \cup \{e(a)\}$. It is *weakly feasible* if $\int f \leq \int e$. It is *strictly feasible* if the inequality is strict.

$S \in \mathcal{A}$ is a *coalition* if $\lambda(S) > 0$. Allocation f is said to be *weakly feasible for coalition* S if $\int_S f \leq \int_S e$. When the weak inequality is replaced by the strict inequality $<$ in this definition, f is said to be *strictly feasible for* S. A pair (S, f) is a (*weakly* or *strictly*) *feasible pair* when f is (weakly or strictly) feasible for S.

An *objection* to an allocation f is a feasible pair (S, g) satisfying

(1) $g(a) \succeq_a f(a)$ for a.e. $a \in S$ and $\lambda(\{a \in S \mid g(a) \succ_a f(a)\}) > 0$, or

(2) $g(a) \succ_a f(a)$ for a.e. $a \in S$.

Objection is said to be *weakly preferred* in the first case and is called *strictly preferred* in the second case.

The *allocation induced by the objection* (S,g) is the mapping h from A to R^ℓ defined by

$$h(a) = \begin{cases} g(a) & \text{if} \quad a \in S \\ f(a) & \text{otherwise.} \end{cases}$$

If (S,g) is an objection to f and h the allocation induced by (S,g), a *counterobjection* to (S,g) is a feasible pair (T,k) satisfying

(1) $k(a) \succeq_a h(a)$ for a.e. $a \in S$ and $\lambda(\{a \in S \cap T \mid k(a) \succ_a h(a)\}) > 0$, or

(2) $k(a) \succ_a h(a)$ for a.e. $a \in S$ and $\lambda(S \cap T) > 0$.

As in the definition of objections, the first and the second cases distinguish between *weakly* and *strictly preferred* couterobjections.

An objection (S, g) is said to be *justified* if there is no counterobjection to it. A ***bargaining set*** is the set of all weakly feasible allocations that do not admit justified objections. This definition gives rise to various notion of bargaining sets depending upon whether objections and counterobjections are weakly or strictly preferred, and whether they are weakly or strictly feasible.

Let us denote by $\mathcal{B}_{1234}(\mathcal{E})$ or simply by $\mathcal{B}_{1234}$ the bargaining set that is composed of all weakly feasible allocations to which every objection with characteristics 1 and 2 has a counterobjection with characteritics 3 and 4. The two characteristics refer respectively to weak or strict preference and to weak or strict feasibility. For example, $\mathcal{B}_{wssw}$ denotes the set of all allocations to which every *weakly preferred strict* objection has a *strictly preferred weak* counterobjection. Although sixteen different bargainig sets may be defined in this manner, we consider only a few of them. The following inclusion relations illustrate some of the barganing sets:

$$\mathcal{B}_{wwss} \subset \left\{ \begin{array}{c} \mathcal{B}_{wwsw} \\ \\ \mathcal{B}_{wwws} \end{array} \right\} \subset \mathcal{B}_{wwww} \subset \left\{ \begin{array}{c} \mathcal{B}_{swww} \\ \\ \mathcal{B}_{wsww} \end{array} \right\} \subset \mathcal{B}_{ssww}$$

The above inclusion relations follow immediately from their definitions. For example, $\mathcal{B}_{wwss}$ is contained in $\mathcal{B}_{wwsw}$ because in $\mathcal{B}_{wwss}$ counterobjections are made more difficult so that objections are easier than in $\mathcal{B}_{wwsw}$.

The *core* is the set of all weakly feasible allocations that do not admit any objections, justified or not. Depending upon the characterisitics of objections, we have four concepts of the core. We denote them by $\mathcal{C}_{ww}$, $\mathcal{C}_{sw}$, $\mathcal{C}_{ws}$, and $\mathcal{C}_{ss}$. The following inclusions are immediate:

$$\mathcal{C}_{ww} \subset \left\{ \begin{array}{c} \mathcal{C}_{sw} \\ \\ \mathcal{C}_{ws} \end{array} \right\} \subset \mathcal{C}_{ss}.$$

Let us denote by α and β the two characteristics, such as ww, sw, *etc.* of objections and counterobjections respectively. Since objections may not be justified in case of the core, for any α and β we have

$$\mathcal{C}_{\alpha} \subset \mathcal{B}_{\alpha\beta}.$$

For example, we have $\mathcal{C}_{ww} \subset \mathcal{B}_{wwss}$, $\mathcal{C}_{sw} \subset \mathcal{B}_{swss}$, *etc.*

Remark 2.1 In the original definition of the Mas-Colell's bargaining set (Mas-Colell [9, p.131]), he did not require that $\lambda(S \cap T) > 0$ in defining a couterobjection (T, k) to an objection (S, g). Since counterobjections are made easier, objections become more difficult. In other words, the set of justified objections become smaller, thus making a bargaining set larger. For the purpose of establishing the equivalence between a bargaining set and the set of Walrasian allocations, one can get a mathematically stronger result by making a bargaining set larger. However, if we consider the meaning of a counterobjection, it appears more appealing to define it as we did here or elsewhere ([16]). Anderson and Trockel [1] also find it appropriate to require the nonemptiness of the intersection of objecting group and couterobjecting group in establishihg a convergence result in finite economies.

3. Walrasian Allocations and Objections

Allocation f is *price-supported* if there is a price vector $p\ (\neq 0) \in R^{\ell}_{+}$ satisfying

(1) $p \cdot x \geq p \cdot e(a)$ whenever $x \succ_a f(a)$, or

(2) $p \cdot x > p \cdot e(a)$ whenever $x \succ_a f(a)$

for a.e. $a \in A$. f is *weakly* or *strictly price-supported* depending upon whether it is the first or the second case.

Weakly feasible allocation f is *Walrasian* if it is price-supported, and *budget feasible* for A, i.e.,

$$p \cdot f(a) \leq p \cdot e(a), \ a.e.\ a \in A.$$

The set of Walrasian allocations are denoted by $\mathcal{W}_w$ or $\mathcal{W}_s$ depending on whether allocations are price-supported weakly or strictly. It is a result of a classical theorem (see, e.g., [7, Proposition 1, p.131]) that one has:

Fact 3.1 $\quad \mathcal{W}_s \subset \mathcal{C}_{ww} \subset \mathcal{C}_{sw}$

Using the similar argument that proves Fact 3.1 it is easy to show:

Fact 3.2 $\quad \mathcal{W}_w \subset \mathcal{C}_{ws} \subset \mathcal{C}_{ss}$

Then, Lemma 3.2 to be stated below implies that:

Fact 3.3 $\quad \mathcal{W}_w = \mathcal{C}_{ws} = \mathcal{C}_{ss}$.

An objection (S, g) to an allocation f is *Walrasian* if there is a price vector $p\ (\neq 0) \in R^{\ell}_{+}$ such that the allocation h induced by (S, g) is ***price-supported by p***, i.e.,

(1) $p \cdot x \geq p \cdot e(a)$ whenever $x \succ_a h(a)$, a.e. $a \in A$, or

(2) $p \cdot x > p \cdot e(a)$ whenever $x \succ_a h(a)$, a.e. $a \in A$,

and is ***budget feasible for S, i.e.,***

$$p \cdot h(a) \leq p \cdot e(a), \text{ a.e. } a \in S.$$

Approximately speaking, a Walrasian objection (S, g) to f is formed by those who prefer to trade the bundle assigned to them by f for the bundle given by g at the market price vector p. $\mathcal{WO}[f]_{\alpha,w}$ and $\mathcal{WO}[f]_{\alpha,s}$ denote the set of Walrasian objections to f, with characteristics α, that are respectively price-supported weakly and strictly.

Similarly, the sets of all justified objections to f are denoted by $\mathcal{JO}[f]_{\alpha\beta}$ where α and β respectively represent the two characterisitics of objections and counterobjections. For example, for $\alpha = sw$ and $\beta = ws$, $\mathcal{JO}[f]_{swws}$ is the set of all strictly preferred weak objections that do not admit weakly preferred strict counterobjections.

Proposition 3.1 *Let f be a weakly feasible allocation. Then, one has:*

(1) $\mathcal{WO}[f]_{\alpha,w} = \mathcal{JO}[f]_{\alpha ws}$

(2) $\mathcal{WO}[f]_{\alpha,s} \subset \mathcal{JO}[f]_{\alpha ww} \subset \mathcal{JO}[f]_{\alpha ss}$

This proposition follows from the following two lemmas:

Lemma 3.1 *Let h be an allocation which is weakly price-supported by p ($\neq 0$) $\in R^{\ell}_{+}$. Then, there is no strictly feasible pair (S, g) satisfying $\lambda(S) > 0$ and $g(a) \succsim_a h(a)$, a.e. $a \in S$. If h is strictly price-supported by p, then there is no weakly feasible pair (S, g) satisfying $g(a) \succsim_a h(a)$, a.e. $a \in S$, and $\lambda(\{a \in S \mid g(a) \succ_a h(a)\}) > 0$.*

Lemma 3.2 *Let h be an allocation. If there is no strictly feasible pair (S, g) satisfying $g(a) \succsim_a h(a)$, a.e. $a \in S$, and $\lambda(\{a \in S \mid g(a) \succ_a h(a)\}) > 0$, then h is weakly price-supported by a p ($\neq 0$) $\in R^{\ell}_{+}$.*

Lemmas 3.1 and 3.2 are polished and modified versions of the First and the Second Fundamental Theorem of welfare economics, modified so as to account for the lack of convexity of consumption sets in terms of the weak price-supportability and for the possibility of improvement by various coalitions. (For a proof, see [16].) By defining h to be the allocation induced by the objection f, Lemma 3.1 implies $\mathcal{WO}[f]_{\alpha,w} \subset \mathcal{JO}[f]_{\alpha ws}$ and $\mathcal{WO}[f]_{\alpha,s} \subset \mathcal{JO}[f]_{\alpha ww}$. Similarly, Lemma 3.2 implies $\mathcal{JO}[f]_{\alpha ws} \subset \mathcal{WO}[f]_{\alpha,w}$. Since $\mathcal{JO}[f]_{\alpha ww} \subset \mathcal{JO}[f]_{\alpha ws} \subset \mathcal{JO}[f]_{\alpha ss}$, the conclusion of Proposition 3.1 follows.

Proposition 3.1 characterizes weakly price-supported justified objections with weakly-preferred strict counterobjections. Mas-Colell [9, Proposition 1, p.132, and Proposition 3, p.133] first introduced the idea of characterizing justified objections with Walrasian objections. He proved the following:

Proposition 3.2 *If all the commodities are perfectly divisible and if preferences of individual agents are strictly monotone with convex consumption sets, then one has:*

$$\mathcal{WO}[f]_{\alpha,s} = \mathcal{JO}[f]_{\alpha ww}.$$

This proposition follows from Proposition 3.1 and the following Lemma 3.3 which only involves a standard argument.

Lemma 3.3 *If all the commodities are perfectly divisible and if preferences are strictly monotone with convex consumption sets, then the weak price-supportability implies the strict price-supportability.*

Remark 3.1 Mas-Colell's original definition of justified objections (see [9, Definition 2, p.131]) is given by $\mathcal{JO}[f]_{asw}$. But if all the commodities are perfectly divisible and if preferences of individual agents are strictly monotone with convex consumption sets, then one has $\mathcal{JO}[f]_{asw} = \mathcal{JO}[f]_{aww}$.

4. Equivalence Theorems

The fundamental theorem concerning the bargaining set is the following Mas-Colell's equivalence theorem.

Theorem 4.1 *(Mas-Colell [9]) If all the commodities are perfectly divisible and if preferences of individual agents are strictly monotone with convex consumption sets, then one has:*

$$\mathcal{W}_s = \mathcal{B}_{wwww}.$$

The Mas-Colell's theorem essentially shows that even if one enlarges the core by taking account of reactions of others against those who engage on an oposing move to a proposed allocation, the equivalence theorem *à la* Aumann [2] holds. However, this theorem need not be true without perfect divisibility of commodities.

Example 4.1 Consider an economy with identical agents having endowments $e = (2,\ 1)$. Commodity 1 is perfectly divisible while commodity 2 is consumable only in integer amounts. Let $X \equiv [\,0,\ \infty\,) \times \{0,\ 1,\ 2, \ldots\}$ be the consumption set. Preference relation is given by indifference curves which are intersections of broken curves with the consumption set in Figure 1. Let $\{A_1,\ A_2\}$ be a measurable partition of the population A such that $\lambda(A_1) = \frac{1}{3}, \lambda(A_2) = \frac{2}{3}$. Define a feasible allocation f by

$$f(a) = \begin{cases} (0,\ 3) & \text{if} \quad a \in A_1 \\ \\ (3,\ 0) & \text{if} \quad a \in A_2 \end{cases}$$

One can check that the allocation f is not strictly price-supported but there are no justified objections in $\mathcal{JO}[f]_{wwww}$. Thus, $\mathcal{W}_s \neq \mathcal{B}_{wwww}$, showing that the conclusion of the Mas-Collel's fundamental theorem as stated above need not be true when we drop the assumption of the perfect divisibility of commodities. Indeed, one can check that $\mathcal{W}_s = \mathcal{C}_{sw} = \mathcal{B}_{wwsw} = \emptyset$ while $f \in \mathcal{B}_{wwww}$. (Strictly speaking, in [9, Theorem 1, p.131] the theorem states $\mathcal{W}_s = \mathcal{B}_{wwsw}$ so that the example here is not entirely appropriate. But the point raised here still stands as one has $\mathcal{B}_{wwsw} = \mathcal{B}_{wwww}$ in the Mas-Colell's

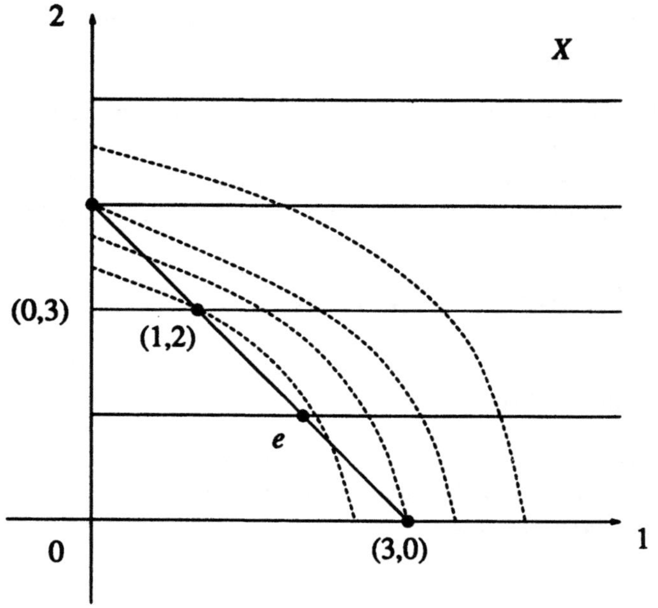

Figure 1

framework. See our remark 3..1.) Note also that f is a weakly price-supported Walrasian allocation. Partition A_2 measurably into $\{S_1, S_2\}$ so that $\lambda(S_1) = \frac{2}{9}$ and $\lambda(S_2) = \frac{4}{9}$. Define g by

$$g(a) = \begin{cases} (0,\ 3) & \text{if} \quad a \in S_1 \\ (3,\ 0) & \text{if} \quad a \in S_2 \\ (2,\ 1) & \text{otherwise.} \end{cases}$$

Then, g is an allocation feasible for A_2. One can check that (A_2, g) belongs to $\mathcal{JO}[f]_{wwws}$. Thus, $f \in \mathcal{W}_w$ but $f \notin \mathcal{B}_{wwws}$.

We now turn to the general case when commodities need not be perfectly divisible so that consumption sets are closed nonconvex sets in R^ℓ. By Fact 3.1 one has $\mathcal{W}_s \subset \mathcal{B}_{wwww}$, but by Example 4.1 $\mathcal{B}_{wwww}$ is not, in general, contained in $\mathcal{W}_s$. Hence, to establish the reverse inclusion relation, one needs to make the bargaining set slightly smaller and the set of Walrasian allocations slightly larger. Indeed, one can show:

Theorem 4.2 $\qquad \mathcal{B}_{wwws} \subset \mathcal{W}_w \subset \mathcal{B}_{wsss}$.

The proof of this theorem, which uses the method introduced by [9], is given in [16]. It involves an existential argument invoking the Kakutani's fixed point theorem. This existential argument is summarized in the following theorem:

Theorem 4.3 *If an allocation f is weakly feasible and if $f \notin \mathcal{W}_w$, then one has:*

$$\mathcal{WO}[f]_{ww,w} \neq \emptyset.$$

The second inclusion relaltion of Theorem 4.2 follows from Fact 3.3. The first inclusion relation follows from Theorem 4.3 and Proposition 3.1.

Let us proceed to give a condition on economic environments under which the strict equivalence between the bargaining set $\mathcal{B}_{wwwww}$ and the set $\mathcal{W}_s$ of Walrasian allocations holds true. Let us begin by defining critical sets.

Given a p $(\neq 0) \in R^\ell$ and a consumption set X, we say $x \in X$ has *local cheaper points* $(x_n)_n$ if the sequence x_n converges to x in X and satisfies $p \cdot x_n < p \cdot x$ for each n. x is a *critical point* in X if x does not have local cheaper points. The set of all critical points in X is denoted by $C_p(X)$. Define, for $\delta > 0$, an *open half ball* by $HN_p(x, \delta) = \{z \in R^\ell \mid \| x - z \| < \delta,\ p \cdot z < p \cdot x\}$. Then we have

$$C_p(X) = \{x \in X \mid (\exists \delta > 0) HN_p(x, \delta) \cap X = \emptyset\}.$$

The *critical set in the wealth space*, $J_p(x)$, is defined by $J_p(X) = \{w \in R \mid (\exists x \in C_p(X)) w = p \cdot x\}$. The *critical set of agents*, A_p, is defined by

$$A_p = \{a \in A \mid p \cdot e(a) \in J_p(X_a)\}.$$

Let $D(a,p)$ and $D_w(a,p)$ denote the demand and the weak demand set of agent a at price p respectively, i.e., $D(a,p) = \{x \in X_a \mid p \cdot x \leq p \cdot e(a)$ and $p \cdot z > p \cdot e(a)$ whenever $z \succ_a x\}$, $D_w(a,p) = \{x \in X_a \mid p \cdot x \leq p \cdot e(a)$ and $p \cdot z \geq p \cdot e(a)$ whenever $z \succ_a x\}$. It is well-known that relations $D(a, \cdot)$ may fail to be upper hemicontinuous when $p \cdot e(a) \in J_p(X_a)$. Note that in this paper consumption sets are simply nonempty closed subsets of R^ℓ_+ which are not necessarily convex. Therefore, the *minimum income condition*, $p \cdot e(a) > \min\ p \cdot X_a$ for a.e. $a \in A$, dose not guarantee the upper hemicontinuity of demand relations nor the existence of Walrasian allocations. However, the upper hemicontinuity of the weak demand relations $D_w(a, \cdot) : R^\ell_{++} \to R^\ell$ always hold true regardless of whether consumption sets are convex or not. This is precisely the point of using weak demand relations instead of usual demand relations.

Let us now define the *strict critical set* of agents A^*_p by $A^*_p = A_p \backslash \{a \in A \mid (\forall x,\ y \in D_w(a,p)) x \sim_a y\}$. Thus, A^*_p consists of economic agents whose budget hyperplanes contain a critical point in their consumption sets, and whose weak demand sets contain consumption bundles which are not indifferent.

By imposing a straightforward condition on the strict critical set of agents at weakly price-supported Walrasian allocations, we obtain an exact equivalence result between the bargaining set and the set of Walrasian allocations.

Given that f is a weakly price-supported Walrasian allocation, $\Pi(f)$ denotes the set of all price vectors by which f is weakly price-supported. Let $\Pi = \cup_f \Pi(f)$, i.e., the set of all price vectors that weakly support Walrasian allocations. The following theorem demonstrates an exact equivalence result.

Theorem 4.4 *(1) If there is $p \in \Pi$ such that A_p^* is λ-null, then* $\mathcal{W}_s \neq \emptyset$.
(2) If, for each weakly price-supported Walrasian allocation f, there is $p \in \Pi(f)$ such that A_p^* is λ-null, then one has:

$$\mathcal{W}_s = \mathcal{B}_{wwww}.$$

The proof of the theorem uses the next lemma.

Lemma 4.1 *Let A_p^* be λ-null. Then, an allocation h is weakly price-supported by p if and only if it is strictly price-supported by p.*

Under the assumption of Theorem 4.4, weakly price-supported Walrasian allocations (resp. obejctions) do coincide with strictly price-supported Walrasian allocations (resp. objections) by Lemma 4.1. Therefore, the combination of Theorem 4.3 and Proposition 3.1 together with Facts 3.1 and 3.2 yields Theorem 4.4.
Remark 4.1 Under the assumption of Theorem 4.4.(2), the following equivalence relation holds:

$$\mathcal{W}_s = \mathcal{C}_{ww} = \mathcal{B}_{wwws} = \mathcal{B}_{wwww} = \mathcal{W}_w = \mathcal{C}_{ws}.$$

Very roughly speaking, one may interpret the above relation to say that the nonemptiness of strictly price-supported Walrasian allocations is necessary and "almost" sufficient for the exact equivalence theorem.
Remark 4.2 Examples of economies satisfying that the strict critical set of agents A_p^* at a price vector p, which weakly price-supports a Walrasian allocation, is λ–null, are presented in [16].

References

[1] R.M. Anderson and Trockel W., *Convergence of the Bargaining Set,* Working Paper, Department of Economics and Department of Mathematics, University of California, Berkeley, 1993.

[2] R. Aumann, *Markets with a continuum of traders,* Econometrica **32** (1964), 39–50.

[3] R. Aumann and Maschler M. , The bargaining set for cooperative games, in *Advances in Game Theory,* (M. Dresher, L.S. Shapley and A.W. Tucker, Eds.), 443–447, Princeton Univ. Press, Princeton, 1964.

[4] B. Dutta, Ray D., Sengupta K. and Vohra R., *A consistent bargaining set,* J. Econ. Theory **49** (1989), 93–112.

[5] J. Geanakoplos, *The Bargaining Set and Nonstandard Analysis,* Chapter 3 of Ph.D. Dissertation, Harvard Univ., Cambridge, 1978.

[6] B. Grodal, *Bargaining Sets and Walrasian Allocations for Atomless Economies with Incomplete Preferences,* MSRI, Berkeley, 1986.

[7] W. Hildenbrand, *Core and Equilibria of a Large Economy,* Princeton Univ. Press, Princeton, 1974.

[8] M.A. Khan and Yamazaki A., *On the cores of economies with indivisible commodities and a continuum of traders,* J. Econ. Theory **24** (1981), 218–225.

[9] A. Mas-Colell, *An equivalence theorem for a bargaining set,* J. Math. Econ. **18** (1989), 129–139.

[10] Y. Nomura, *Elementary Equivalence Theorems for Bargaining Sets with Infinitely Many Commodities,* Working Paper, Chiba University, Chiba, 1993.

[11] L.S. Shapley and Shubik M., Convergence of the bargaining set for differentiable market games, Appendix B, in *A Game Theoretic Approach to Political Economy,* (M. Shubik), 683–692, MIT Press, Cambridge, 1984.

[12] B. Shitovitz, *The bargaining set and the core in mixed markets with atoms and an atomless sector,* J. Math. Econ. **18** (1989), 377–383.

[13] K. Vind, *Two characterizations of bargaining sets,* J. Math. Econ. **21** (1992), 89–97.

[14] R. Vohra, *An existence theorem for a bargaining set,* J. Math. Econ. **20** (1991), 19–34.

[15] A. Yamazaki, *Diversified consumption characteristics and conditionally dispersed endowment distribution: Regularizaing effect and existence of equilibria,* Econometrica **49** (1981), 639–654.

[16] A. Yamazaki, *Bargaining Sets and Equilibria without Perfect Divisibility of Commodities,* RUEE Working Paper, Department of Economics, Hitotsubashi University, Kunitachi, Tokyo,1993.

Nonlinear and Convex Analysis in Economic Theory

Organized by :

Tokyo Institute of Technology (T.I.Tech.)
Keio Economic Society (K.E.S.)

Supported by :

Nihon Keizai Shimbun Inc.

Scientific Committee :

G.Debreu(U.C.Berkeley) K.Fan(U.C.Santa Barbara)
M.Fukuoka(Keio U.) K.Itô(Kyoto U.)

Programme Committee :

T.Fujimoto(Okayama U.) K.Kawamata(Keio U.) M.Kojima(T.I.Tech.)
H.Komiya(Keio U.) S.Kusuoka(U.Tokyo) T.Maruyama(Keio U.)
M.Nakayama(Hosei U.) K.Nishimura(Kyoto U.) M.Nisio(Osaka E-C U.)
I.Sawashima(Ochanomizu U.) S.Simons(U.C.S.B) W.Takahashi(T.I.Tech.)
K.Tanaka(Niigata U.) S.Ukai(T.I.Tech.) A.Yamazaki(Hitotsubashi U.)

Sponsors :

Mr. Amikura, Katsuhiko
Daiwa Institute of Research Ltd.
Daiyuu Kensetsu Co., Ltd
Egusa Foundation for International Cooperation in the Social Sciences
Hashimoto Sangyo Co., Ltd.
Japan Airlines
Japan Society for the Promotion of Sciences
Mr. Kataoka, Takashi
Kindai Kagaku sha Co., Ltd.
The Nomura Securities Investment Trust Management Co. Ltd.
Saiken Bisou Co., Ltd.
SEIBUNSHA Publishing Co., Ltd.
Shonan Institute of Technology
WACOM CO., LTD.

Programme of the Conference

July 2 (Friday)

Plenary Talk

KY FAN, Univ. of California, Santa Barbara (USA)
A Survey of Existence Theorems for Certain Convex Inequalities and Related Results
chaired by Aderzej Granas

KIYOSI ITÔ, Kyoto Univ. (Japan)
A Survey of Stochastic Differential Equations
chaired by Hans Föllmer

Session 1 Stochastic Analysis and Financial Market

HANS FÖLLMER, Univ. of Bonn (Germany)
A Microeconomic Approach to Diffusion Models for Stock Prices
SHIGEO KUSUOKA, Univ. of Tokyo (Japan)
Option Replication with Transaction Costs
chaired by Makiko Nisio

DARRELL J. DUFFIE, Stanford Univ. (USA)
A Yield-Factor Model of Interest Rates
RYOZO MIURA, Hitotsubashi Univ. (Japan)
Pricing of Bonds and their Derivatives with Multifactor Stochastic Interest Rates
chaired by Kiyosi Itô

Session 2 Fixed Point Theory

ANTHONY TO-MING LAU, Univ. of Alberta (Canada)
Fixed Point Property for Semigroup of Non-expansive Mappings and Amenability
HIDETOSHI KOMIYA, Keio Univ. (Japan)
On Covering Theorems of a Simplex and Their Generalizations
chaired by Ikuko Sawashima

STEPHEN SIMONS, Univ. of California, Santa Barbara (USA)
Swimming below Icebergs
MASAKAZU KOJIMA, Tokyo Institute of Technology (Japan)
Interior Point Methods for Linear Complementarity Problems
chaired by Ky Fan

July 3 (Saturday)

Plenary Talk

ADERZEJ GRANAS, Nicolas Copernicus Univ. (Poland)
Some Geometric Principles of Convex Analysis
chaired by Stephen Simons

Session 3 Convex Analysis and Optimization

GIORGIO GIORGI, Univ. Degli Studi di Pavia (Italy)
On First Order Sufficient Conditions for Constrained Optima
KENSUKE TANAKA, Niigata Univ.(Japan)
On a Perturbation by Dynamic Programming
chaired by Mikio Nakayama

CHARLES CASTAING, Univ. Montpellier II (France)
Integral Functionals on the Space of Vector Measures with Applications to the Sweeping Process
NORIO KIKUCHI, Keio Univ. (Japan)
A Construction Method of Morse Flows to Variational Functionals
chaired by Seiji Ukai

Session 4 Nonlinear Dynamical Systems and Business Fluctuations

JEAN-MICHEL GRANDMONT, CEPREMAP (France)
Nonlinear Endogenous Business Cycles
RICHARD H. DAY, Univ. of Southern California (USA)
Multiple Phase Dynamics and Economic Systems
chaired by Kazuo Nishimura

MAKOTO YANO, Yokohama National Univ. (Japan)
KAZUO NISHIMURA, Kyoto Univ.(Japan)
Non-Linear Dynamics and Chaos in Optimal Growth: An Example
YOICHIRO TAKAHASHI, Univ. of Tokyo (Japan)
Classification of Chaos and a Large Deviation Theory for Compact Dynamical Systems
chaired by Kyuya Masuda

Plenary Talk

KUNIO KAWAMATA, Keio Univ. (Japan)
Stability and Osclations in the Kaldorian Dynamics Process
—Internal and External Cycles—
chaired by Akira Yamazaki

July 4 (Sunday)

Plenary Talk

STEPHEN SMALE, Univ. of California, Berkeley (USA)
On Solving the Equations for Supply = Demand
chaired by Tosiya Saito

Session 5 Eigenvalue of Positive Operators

ROGER D. NUSSBAUM, Rutgers Univ. (USA)
Differential-Delay Equations with State-Dependent Time Lags
ULRICH KRAUSE, Univ. of Bremen (Germany)
Positive Nonlinear Systems in Economics
chaired by Anthony To-Ming Lau

IKUKO SAWASHIMA, Ochanomizu Univ. (Japan)
Method of Duals in Nonlinear Analysis
—Lipschitz Duals of Banach Spaces and Some Applications—
TAKAO FUJIMOTO, Okayama Univ. (Japan)
Stabilizing Unstable Systems
chaired by Tsuyoshi Ando

Session 6 General Equilibrium Analysis

M. ALI KHAN, The Johns Hopkins Univ. (USA)
Economies with Imperfect Information
ROBERT ANDERSON, Univ. of California, Berkeley
Convergence of the Bargaining Set
chaired by Tatsuro Ichiishi

TATSURO ICHIISHI, Ohio State Univ. (USA)
Cooperative Processing of Information
AKIRA YAMAZAKI, Hitotsubashi Univ. (Japan)
Bargaining Sets and Equilibria without Perfect Divisibility of Commodities
chaired by Shinichi Takekuma

Plenary Talk

LEONID HURWICZ, Univ. of Minnesota (USA)
Cores, Optima, and Almost Competitive Prices in Discrete Spaces
chaired by Masao Fukuoka

Participants

Shigeo Akashi (Niigata Univ.)
Robert Anderson (Univ. of California at Berkeley)
Tsuyoshi Ando (Hokkaido Univ.)
Kazuo Anzai (Kagawa Univ.)
Nobuo Aoki (Tokyo Metropolitan Univ.)
Yuji Aruka (Chuo Univ.)
Toichiro Asada (Chuo Univ.)
Serguey Braguinsky (Yokohama City Univ.)
Charles Castaing (Univ. Montpellie II)
Yoshindo Chigusa (Kanto Gakuen Univ.)
Karaku Chou (Hitotsubashi Univ.)
Wen-Tian Cui (Keio Univ.)
Richard H. Day (Univ. of Southern California)
Yihong Du (Univ. of New England)
Darrell J. Duffie (Stanford Univ.)
Ky Fan (Univ. of California at Santa Barbara)
Hans Föllmer (Univ. Bonn)
Yoriaki Fujimori (Waseda Univ.)
Takao Fujimoto (Okayama Univ.)
Daisuke Fujiwara (Gakushuin Univ.)
Hiroshi Fukanoki (Hitotsubashi Univ.)
Shoichi Fukaya (Defence Academy)
Masao Fukuoka (Keio Univ.)
Giorgio Giorgi (Univ. of Pavia)
Motonari Goto (Tokyo Institute of Technology)
Aderzej Granas (Nicolas Copernicus Univ.)
Jean-Michel Grandmont (CEPREMAP)
Tetsuya Hayami (Takugin Credit Co. Ltd)
Norimichi Hirano (Yokohama National Univ.)
Kazuhiko Hirashima (Kyushu Univ.)
Masayoshi Hirota (Science Univ. of Tokyo)
Sin Hitotumatu (Tokyo Denki Univ.)
Toshikatsu Honda (Hitotsubashi Univ.)
Mitsuhiro Hoshino (Niigata Univ.)
Yoshiaki Hoshino (Hitotsubashi Univ.)
Leonid Hurwicz (Univ. of Minnesota)
Tatsuro Ichiishi (Ohio State Univ.)
Atsushi Ihaya (Mitsui Trust & Banking Co. Ltd)
Hiroyasu Iida (Keio Univ.)
Tyuei Iida (Keio Univ.)
Nobuyuki Ikeda (Ritsumeikan Univ.)
Takanobu Ikeda (Takushoku Univ.)
Tadashi Inoue (Hirosaki Univ.)
Motohiro Ishii (Akashi Labo.)
Yasushi Ishikawa (Hirosaki Univ.)
Shiro Ishikawa (Keio Univ.)
Kiyosi Itô (Kyoto Univ.)
Mikio Ito (Keio Univ.)
Nariyasu Ito (Musashi Univ.)
Gyoichi Iwata (Keio Univ.)
Osamu Kada (Tokyo Institute of Technology)
Akio Kagawa (Tokyo Keizai Univ.)
Koichiro Kamada (Bank of Japan)
Denzo Kamiya (Keio Univ.)
Naohisa Kaneda (Univ. of California at Santa Barbara)
Hiroshi Kaneko (Science Univ. of Tokyo)
Hideo Kanemitsu (Sophia Univ.)
Takeaki Kariya (Hitotsubashi Univ.)
Takashi Kataoka (Bear Stearns Real Estate Group Inc.)
Kunio Kawamata (Keio Univ.)
Hidefumi Kawasaki (Kyushu Univ.)
M. Ali Khan (The Johns Hopkins Univ.)
Yōichi Kijima (Yokohama National Univ.)
Norio Kikuchi (Keio Univ.)
Gang-Eun Kim (Tokyo Institute of Technology)
Mika Kishino (Waseda Univ.)
Tadashi Kiritani (Tokyo Metropolitan Univ.)
Hirobumi Kiuchi (Tokyo Institute of Technology)
Sayuri Koga (Kyushu Univ.)
Masakazu Kojima (Tokyo Institute of Technology)
Hidetoshi Komiya (Keio Univ.)
Hitoshi Kon (Niigata Univ.)
Shōzo Koshi (Utsunomiya Univ.)
Ulrich Krause (Univ. Bremen)
Yoshiaki Kumafuyu (Keio Univ.)
Naoto Kunitomo (Univ. of Tokyo)
Yukio Kuribayashi (Tottori Univ.)
ShigeToshi Kuroda (Gakushuin Univ.)
Daishi Kuroiwa (Niigata Univ.)
Shigeo Kusuoka (Univ. of Tokyo)
Anthony To-Ming Lau (Univ. of Alberta)
Nae-Chan Lee (Keio Univ.)
Jun-ichi Maeda (Hiroshima Shudo Univ.)
Makoto Maejima (Keio Univ.)
Toru Maruyama (Keio Univ.)

Yoshihiro Maruyama (Univ. of Tsukuba)
Kyuya Masuda (Rikkyo Univ.)
Hiroshi Matano (Univ. of Tokyo)
Minoru Matsuda (Shizuoka Univ.)
Akio Matsumoto (Niigata Univ.)
Ryozo Miura (Hitotsubashi Univ.)
Yoshio Miyahara (Nagoya City Univ.)
Shizuo Miyajima (Science Univ. of Tokyo)
Takashi Miyamoto (Keio Univ.)
Kiyoshi Mizohata (Tokyo Institute of Technology)
Masayoshi Mizutani (Keio Univ.)
Kiyoshi Murakami (Ministry of Education)
Akiko Murase (Univ. of Tokyo)
Isao Muto (Defence Academy)
Kazuhide Nakajo (Tokyo Institute of Technology)
Shinichiro Nakamura (Waseda Univ.)
Shinsuke Nakamura (Keio Univ.)
Hiroharu Nakano (Shiga Univ.)
Mayumi Nakatani (Keio Univ.)
Mikio Nakayama (Hosei Univ.)
Hukukane Nikaido (Tokyo International Univ.)
Kazuo Nishimura (Kyoto Univ.)
Hisakazu Nishino (Keio Univ.)
Kiyoko Nishizawa (Sophia Univ.)
Makiko Nisio (Osaka Electro-Communication Univ.)
Hideo Noda (Mitsui Trust & Banking Co. Ltd)
Roger D. Nussbaum (Rutgers Univ.)
Kinichi Obi (Tokyo Institute of Technology)
Toshiko Ogiwara (Univ. of Tokyo)
Kazuhiko Ohashi
Sachio Ohba (Kanagawa Univ.)
Michihiro Ohyama (Keio Univ.)
Masahiro Okuno (Univ. of Tokyo)
Setsuo Ōnari (Hitotsubashi Univ.)
Hiroaki Osana (Keio Univ.)
Makoto Ohsawa (Bank of Japan)
Yorimasa Oshime (Doshisha Univ.)
Keisuke Osumi (Kyushu Univ.)
Tatsuji Owase(Waseda Univ.)
Yujiro Ohya (Kyoto Univ.)
Yoshihiko Oyoshi (Bank of Nona Scatia)
Hiroyuki Ozaki (Univ. of Wisconsin)
Taro Ozawa (Hosei Univ.)
Jong-Yeoul Park (Pusan National Univ.)
Nobusumi Sagara (Hitotsubashi Univ.)
Mitsuaki Saito (Tokyo Metropolitan Univ.)
Tosiya Saito (Kawai juku)
Tomoya Sakagami (Kumamoto Univ. of Commerce)
Yasuhiro Sakai (Univ. of Tsukuba)
Yoshikiyo Sakai (Yokohama City Univ.)
Yūji Sakai (Shinshu Univ.)
Yukiyoshi Sakai (Kagoshima Univ.)
Hitoshi Sakaide (Kajimaroad)
Shigekazu Sato
Motohiro Sato (Hitotsubashi Univ.)
Yuji Sato (Keio Univ.)
Ikuko Sawashima (Ochanomizu Univ.)
Tadashi Shigoka (Kyoto Univ.)
Tomoo Shimizu (The Nishi-Tokyo Univ.)
Naoki Shioji (Tamagawa Univ.)
Shuhei Shiozawa (Keio Univ.)
Shunsuke Shiraishi (Toyama Univ.)
Kenichi Shiraiwa (Science Univ. of Tokyo)
Stephen Simons (Univ. of California at Santa Barbara)
Stephen Smale (Univ. of California at Berkeley)
Takuya Sobukawa (Okayama Univ.)
Makashi Sueki (Sanwa Research Institute Co.)
Shinichi Suda (Keio Univ.)
Akihiro Sugawara (Tamagawa Univ.)
Yeneng Sun (National Univ. of Singapore)
Yoji Sunada (Keio Univ.)
Takashi Suzuki (Hitotsubashi Univ.)
Yuhsuke Tabayashi (Aoyama Gakuin Univ.)
Harutaka Takahashi (Meiji Gakuin Univ.)
Shuichi Takahashi (Tokyo Univ. of Science)
Wataru Takahashi (Tokyo Institute of Technology)
Yoichiro Takahashi (Univ. of Tokyo)
Koichiro Takaoka (Univ. of Tokyo)
Jinryo Takasaki (Meiji Gakuin Univ.)
Shinichi Takekuma (Hitotubashi Univ.)
Fukiko Takeo (Ochanomizu Univ.)
Yasuyuki Tanabe (Hitotsubashi Univ.)
Eiji Tanaka (Tokyo Institute of Technology)
Kensuke Tanaka (Niigata Univ.)
Tamaki Tanaka (Hirosaki Univ.)
Hiroshi Tateishi (Takushoku Univ.)
Phan Thien Thach (Tokyo Institute of Technology)
Manabu Toda (Tokyo Keizai Univ.)
Tsutomu Tokimasa (Saga Univ.)
Kunio Tsujioka (Saitama Univ.)
Hoang Duong Tuan (Sophia Univ.)

Yoshimichi Tsuno (Sophia Univ.)
Seiji Ukai (Tokyo Institute of Technology)
Nobuo Wakabayashi (Otaru Univ. of Commerce)
Jun Wako (Gakushuin Univ.)
Hisako Watanabe (Ochanomizu Univ.)
Tetsufumi Yamakawa (Bank of Japan)
Yoshihiro Yajima (Univ. of Tokyo)
Kenji Yamamoto (Otaru Univ. of Commerce)
Akira Yamazaki (Hitotsubashi Univ.)
Noriyuki Yanagawa (Keio Univ.)
Masataka Yogo (Kanazawa College of Economics)
Makoto Yano (Yokohama National Univ.)
Kazunori Yokoyama (Niigata Chuoh Junior College)
Jun Yoshida (Keio Univ.)
Mariko Yoshida (Musashi Univ.)
Takeshi Yoshimoto (Toyo Univ.)
Naoyuki Yoshino (Keio Univ.)
Tadaaki Yoshioka (Kanagawa Univ.)
Hiroshi Yoshiwara (Niigata Univ.)

Springer-Verlag and the Environment

We at Springer-Verlag firmly believe that an international science publisher has a special obligation to the environment, and our corporate policies consistently reflect this conviction.

We also expect our business partners – paper mills, printers, packaging manufacturers, etc. – to commit themselves to using environmentally friendly materials and production processes.

The paper in this book is made from low- or no-chlorine pulp and is acid free, in conformance with international standards for paper permanency.

Printing: Weihert-Druck GmbH, Darmstadt
Binding: Theo Gansert Buchbinderei GmbH, Weinheim

Lecture Notes in Economics and Mathematical Systems

For information about Vols. 1–239
please contact your bookseller or Springer-Verlag

Vol. 240: Contributions to Operations Research. Proceedings, 1984. Edited by K. Neumann and D. Pallaschke. V, 190 pages. 1985.

Vol. 241: U. Wittmann, Das Konzept rationaler Preiserwartungen. XI, 310 Seiten. 1985.

Vol. 242: Decision Making with Multiple Objectives. Proceedings, 1984. Edited by Y.Y. Haimes and V. Chankong. XI, 571 pages. 1985.

Vol. 243: Integer Programming and Related Areas. A Classified Bibliography 1981–1984. Edited by R. von Randow. XX, 386 pages. 1985.

Vol. 244: Advances in Equilibrium Theory. Proceedings, 1984. Edited by C.D. Aliprantis, O. Burkinshaw and N.J. Rothman. II, 235 pages. 1985.

Vol. 245: J.E.M. Wilhelm, Arbitrage Theory. VII, 114 pages. 1985.

Vol. 246: P.W. Otter, Dynamic Feature Space Modelling, Filtering and Self-Tuning Control of Stochastic Systems. XIV, 177 pages.1985.

Vol. 247: Optimization and Discrete Choice in Urban Systems. Proceedings, 1983. Edited by B.G. Hutchinson, P. Nijkamp and M. Batty VI, 371 pages. 1985.

Vol. 248: Pural Rationality and Interactive Decision Processes. Proceedings, 1984. Edited by M. Grauer, M. Thompson and A.P. Wierzbicki. VI, 354 pages. 1985.

Vol. 249: Spatial Price Equilibrium: Advances in Theory, Computation and Application. Proceedings, 1984. Edited by P.T. Harker. VII, 277 pages. 1985.

Vol. 250: M. Roubens, Ph. Vincke, Preference Modelling. VIII, 94 pages. 1985.

Vol. 251: Input-Output Modeling. Proceedings, 1984. Edited by A. Smyshlyaev. VI, 261 pages. 1985.

Vol. 252: A. Birolini, On the Use of Stochastic Processes in Modeling Reliability Problems. VI, 105 pages. 1985.

Vol. 253: C. Withagen, Economic Theory and International Trade in Natural Exhaustible Resources. VI, 172 pages. 1985.

Vol. 254: S. Müller, Arbitrage Pricing of Contingent Claims. VIII, 151 pages. 1985.

Vol. 255: Nondifferentiable Optimization: Motivations and Applications. Proceedings, 1984. Edited by V.F. Demyanov and D. Pallaschke. VI, 350 pages. 1985.

Vol. 256: Convexity and Duality in Optimization. Proceedings, 1984. Edited by J. Ponstein. V, 142 pages. 1985.

Vol. 257: Dynamics of Macrosystems. Proceedings, 1984. Edited by J.-P. Aubin, D. Saari and K. Sigmund. VI, 280 pages. 1985.

Vol. 258: H. Funke, Eine allgemeine Theorie der Polypol- und Oligopolpreisbildung. III, 237 pages. 1985.

Vol. 259: Infinite Programming. Proceedings, 1984. Edited by E.J. Anderson and A.B. Philpott. XIV, 244 pages. 1985.

Vol. 260: H.-J. Kruse, Degeneracy Graphs and the Neighbourhood Problem. VIII, 128 pages. 1986.

Vol. 261: Th.R. Gulledge, Jr., N.K. Womer, The Economics of Made-to-Order Production. VI, 134 pages. 1986.

Vol. 262: H.U. Buhl, A Neo-Classical Theory of Distribution and Wealth. V, 146 pages. 1986.

Vol. 263: M. Schäfer, Resource Extraction and Market Struucture. XI, 154 pages. 1986.

Vol. 264: Models of Economic Dynamics. Proceedings, 1983. Edited by H.F. Sonnenschein. VII, 212 pages. 1986.

Vol. 265: Dynamic Games and Applications in Economics. Edited by T. Basar. IX, 288 pages. 1986.

Vol. 266: Multi-Stage Production Planning and Inventory Control. Edited by S. Axsäter, Ch. Schneeweiss and E. Silver. V, 264 pages.1986.

Vol. 267: R. Bemelmans, The Capacity Aspect of Inventories. IX, 165 pages. 1986.

Vol. 268: V. Firchau, Information Evaluation in Capital Markets. VII, 103 pages. 1986.

Vol. 269: A. Borglin, H. Keiding, Optimality in Infinite Horizon Economies. VI, 180 pages. 1986.

Vol. 270: Technological Change, Employment and Spatial Dynamics. Proceedings, 1985. Edited by P. Nijkamp. VII, 466 pages. 1986.

Vol. 271: C. Hildreth, The Cowles Commission in Chicago, 1939–1955. V, 176 pages. 1986.

Vol. 272: G. Clemenz, Credit Markets with Asymmetric Information. VIII,212 pages. 1986.

Vol. 273: Large-Scale Modelling and Interactive Decision Analysis. Proceedings, 1985. Edited by G. Fandel, M. Grauer, A. Kurzhanski and A.P. Wierzbicki. VII, 363 pages. 1986.

Vol. 274: W.K. Klein Haneveld, Duality in Stochastic Linear and Dynamic Programming. VII, 295 pages. 1986.

Vol. 275: Competition, Instability, and Nonlinear Cycles. Proceedings, 1985. Edited by W. Semmler. XII, 340 pages. 1986.

Vol. 276: M.R. Baye, D.A. Black, Consumer Behavior, Cost of Living Measures, and the Income Tax. VII, 119 pages. 1986.

Vol. 277: Studies in Austrian Capital Theory, Investment and Time. Edited by M. Faber. VI, 317 pages. 1986.

Vol. 278: W.E. Diewert, The Measurement of the Economic Benefits of Infrastructure Services. V, 202 pages. 1986.

Vol. 279: H.-J. Büttler, G. Frei and B. Schips, Estimation of Disequilibrium Modes. VI, 114 pages. 1986.

Vol. 280: H.T. Lau, Combinatorial Heuristic Algorithms with FORTRAN. VII, 126 pages. 1986.

Vol. 281: Ch.-L. Hwang, M.-J. Lin, Group Decision Making under Multiple Criteria. XI, 400 pages. 1987.

Vol. 282: K. Schittkowski, More Test Examples for Nonlinear Programming Codes. V, 261 pages. 1987.

Vol. 283: G. Gabisch, H.-W. Lorenz, Business Cycle Theory. VII, 229 pages. 1987.

Vol. 284: H. Lütkepohl, Forecasting Aggregated Vector ARMA Processes. X, 323 pages. 1987.

Vol. 285: Toward Interactive and Intelligent Decision Support Systems. Volume 1. Proceedings, 1986. Edited by Y. Sawaragi, K. Inoue and H. Nakayama. XII, 445 pages. 1987.

Vol. 286: Toward Interactive and Intelligent Decision Support Systems. Volume 2. Proceedings, 1986. Edited by Y. Sawaragi, K. Inoue and H. Nakayama. XII, 450 pages. 1987.

Vol. 287: Dynamical Systems. Proceedings, 1985. Edited by A.B. Kurzhanski and K. Sigmund. VI, 215 pages. 1987.

Vol. 288: G.D. Rudebusch, The Estimation of Macroeconomic Disequilibrium Models with Regime Classification Information. VII,128 pages. 1987.

Vol. 289: B.R. Meijboom, Planning in Decentralized Firms. X, 168 pages. 1987.

Vol. 290: D.A. Carlson, A. Haurie, Infinite Horizon Optimal Control. XI, 254 pages. 1987.

Vol. 291: N. Takahashi, Design of Adaptive Organizations. VI, 140 pages. 1987.

Vol. 292: I. Tchijov, L. Tomaszewicz (Eds.), Input-Output Modeling. Proceedings, 1985. VI, 195 pages. 1987.

Vol. 293: D. Batten, J. Casti, B. Johansson (Eds.), Economic Evolution and Structural Adjustment. Proceedings, 1985. VI, 382 pages.

Vol. 294: J. Jahn, W. Knabs (Eds.), Recent Advances and Historical Development of Vector Optimization. VII, 405 pages. 1987.

Vol. 295. H. Meister, The Purification Problem for Constrained Games with Incomplete Information. X, 127 pages. 1987.

Vol. 296: A. Börsch-Supan, Econometric Analysis of Discrete Choice. VIII, 211 pages. 1987.

Vol. 297: V. Fedorov, H. Läuter (Eds.), Model-Oriented Data Analysis. Proceedings, 1987. VI, 239 pages. 1988.

Vol. 298: S.H. Chew, Q. Zheng, Integral Global Optimization. VII, 179 pages. 1988.

Vol. 299: K. Marti, Descent Directions and Efficient Solutions in Discretely Distributed Stochastic Programs. XIV, 178 pages. 1988.

Vol. 300: U. Derigs, Programming in Networks and Graphs. XI, 315 pages. 1988.

Vol. 301: J. Kacprzyk, M. Roubens (Eds.), Non-Conventional Preference Relations in Decision Making. VII, 155 pages. 1988.

Vol. 302: H.A. Eiselt, G. Pederzoli (Eds.), Advances in Optimization and Control. Proceedings, 1986. VIII, 372 pages. 1988.

Vol. 303: F.X. Diebold, Empirical Modeling of Exchange Rate Dynamics. VII, 143 pages. 1988.

Vol. 304: A. Kurzhanski, K. Neumann, D. Pallaschke (Eds.), Optimization, Parallel Processing and Applications. Proceedings, 1987. VI, 292 pages. 1988.

Vol. 305: G.-J.C.Th. van Schijndel, Dynamic Firm and Investor Behaviour under Progressive Personal Taxation. X, 215 pages.1988.

Vol. 306: Ch. Klein, A Static Microeconomic Model of Pure Competition. VIII, 139 pages. 1988.

Vol. 307: T.K. Dijkstra (Ed.), On Model Uncertainty and its Statistical Implications. VII, 138 pages. 1988.

Vol. 308: J.R. Daduna, A. Wren (Eds.), Computer-Aided Transit Scheduling. VIII, 339 pages. 1988.

Vol. 309: G. Ricci, K. Velupillai (Eds.), Growth Cycles and Multisectoral Economics: the Goodwin Tradition. III, 126 pages. 1988.

Vol. 310: J. Kacprzyk, M. Fedrizzi (Eds.), Combining Fuzzy Imprecision with Probabilistic Uncertainty in Decision Making. IX, 399 pages. 1988.

Vol. 311: R. Färe, Fundamentals of Production Theory. IX, 163 pages. 1988.

Vol. 312: J. Krishnakumar, Estimation of Simultaneous Equation Models with Error Components Structure. X, 357 pages. 1988.

Vol. 313: W. Jammernegg, Sequential Binary Investment Decisions. VI, 156 pages. 1988.

Vol. 314: R. Tietz, W. Albers, R. Selten (Eds.), Bounded Rational Behavior in Experimental Games and Markets. VI, 368 pages. 1988.

Vol. 315: I. Orishimo, G.J.D. Hewings, P. Nijkamp (Eds), Information Technology: Social and Spatial Perspectives. Proceedings 1986. VI, 268 pages. 1988.

Vol. 316: R.L. Basmann, D.J. Slottje, K. Hayes, J.D. Johnson, D.J. Molina, The Generalized Fechner-Thurstone Direct Utility Function and Some of its Uses. VIII, 159 pages. 1988.

Vol. 317: L. Bianco, A. La Bella (Eds.), Freight Transport Planning and Logistics. Proceedings, 1987. X, 568 pages. 1988.

Vol. 318: T. Doup, Simplicial Algorithms on the Simplotope. VIII, 262 pages. 1988.

Vol. 319: D.T. Luc, Theory of Vector Optimization. VIII, 173 pages. 1989.

Vol. 320: D. van der Wijst, Financial Structure in Small Business. VII, 181 pages. 1989.

Vol. 321: M. Di Matteo, R.M. Goodwin, A. Vercelli (Eds.), Technological and Social Factors in Long Term Fluctuations. Proceedings. IX, 442 pages. 1989.

Vol. 322: T. Kollintzas (Ed.), The Rational Expectations Equilibrium Inventory Model. XI, 269 pages. 1989.

Vol. 323: M.B.M. de Koster, Capacity Oriented Analysis and Design of Production Systems. XII, 245 pages. 1989.

Vol. 324: I.M. Bomze, B.M. Pötscher, Game Theoretical Foundations of Evolutionary Stability. VI, 145 pages. 1989.

Vol. 325: P. Ferri, E. Greenberg, The Labor Market and Business Cycle Theories. X, 183 pages. 1989.

Vol. 326: Ch. Sauer, Alternative Theories of Output, Unemployment, and Inflation in Germany: 1960–1985. XIII, 206 pages. 1989.

Vol. 327: M. Tawada, Production Structure and International Trade. V, 132 pages. 1989.

Vol. 328: W. Güth, B. Kalkofen, Unique Solutions for Strategic Games. VII, 200 pages. 1989.

Vol. 329: G. Tillmann, Equity, Incentives, and Taxation. VI, 132 pages. 1989.

Vol. 330: P.M. Kort, Optimal Dynamic Investment Policies of a Value Maximizing Firm. VII, 185 pages. 1989.

Vol. 331: A. Lewandowski, A.P. Wierzbicki (Eds.), Aspiration Based Decision Support Systems. X, 400 pages. 1989.

Vol. 332: T.R. Gulledge, Jr., L.A. Litteral (Eds.), Cost Analysis Applications of Economics and Operations Research. Proceedings. VII, 422 pages. 1989.

Vol. 333: N. Dellaert, Production to Order. VII, 158 pages. 1989.

Vol. 334: H.-W. Lorenz, Nonlinear Dynamical Economics and Chaotic Motion. XI, 248 pages. 1989.

Vol. 335: A.G. Lockett, G. Islei (Eds.), Improving Decision Making in Organisations. Proceedings. IX, 606 pages. 1989.

Vol. 336: T. Puu, Nonlinear Economic Dynamics. VII, 119 pages. 1989.

Vol. 337: A. Lewandowski, I. Stanchev (Eds.), Methodology and Software for Interactive Decision Support. VIII, 309 pages. 1989.

Vol. 338: J.K. Ho, R.P. Sundarraj, DECOMP: an Implementation of Dantzig-Wolfe Decomposition for Linear Programming. VI, 206 pages.

Vol. 339: J. Terceiro Lomba, Estimation of Dynamic Econometric Models with Errors in Variables. VIII, 116 pages. 1990.

Vol. 340: T. Vasko, R. Ayres, L. Fontvieille (Eds.), Life Cycles and Long Waves. XIV, 293 pages. 1990.

Vol. 341: G.R. Uhlich, Descriptive Theories of Bargaining. IX, 165 pages. 1990.

Vol. 342: K. Okuguchi, F. Szidarovszky, The Theory of Oligopoly with Multi-Product Firms. V, 167 pages. 1990.

Vol. 343: C. Chiarella, The Elements of a Nonlinear Theory of Economic Dynamics. IX, 149 pages. 1990.

Vol. 344: K. Neumann, Stochastic Project Networks. XI, 237 pages. 1990.

Vol. 345: A. Cambini, E. Castagnoli, L. Martein, P Mazzoleni, S. Schaible (Eds.), Generalized Convexity and Fractional Programming with Economic Applications. Proceedings, 1988. VII, 361 pages. 1990.

Vol. 346: R. von Randow (Ed.), Integer Programming and Related Areas. A Classified Bibliography 1984–1987. XIII, 514 pages. 1990.

Vol. 347: D. Ríos Insua, Sensitivity Analysis in Multiobjective Decision Making. XI, 193 pages. 1990.

Vol. 348: H. Störmer, Binary Functions and their Applications. VIII, 151 pages. 1990.

Vol. 349: G.A. Pfann, Dynamic Modelling of Stochastic Demand for Manufacturing Employment. VI, 158 pages. 1990.

Vol. 350: W.-B. Zhang, Economic Dynamics. X, 232 pages. 1990.

Vol. 351: A. Lewandowski, V. Volkovich (Eds.), Multiobjective Problems of Mathematical Programming. Proceedings, 1988. VII, 315 pages. 1991.

Vol. 352: O. van Hilten, Optimal Firm Behaviour in the Context of Technological Progress and a Business Cycle. XII, 229 pages. 1991.

Vol. 353: G. Ricci (Ed.), Decision Processes in Economics. Proceedings, 1989. III, 209 pages 1991.

Vol. 354: M. Ivaldi, A Structural Analysis of Expectation Formation. XII, 230 pages. 1991.

Vol. 355: M. Salomon. Deterministic Lotsizing Models for Production Planning. VII, 158 pages. 1991.

Vol. 356: P. Korhonen, A. Lewandowski, J . Wallenius (Eds.), Multiple Criteria Decision Support. Proceedings, 1989. XII, 393 pages. 1991.

Vol. 357: P. Zörnig, Degeneracy Graphs and Simplex Cycling. XV, 194 pages. 1991.

Vol. 358: P. Knottnerus, Linear Models with Correlated Disturbances. VIII, 196 pages. 1991.

Vol. 359: E. de Jong, Exchange Rate Determination and Optimal Economic Policy Under Various Exchange Rate Regimes. VII, 270 pages. 1991.

Vol. 360: P. Stalder, Regime Translations, Spillovers and Buffer Stocks. VI, 193 pages . 1991.

Vol. 361: C. F. Daganzo, Logistics Systems Analysis. X, 321 pages. 1991.

Vol. 362: F. Gehrels, Essays In Macroeconomics of an Open Economy. VII, 183 pages. 1991.

Vol. 363: C. Puppe, Distorted Probabilities and Choice under Risk. VIII, 100 pages . 1991

Vol. 364: B. Horvath, Are Policy Variables Exogenous? XII, 162 pages. 1991.

Vol. 365: G. A. Heuer, U. Leopold-Wildburger. Balanced Silverman Games on General Discrete Sets. V, 140 pages. 1991.

Vol. 366: J. Gruber (Ed.), Econometric Decision Models. Proceedings, 1989. VIII, 636 pages. 1991.

Vol. 367: M. Grauer, D. B. Pressmar (Eds.), Parallel Computing and Mathematical Optimization. Proceedings. V, 208 pages. 1991.

Vol. 368: M. Fedrizzi, J. Kacprzyk, M. Roubens (Eds.), Interactive Fuzzy Optimization. VII, 216 pages. 1991.

Vol. 369: R. Koblo, The Visible Hand. VIII, 131 pages.1991.

Vol. 370: M. J. Beckmann, M. N. Gopalan, R. Subramanian (Eds.), Stochastic Processes and their Applications. Proceedings, 1990. XLI, 292 pages. 1991.

Vol. 371: A. Schmutzler, Flexibility and Adjustment to Information in Sequential Decision Problems. VIII, 198 pages. 1991.

Vol. 372: J. Esteban, The Social Viability of Money. X, 202 pages. 1991.

Vol. 373: A. Billot, Economic Theory of Fuzzy Equilibria. XIII, 164 pages. 1992.

Vol. 374: G. Pflug, U. Dieter (Eds.), Simulation and Optimization. Proceedings, 1990. X, 162 pages. 1992.

Vol. 375: S.-J. Chen, Ch.-L. Hwang, Fuzzy Multiple Attribute Decision Making. XII, 536 pages. 1992.

Vol. 376: K.-H. Jöckel, G. Rothe, W. Sendler (Eds.), Bootstrapping and Related Techniques. Proceedings, 1990. VIII, 247 pages. 1992.

Vol. 377: A. Villar, Operator Theorems with Applications to Distributive Problems and Equilibrium Models. XVI, 160 pages. 1992.

Vol. 378: W. Krabs, J. Zowe (Eds.), Modern Methods of Optimization. Proceedings, 1990. VIII, 348 pages. 1992.

Vol. 379: K. Marti (Ed.), Stochastic Optimization. Proceedings, 1990. VII, 182 pages. 1992.

Vol. 380: J. Odelstad, Invariance and Structural Dependence. XII, 245 pages. 1992.

Vol. 381: C. Giannini, Topics in Structural VAR Econometrics. XI, 131 pages. 1992.

Vol. 382: W. Oettli, D. Pallaschke (Eds.), Advances in Optimization. Proceedings, 1991. X, 527 pages. 1992.

Vol. 383: J. Vartiainen, Capital Accumulation in a Corporatist Economy. VII, 177 pages. 1992.

Vol. 384: A. Martina, Lectures on the Economic Theory of Taxation. XII, 313 pages. 1992.

Vol. 385: J. Gardeazabal, M. Regúlez, The Monetary Model of Exchange Rates and Cointegration. X, 194 pages. 1992.

Vol. 386: M. Desrochers, J.-M. Rousseau (Eds.), Computer-Aided Transit Scheduling. Proceedings, 1990. XIII, 432 pages. 1992.

Vol. 387: W. Gaertner, M. Klemisch-Ahlert, Social Choice and Bargaining Perspectives on Distributive Justice. VIII, 131 pages. 1992.

Vol. 388: D. Bartmann, M. J. Beckmann, Inventory Control. XV, 252 pages. 1992.

Vol. 389: B. Dutta, D. Mookherjee, T. Parthasarathy, T. Raghavan, D. Ray, S. Tijs (Eds.), Game Theory and Economic Applications. Proceedings, 1990. IX, 454 pages. 1992.

Vol. 390: G. Sorger, Minimum Impatience Theorem for Recursive Economic Models. X, 162 pages. 1992.

Vol. 391: C. Keser, Experimental Duopoly Markets with Demand Inertia. X, 150 pages. 1992.

Vol. 392: K. Frauendorfer, Stochastic Two-Stage Programming. VIII, 228 pages. 1992.

Vol. 393: B. Lucke, Price Stabilization on World Agricultural Markets. XI, 274 pages. 1992.

Vol. 394: Y.-J. Lai, C.-L. Hwang, Fuzzy Mathematical Programming. XIII, 301 pages. 1992.

Vol. 395: G. Haag, U. Mueller, K. G. Troitzsch (Eds.), Economic Evolution and Demographic Change. XVI, 409 pages. 1992.

Vol. 396: R. V. V. Vidal (Ed.), Applied Simulated Annealing. VIII, 358 pages. 1992.

Vol. 397: J. Wessels, A. P. Wierzbicki (Eds.), User-Oriented Methodology and Techniques of Decision Analysis and Support. Proceedings, 1991. XII, 295 pages. 1993.

Vol. 398: J.-P. Urbain, Exogeneity in Error Correction Models. XI, 189 pages. 1993.

Vol. 399: F. Gori, L. Geronazzo, M. Galeotti (Eds.), Nonlinear Dynamics in Economics and Social Sciences. Proceedings, 1991. VIII, 367 pages. 1993.

Vol. 400: H. Tanizaki, Nonlinear Filters. XII, 203 pages. 1993.

Vol. 401: K. Mosler, M. Scarsini, Stochastic Orders and Applications. V, 379 pages. 1993.

Vol. 402: A. van den Elzen, Adjustment Processes for Exchange Economies and Noncooperative Games. VII, 146 pages. 1993.

Vol. 403: G. Brennscheidt, Predictive Behavior. VI, 227 pages. 1993.

Vol. 404: Y.-J. Lai, Ch.-L. Hwang, Fuzzy Multiple Objective Decision Making. XIV, 475 pages. 1994.

Vol. 405: S. Komlósi, T. Rapcsák, S. Schaible (Eds.), Generalized Convexity. Proceedings, 1992. VIII, 404 pages. 1994.

Vol. 406: N. M. Hung, N. V. Quyen, Dynamic Timing Decisions Under Uncertainty. X, 194 pages. 1994.

Vol. 407: M. Ooms, Empirical Vector Autoregressive Modeling. XIII, 380 pages. 1994.

Vol. 408: K. Haase, Lotsizing and Scheduling for Production Planning. VIII, 118 pages. 1994.

Vol. 409: A. Sprecher, Resource-Constrained Project Scheduling. XII, 142 pages. 1994.

Vol. 410: R. Winkelmann, Count Data Models. XI, 213 pages. 1994.

Vol. 411: S. Dauzère-Péres, J.-B. Lasserre, An Integrated Approach in Production Planning and Scheduling. XVI, 137 pages. 1994.

Vol. 412: B. Kuon, Two-Person Bargaining Experiments with Incomplete Information. IX, 293 pages. 1994.

Vol. 413: R. Fiorito (Ed.), Inventory, Business Cycles and Monetary Transmission. VI, 287 pages. 1994.

Vol. 414: Y. Crama, A. Oerlemans, F. Spieksma, Production Planning in Automated Manufacturing. X, 210 pages. 1994.

Vol. 415: P. C. Nicola, Imperfect General Equilibrium. XI, 167 pages. 1994.

Vol. 416: H. S. J. Cesar, Control and Game Models of the Greenhouse Effect. XI, 225 pages. 1994.

Vol. 417: B. Ran, D. E. Boyce, Dynamic Urban Transportation Network Models. XV, 391 pages. 1994.

Vol. 418: P. Bogetoft, Non-Cooperative Planning Theory. XI, 309 pages. 1994.

Vol. 419: T. Maruyama, W. Takahashi (Eds.), Nonlinear and Convex Analysis in Economic Theory. VIII, 306 pages. 1995.

GPSR Compliance

The European Union's (EU) General Product Safety Regulation (GPSR) is a set of rules that requires consumer products to be safe and our obligations to ensure this.

If you have any concerns about our products, you can contact us on ProductSafety@springernature.com

In case Publisher is established outside the EU, the EU authorized representative is:

Springer Nature Customer Service Center GmbH
Europaplatz 3
69115 Heidelberg, Germany

Batch number: 10845482

Printed by Printforce, the Netherlands